T0222275

This text provides a self-contained introduction to applications of loop representations and knot theory in particle physics and quantum gravity.

Loop representations (and the related topic of knot theory) are of considerable current interest because they provide a unified arena for the study of the gauge invariant quantization of Yang-Mills theories and gravity, and suggest a promising approach to the eventual unification of the four fundamental forces. This text begins with a review of calculus in loop space and the fundamentals of loop representations. It then goes on to describe loop representations in Maxwell theory and Yang-Mills theories as well as lattice techniques. Applications in quantum gravity are then discussed in detail. Following chapters move on to consider knot theories, the braid algebra and extended loop representations in quantum gravity. A final chapter assesses the current status of the theory and points out possible directions for future research.

This self-contained introduction will be of interest to graduate students and researchers in theoretical physics and applied mathematics.

CAMBRIDGE MONOGRAPHS ON MATHEMATICAL PHYSICS

General editors: P. V. Landshoff, D. R. Nelson, D. W. Sciama, S. Weinberg

LOOPS, KNOTS, GAUGE THEORIES AND QUANTUM GRAVITY

Cambridge Monographs on Mathematical Physics

J. Ambjørn, B. Durhuus and T. Jonsson *Quantum Geometry: A Statistical Field Theory Approach*

A. M. Anile *Relative Fluids and Magneto-Fluids*

J. A. de Azcárraga and J. M. Izquierdo *Lie Groups, Lie Algebras, Cohomology and Some Applications in Physics*[†]

J. Bernstein *Kinetic Theory in the Early Universe*

G. F. Bertsch and R. A. Broglia *Oscillations in Finite Quantum Systems*

N. D. Birrell and P. C. W. Davies *Quantum Fields in Curved Space*[†]

D. M. Brink *Semiclassical Methods in Nucleus-Nucleus Scattering*

S. Carlip *Quantum Gravity in $2 + 1$ Dimensions*

J. C. Collins *Renormalization*[†]

P. D. B. Collins *An Introduction to Regge Theory and High Energy Physics*[†]

M. Creutz *Quarks, Gluons and Lattices*[†]

P. D. D'eath *Supersymmetric Quantum Cosmology*

F. de Felice and C. J. S. Clarke *Relativity on Curved Manifolds*[†]

B. De Witt *Supermanifolds, second edition*[†]

P. G. O. Freund *Introduction to Supersymmetry*[†]

F. G. Friedlander *The Wave Equation on a Curved Space-Time*

J. Fuchs *Affine Lie Algebras and Quantum Groups*[†]

J. A. H. Futterman, F. A. Handler and R. A. Matzner *Scattering from Black Holes*

R. Gambini and J. Pullin *Loops, Knots, Gauge Theories and Quantum Gravity*[†]

M. Göckeler and T. Schücker *Differential Geometry, Gauge Theories and Gravity*[†]

C. Gómez, M. Ruiz Altaba and G. Sierra *Quantum Groups in Two-dimensional Physics*

M. B. Green, J. H. Schwarz and E. Witten *Superstring Theory, volume 1:* INTRODUCTION[†]

M. B. Green, J. H. Schwarz and E. Witten *Superstring Theory, volume 2:* LOOP AMPLITUDES, ANOMALIES AND PHENOMENOLOGY[†]

S. W. Hawking and G. F. R. Ellis *The Large-Scale Structure of Space-Time*[†]

F. Iachello and A. Arima *The Interacting Boson Model*[†]

F. Iachello and P. van Isacker *The Interacting Boson–Fermion Model*

C. Itzykson and J.-M. Drouffe *Statistical Field Theory, volume 1:* FROM BROWNIAN MOTION TO RENORMALIZATION AND LATTICE GAUGE THEORY[†]

C. Itzykson and J.-M. Drouffe *Statistical Field Theory, volume 2:* STRONG COUPLING, MONTE CARLO METHODS, CONFORMAL FIELD THEORY, AND RANDOM SYSTEMS[†]

J. I. Kapusta *Finite-Temperature Field Theory*[†]

V. E. Korepin, A. G. Izergin and N. M. Boguliubov *The Quantum Inverse Scattering Method and Correlation Functions*[†]

D. Kramer, H. Stephani, M. A. H. MacCallum and E. Herlt *Exact Solutions of Einstein's Field Equations*

M. Le Bellac *Thermal Field Theory*[†]

N. H. March *Liquid Metals: Concepts and Theory*

I. M. Montvay and G. Münster *Quantum Fields on a Lattice*[†]

L. O'Raifeartaigh *Group Structure of Gauge Theories*[†]

A. Ozorio de Almeida *Hamiltonian Systems: Chaos and Quantization*[†]

R. Penrose and W. Rindler *Spinors and Space-time, volume 1:* TWO-SPINOR CALCULUS AND RELATIVISTIC FIELDS[†]

R. Penrose and W. Rindler *Spinors and Space-time, volume 2:* SPINOR AND TWISTOR METHODS IN SPACE-TIME GEOMETRY[†]

S. Pokorski *Gauge Field Theories*[†]

J. Polchinski *String Theory, volume 1*

J. Polchinski *String Theory, volume 2*

V. N. Popov *Functional Integrals and Collective Excitations*[†]

R. Rivers *Path Integral Methods in Quantum Field Theory*[†]

R. G. Roberts *The Structure of the Proton*[†]

W. C. Saslaw *Gravitational Physics of Stellar and Galactic Systems*[†]

J. M. Stewart *Advanced General Relativity*

A. Vilenkin and E. P. S. Shellard *Cosmic Strings and other Topological Defects*[†]

R. S. Ward, R. O. Wells Jr *Twistor Geometry and Field Theories*[†]

[†] Issued as paperback

Loops, Knots, Gauge Theories and Quantum Gravity

RODOLFO GAMBINI

Universidad de la Republica, Montevideo

JORGE PULLIN

Pennsylvania State University

CAMBRIDGE
UNIVERSITY PRESS

PUBLISHED BY THE PRESS SYNDICATE OF THE UNIVERSITY OF CAMBRIDGE
The Pitt Building, Trumpington Street, Cambridge, United Kingdom

CAMBRIDGE UNIVERSITY PRESS
The Edinburgh Building, Cambridge CB2 2RU, UK   http://www.cup.cam.ac.uk
40 West 20th Street, New York, NY 10011–4211, USA   http://www.cup.org
10 Stamford Road, Oakleigh, Melbourne 3166, Australia
Ruiz de Alarcón 13, 28014 Madrid, Spain

First published 1996
First paperback edition 2000

Typeset in 11/13pt Computer Modern

A catalogue record for this book is available from the British Library

Library of Congress Cataloguing in Publication data

Gambini, Rodolfo.
Loops, knots, gauge theories and quantum gravity / Rodolfo Gambini
and Jorge Pullin; [foreword by Abhay Ashtekar].
p. cm. – (Cambridge monographs on mathematical physics)
Includes bibliographical references and index.
ISBN 0 521 47332 2 (hc)
1. Quantum gravity – Mathematics. 2. Loops (Group theory).
3. Knot theory. 4. Gauge fields (Physics). 5. Quantum field theory.
I. Pullin, Jorge. II. Title. III. Series.
QC178.G25   1996
530.1'43–dc20   95-35159   CIP

ISBN 0 521 47332 2 hardback
ISBN 0 521 65475 0 paperback

Transferred to digital printing 2002

A Gabriela y Martha

Contents

Foreword

For about twenty years after its invention, quantum electrodynamics remained an isolated success in the sense that the underlying ideas seemed to apply only to the electromagnetic force. In particular, its techniques did not seem to be useful in dealing with weak and strong interactions. These interactions seemed to lie outside the scope of the framework of local quantum field theory and there was a wide-spread belief that the best way to handle them would be via a more general, abstract S-matrix theory. All this changed dramatically with the discovery that non-Abelian gauge theories were renormalizable. Once the power of the gauge principle was fully recognized, local quantum field theory returned to the scene and, by now, dominates our thinking. Quantum gauge theories provide not only the most natural but also the only viable candidates we have for the description of electroweak and strong forces.

The basic dynamical variables in these theories are represented by non-Abelian connections. Since all the gauge invariant information in a connection is contained in the Wilson loops variables (i.e., traces of holonomies), it is natural to try to bring them to the forefront. This is precisely what is done in the lattice approaches which are the most successful tools we have to probe the non-perturbative features of quantum gauge theories. In the continuum, there have also been several attempts to formulate the theory in terms of Wilson loops. In the perturbative approach, it is known that Wilson loop "Schwinger functions" are finite to all orders after renormalization. This is a strong indication that they may be also mathematically meaningful in a non-perturbative treatment of the continuum theory. Since these are functions on an appropriate space of loops, one can derive differential equations they satisfy on that loop space. The hope is that once a complete set of equations is obtained, physical "boundary conditions" will lead to unique solutions which in turn will determine the theory. Thus, the space of loops offers a natural arena for the quantum theories of connections.

In the last few years, this relation between connections and loops has acquired another dimension. In these developments, the emphasis is on the Hamiltonian formulation. It turns out that there is a remarkable mathematical interplay between measures on the spaces of connections and functions on the loop space, which gives rise to a generalization of the Fourier transform, called the loop transform. This transform can be defined rigorously. As a result, quantum states can be regarded either as gauge invariant functions of connections or as suitable functions of loops. The loop picture suggests new strategies for defining operators and provides a number of new insights.

Quite surprisingly, it turns out that these insights are especially useful while dealing with a force that one does not, normally, associate with theories of connections: gravity. General relativity is usually thought of as a theory of metrics and, therefore, quite removed from theories of other interactions. One can, however, think of it also as a dynamical theory of connections. This idea is not new. Indeed, such a reformulation was obtained already by Einstein and Schrödinger. In their new version, the Levi Civita connection is regarded as the basic variable; metric is a secondary, derived object. The problem was that the equations of the theory became more complicated. It turns out, however, that if one uses chiral connections in place of Levi Civita, the equations actually simplify. With this observation, general relativity moves closer to theories governing other fundamental forces. As in other theories, one can now represent states of quantum gravity as functions of (chiral) connections or, via loop transform, of loops. Thus, the loop representation offers a unified arena for the quantum description of all four fundamental interactions. In the case of general relativity, further structures arise because physical states are required to be invariant under the action of diffeomorphisms. In the loop representation, they depend not on individual loops but also on the (generalized) knot to which the loop belongs. There is thus an unexpected interplay between loops, knots, gauge fields and gravity.

This monograph is devoted to this interplay. The authors are eminently qualified to unfold this saga as they are among the leaders in the field. Indeed, many of the developments that I have alluded to are due to them and their close colleagues. They provide not only a comprehensive summary of the entire subject, but, in the last few chapters, also a glimpse of two frontier areas of active research. Graduate students would find this unified treatment of a large subject extremely useful. More advanced researchers would be able to appreciate the fascinating confluence of ideas from particle physics general relativity and contemporary mathematics.

University Park, 1996   Abhay Ashtekar

Preface

Loops have been used as a tool to study classical and quantum Yang–Mills theory since the work of Mandelstam in the early 1960s. They have led to many insights concerning the non-perturbative dynamics of the theory including the issue of confinement and the lattice formulation. Since the inception of the Asthekar new variables, loop techniques have also found important applications in quantum gravity. Due to the diffeomorphism invariance of the theory they have led to surprising connections with knot theory and topological field theories.

The intention in this book is to present several of these results in a common framework and language. In particular it is an attempt to combine ideas developed some time ago in the context of Yang–Mills theories with the recent applications in quantum gravity. It should be emphasized that our treatment of Yang–Mills theories only covers a small part of all results obtained with loops: that which seems of most relevance for applications in gravity.

This book should allow people from outside the field to gain access in a pedagogical way to the current state of the art. Moreover, it allows experts within this wide field with heterogeneous backgrounds to learn about specific results outside their main area of expertise and as a reference volume. It should be well suited as an introductory guide for graduate students who want to get started in the subject.

Subjects in this area are being developed at two different levels: one more "mathematical", related to constructive quantum field theory and the other more "physical" in which several subtleties are ignored in order to gain rapid insight into the theory. This book will largely concentrate on this latter approach. We will present in some detail the mathematics underlying loops but only at the level needed for a physicist to operate with the resulting formalism.

Due to the rapid development of these ideas in recent times we wrote this book with expediency in mind in order to offer it to the public as soon as possible. We ask readers to forgive any notational and conventions errors that might have arisen in this process. In spite of our efforts to deliver the book expediently, during the time it took to write it the field has already evolved. Although we have kept the manuscript updated, some aspects of the presentation may not be completely in line with current thinking. For instance, the recently pinpointed convergence issues of extended loops and their possible solutions are only briefly mentioned. The rapid development over the last few months of a measure theory in infinite-dimensional spaces and its impact on a rigorous definition of a loop representation is only briefly discussed in chapter 3 and its implications for later chapters are ignored. A similar remark applies to the recent developments concerning the solution of the Mandelstam constraints in terms of spin network states and the possibility of having a basis of independent loop states. We have tried to present a current outlook from our perspective in chapter 12, including very recent references to work in progress.

Many people have contributed to make this book possible. We cannot attempt to exhaustively list all the colleagues from which we have benefited through discussions and interactions. We like to thank the colleagues with which we have had a closer interaction and whose ideas have certainly left an imprint on this book: Abhay Ashtekar, Bernd Brügmann, Cayetano Di Bartolo, Jorge Griego, Jerzy Lewandowski, Jose Mourão, Carlo Rovelli, Lee Smolin and Antoni Trias. We also wish to thank Abhay Ashtekar, John Baez, John Baker, José Mourão, Peter Peldán, Carlo Rovelli, Thorsten Schwander, José Antonio Zapata and especially Bernd Brügmann and Jerzy Lewandowski for many useful comments about the manuscript.

This work was supported in part by grants NSF-INT-94-06269, NSF-PHY-94-06269, NSF-PHY93-96246, NSF-PHY-92-07225, by an Alfred P. Sloan fellowship to JP and by research funds of the University of Utah, The Pennsylvania State University, the Eberly Family research fund, the PSU Office for Minority Faculty Development, Conicyt and PEDECIBA.

State College, 1996   Rodolfo Gambini, Jorge Pullin

1

Holonomies and the group of loops

1.1 Introduction

In this chapter we will introduce holonomies and some associated concepts which will be important in the description of gauge theories to be presented in the following chapters. We will describe the group of loops and its infinitesimal generators, which will turn out to be a fundamental tool in describing gauge theories in the loop language.

Connections and the associated concept of parallel transport play a key role in locally invariant field theories like Yang–Mills and general relativity. All the fundamental forces in nature that we know of may be described in terms of such fields. A connection allows us to compare points in neighboring fibers (vectors or group elements depending on the description of the particular theory) in an invariant form. If we know how to parallel transport an object along a curve, we can define the derivative of this object in the direction of the curve. On the other hand, given a notion of covariant derivative, one can immediately introduce a notion of parallel transport along any curve.

For an arbitrary closed curve, the result of a parallel transport in general depends on the choice of the curve. To each closed curve γ in the base manifold with origin at some point o the parallel transport will associate an element H of the Lie group G associated to the fiber bundle. The parallel transported element of the fiber is obtained from the original one by the action of the group element H. The path dependent object $H(\gamma)$ is usually called the *holonomy*. It has been considered in various contexts in physics and given different names. For instance, it is known as the *Wu–Yang phase factor* in particle physics.

Curvature is related to the failure of an element of the fiber to return to its original value when parallel transported along a small closed curve. When evaluated on an infinitesimal closed curve with basepoint o, the

1

holonomy has the same information as the curvature at o. Knowledge of the holonomy for any closed curve with a base point o allows one, under very general hypotheses, to reconstruct the connection at any point of the base manifold up to a gauge transformation. An important fact about holonomies is their invariance under the set of gauge transformations which act trivially at the base point. We will later show that this will imply that the physical configurations of any gauge theory can be faithfully and uniquely (up to transformations at the base point) represented by their holonomies. They can therefore be used to encode all the kinematical information about the theory in question.

Since the early 1960s several descriptions of gauge theories in terms of holonomies have been considered. They seem to be particularly well suited to study the non-perturbative features at the quantum level. In recent years interest in the non-local descriptions of gauge theories has been greatly increased by the introduction of a new set of canonical variables that allow one to describe the phase space of general relativity in a manner that resembles an $SU(2)$ Yang–Mills theory. In fact, holonomies may well provide a common geometrical framework for all the fundamental forces in nature

A generalization of the notion of holonomy may be defined intrinsically without any reference to connections. It will turn out that this point of view has more than a purely mathematical interest and is the origin of important results that are relevant to the physical applications. Holonomies can be viewed as homomorphisms from a group structure defined in terms of equivalence classes of closed curves onto a Lie group G. Each equivalence class of closed curves is what we will technically call a loop and the group structure defined by them is called the group of loops.

The group of loops is the basic underlying structure of all the non-local formulations of gauge theories in terms of holonomies. In particular, when quantizing the theory, wavefunctions in the "loop representation" are really functions dependent on the elements of the group of loops*. This is the physical reason why it is important to understand the structure of the group of loops, since it is the "arena" where the quantum loop representation takes place.

In spite of the fact that the group of loops is not a Lie group, it is possible to define infinitesimal generators for it. When they are represented in the space of functions of loops, they give rise to differential operators in loop space. Some of these operators have appeared in various physical contexts and have been given diverse names such as "area derivative",

---

* In this context the group of loops is usually referred to as "loop space" and we will loosely use this terminology when it does not give rise to ambiguities. Notice that it is not related to the "loop groups" in the main mathematical literature.

"keyboard derivative", "loop derivative". In most of these presentations the group properties of loops were largely ignored and this resulted in various inconsistencies. In the approach we follow in this chapter all these operators arise simply and consistently as representations of the infinitesimal generators of the group of loops.

In many presentations, loop space is formulated with parametrized curves. In this context differential operators are usually written in terms of functional derivatives. The group structure of loops is hidden by these formulations and it is easy to overlook it, again leading to inconsistencies. In this book we will deal with unparametrized loops which allow for a cleaner formulation, only resorting to parametrizations for some particular results.

This chapter is structured in the following way. In section 1.2 we define the group of loops and discuss its topology and its action on open paths. In section 1.3 we introduce the infinitesimal generators of the group and their differential representation. We also introduce differential operators acting on open paths. In section 1.3.3 we introduce the connection derivative, its relation to the loop derivative and to usual notions of gauge theory. In section 1.3.4 we discuss the contact and functional derivatives in loop space and their relations with diffeomorphisms. In section 1.4 we introduce the idea of representations of the group of loops in a Lie group and we retrieve the classical kinematics of gauge theories. We end with a summary of the ideas developed in this chapter.

1.2 The group of loops

We start by considering a set of parametrized curves on a manifold M that are continuous and piecewise smooth. A curve p is a map

$$p : [0, s_1] \cup [s_1, s_2] \cdots [s_{n-1}, 1] \rightarrow M \tag{1.1}$$

smooth in each closed interval $[s_i, s_{i+1}]$ and continuous in the whole domain. There is a natural composition of parametrized curves. Given two piecewise smooth curves p_1 and p_2 such that the end point of p_1 is the same as the beginning point of p_2, we denote by $p_1 \circ p_2$ the curve:

$$p_1 \circ p_2(s) = \begin{cases} p_1(2s), & \text{for } s \in [0, 1/2] \\ p_2(2(s - 1/2)) & \text{for } s \in [1/2, 1]. \end{cases} \tag{1.2}$$

The curve traversed in the opposite orientation ("opposite curve") is given by

$$p^{-1}(s) := p(1 - s). \tag{1.3}$$

In what follows, we will mainly be interested in unparametrized curves. We will therefore define an equivalence relation by identifying the curve

p and $p \circ \phi$ for all orientation preserving differentiable reparametrizations $\phi : [0,1] \rightarrow [0,1]$. It is important to note that the composition of unparametrized curves is well defined and independent of the members of the equivalence classes used in their definition.

We will now consider closed curves $l, m, ...$, that is, curves which start and end at the same point o. We denote by L_o the set of all these closed curves. The set L_o is a semi-group under the composition law $(l, m) \rightarrow l \circ m$. The identity element ("null curve") is defined to be the constant curve $i(s) = o$ for any s and any parametrization. However, we do not have a group structure, since the opposite curve l^{-1} is not a group inverse in the sense that $l \circ l^{-1} \neq i$.

Holonomies are associated with the parallel transport around closed curves. In the case of a trivial bundle the connection is given by a Lie-algebra-valued one form A_a on M. The parallel transport around a closed curve $l \in L_o$ is a map from the fiber over o to itself given by the path ordered exponential (for the definition of path ordered exponential see reference [1]),

$$H_A(l) = P \exp \int_l A_a(y) dy^a. \qquad (1.4)$$

In the general case of a principal fiber bundle $P(M, G)$ with group G over M the holonomy map is defined as follows. We choose a point $\hat{o}$ in the fiber over o and by using the connection A we lift the closed curve l in M to a curve $\hat{l}$ in P such that the beginning point is

$$\hat{l}(0) = \hat{o} \qquad (1.5)$$

and the end point is given by

$$\hat{l}(1) = \hat{l}(0) H_A(l), \qquad (1.6)$$

which defines $H_A(l)$. The holonomy H_A is an element of the group G and the product denotes the right action of G. The main property of H_A is

$$H_A(l \circ m) = H_A(l) H_A(m). \qquad (1.7)$$

A change in the choice of the point on the fiber over o replacing $\hat{o}$ for $\hat{o}' = \hat{o} g$ induces the transformation

$$H'_A(l) = g^{-1} H_A(l) g. \qquad (1.8)$$

In order to transform the set L_o into a group, we need to introduce a further equivalence relation. The rationale for this relation is to try to identify all closed curves leading to the the same holonomy for all smooth connections, since curves with the same holonomy carry the same information towards building the physical quantities of the theory. The classes of equivalence under this relation are what we will from now on call *loops* and we will denote them with Greek letters, to distinguish them from the individual curves which form the equivalence classes. Several definitions of this equivalence relation have been proposed. Each of them sheds some light on the group structure so we will take a minute to consider them in some detail.

Definition 1
Let

$$H_A : L_o \to G \tag{1.9}$$

be the holonomy map of a connection A defined on a bundle $P(M, G)$. Two curves $l, m \in L_o$ are equivalent [2] [4] $l \sim m$ iff

$$H_A(l) = H_A(m) \tag{1.10}$$

for every bundle $P(M, G)$ and smooth connection A.

Definition 2
We start by defining loops which are equivalent to the identity. A closed curve l is called a *tree*[5] or *thin* [6] if there exists a homotopy of l to the null curve in which the image of the homotopy is included in the image of l. This kind of curves does not "enclose any area" of M. Two closed curves $l, m \in L_o$ are equivalent $l \sim m$ iff $l \circ m^{-1}$ is thin. Obviously a thin curve is equivalent to the null curve.

Definition 3 [7]
Given the closed curves l and m and three open curves p_1, p_2 and q such that

$$l = p_1 \circ p_2 \tag{1.11}$$
$$m = p_1 \circ q \circ q^{-1} \circ p_2 \tag{1.12}$$

then $l \sim m$.

There is a fourth definition, due to Chen [7], that requires the use of a set of objects (Chen integrals, which we will call "loop multitangents") that we will define in chapter 2, but we will not discuss it here.

It can be shown that definitions 2 and 3 are equivalent. Moreover, it is also immediate to notice that two curves equivalent under definitions 2 or 3 are also equivalent under definition 1. The reciprocal is not obvious. Partial results can be found in reference [7] and a complete proof for piecewise analytic curves has been presented by Ashtekar and Lewandowski [40].

With any of these definitions one can show that the composition between loops is well defined and is again a loop. In other words if $\alpha \equiv [l]$

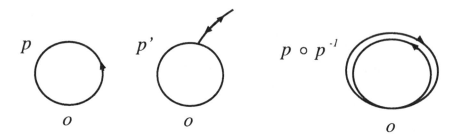

Fig. 1.1.   Curves p and p' differ by a tree. The composition of a curve and its inverse is a tree.

and $\beta \equiv [m]$ then $\alpha \circ \beta = [l \circ m]$ where by $[]$ we denote the equivalence classes. From now on we will denote loops with greek letters, to distinguish them from curves[†].

Notice that with the equivalence relation defined, it makes sense to define an inverse of a loop. Since the composition of a curve with its opposite yields a tree (see figure 1.1) it is natural, given a loop α, to define its inverse α^{-1} by $\alpha \circ \alpha^{-1} = \iota$ where ι is the set of closed curves equivalent to the null curve (thin loops or trees). α^{-1} is the set of curves opposite to the elements of α.

We will denote the set of loops basepointed at o by $\mathcal{L}_o$. Under the composition law given by $\circ$ this set is a non-Abelian group, which is called the group of loops.

A well known result [5] is that any homomorphism,

$$\mathcal{L}_o \to G, \tag{1.13}$$

where G is a Lie group, defines a *holonomy* associated with a "generalized" connection. By generalized we mean that the connection will not, in general, be a smooth function (for instance it could be distributional or worse). One can, by imposing extra smoothness conditions [6, 4] on the homomorphism, ensure that a differentiable principal fiber bundle and a connection are defined such that H is the holonomy of this connection. Recall that under a homomorphism, the composition law of the group of loops is mapped onto the composition law of the Lie group G,

$$H(\alpha \circ \beta) = H(\alpha)H(\beta), \tag{1.14}$$

[†] Notice that in this book we will use the word loop in a very precise sense, denoting the holonomic-equivalent classes of curves. Other equivalences can be considered. The idea of a group of loops has appeared in other unrelated contexts [42]. For this reason some authors have proposed calling the holonomic equivalence classes "hoops" to avoid confusion [3].

and that inverses are mapped to each other,

$$H(\alpha^{-1}) = (H(\alpha))^{-1}. \qquad (1.15)$$

We will come back to this property in section 1.4 when we discuss the infinitesimal generators and their relations to the physical quantities.

From now on we will routinely use functions of loops, such as the holonomy that we just introduced. Obviously, not any function of curves qualifies as a function of loops. An immediate example of this would be to consider the length of a curve, which takes different values on the different curves that form the equivalence class defining a loop.

It is useful to introduce a notion of continuity in loop space, since we will be frequently using functions defined on this space. We will define two loops α and β to be close, in the sense that α in a neighborhood $U_\epsilon(\beta)$ if there exist at least two parametrized curves $a(s) \in \alpha$ and $b(s) \in \beta$ such that $a(s) \in U_\epsilon(b(s))$ with the usual topology of curves in the manifold[‡]. With this topology, the group of loops is a topological group.

It is convenient for future use to introduce an equivalence relation for open curves similar to the one we introduced for closed curves. We will call the equivalence classes of open curves "paths". Given two open curves p_o^x and q_o^x from the basepoint to a point x in the manifold, we will define these curves to be equivalent iff $p_o^x q^{-1}{}_o^x$ is a tree[§]. We will denote paths with Greek letters as we do for loops, but indicating the origin and end points, as in α_o^x. Given two different paths starting and ending at the same points, it is immediate to see that the composition of one with the opposite of the other is a loop. Analogously one can compose loops with paths to produce new paths with the same end points. Furthermore, the notion of topology introduced for loops can immediately be generalized to paths. However, paths cannot be structured into a group, since it is not possible to compose, in general, two paths to form a new path (the end of one of them has to coincide with the beginning of the other in order to do this).

1.3   Infinitesimal generators of the group of loops

We will now consider a representation of the group of loops given by operators acting on continuous functions under the topology introduced in the previous section. We will introduce a set of differential operators

---

[‡] Lewandowski [4], elaborating on a suggestion by Barrett [6] has introduced a topology defined in terms of homotopies of loops. The group of loops endowed with this topology is a topological Haussdorff group.

[§] From now on we will interchangeably use the notations $q^{-1}{}_o^x$ and q_x^o to designate the same object, the curve q traversed from x to o. A similar convention will be adopted for paths.

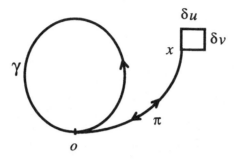

Fig. 1.2.   The infinitesimal loop that defines the loop derivative.

acting on these functions that are related to the infinitesimal generators of the group of loops, in terms of which one can construct the elements of the group. In later chapters we will show that these operators are related to physical quantities of gauge theories. Although the explicit introduction of the differential operators will be made in a coordinate chart, we will show that the definitions do not depend on the particular chart chosen. A more intrinsic definition, also making use of the properties of the group of loops has been proposed by Tavares [43].

1.3.1   The loop derivative

Given $\Psi(\gamma)$ a continuous, complex-valued function of $\mathcal{L}_o$ we want to consider its variation when the loop γ is changed by the addition of an infinitesimal loop $\delta\gamma$ basepointed at a point x connected by a path π_o^x to the basepoint of γ, as shown in figure 1.2. That is, we want to evaluate the change in the function when changing its argument from γ to $\pi_o^x \circ \delta\gamma \circ \pi_x^o \circ \gamma$. In order to do this we will consider a two-parameter family of infinitesimal loops $\delta\gamma$ that contain in a particular coordinate chart the curve obtained by traversing the vector u^a from x^a to $x^a + \epsilon_1 u^a$, the vector v^a from $x^a + \epsilon_1 u^a$ to $x^a + \epsilon_1 u^a + \epsilon_2 v^a$, the vector $-u^a$ from $x^a + \epsilon_1 u^a + \epsilon_2 v^a$ to $x^a + \epsilon_2 v^a$ and the vector $-v^a$ from $x^a + \epsilon_2 v^a$ back to x^a as shown in figure 1.2. We will denote these kinds of curves with the notation[¶] $\delta u \delta v \delta \bar{u} \delta \bar{v}$.

[¶] In order not to clutter the notation we will not distinguish between curves and paths here. We also drop the ϵ_i dependence of each path. The path $\delta\bar{u} \equiv (\delta u)^{-1}$.

For a given π and γ a loop differentiable function depends only on the infinitesimal vectors $\epsilon_1 u^a$ and $\epsilon_2 v^a$. We will assume it has the following expansion with respect to them,

$$
\begin{aligned}
\Psi(\pi_o^x \circ \delta\gamma \circ \pi_x^o \circ \gamma) =\ & \Psi(\gamma) + \epsilon_1 u^a Q_a(\pi_o^x)\Psi(\gamma) + \epsilon_2 v^a P_a(\pi_o^x)\Psi(\gamma) \\
& + \tfrac{1}{2}\epsilon_1\epsilon_2(u^a v^b + v^a u^b)S_{ab}(\pi_o^x)\Psi(\gamma) \\
& + \tfrac{1}{2}\epsilon_1\epsilon_2(u^a v^b - v^a u^b)\Delta_{ab}(\pi_o^x)\Psi(\gamma).
\end{aligned} \tag{1.16}
$$

where Q, P, S, Δ are differential operators on the space of functions $\Psi(\gamma)$. If ϵ_1 or ϵ_2 vanishes or if u is collinear with v then $\delta\gamma$ is a tree and all the terms of the right-hand side except the first one must vanish. This means that $Q = P = S = 0$. Since the antisymmetric combination $(u^a v^b - v^a u^b)$ vanishes, Δ need not be zero. That is, a function is loop differentiable if for any path π_o^x and vectors u, v, the effect of an infinitesimal deformation is completely contained in the path dependent antisymmetric operator $\Delta_{ab}(\pi_o^x)$,

$$
\Psi(\pi_o^x \circ \delta\gamma \circ \pi_x^o \circ \gamma) = (1 + \tfrac{1}{2}\sigma^{ab}(x)\Delta_{ab}(\pi_o^x))\Psi(\gamma), \tag{1.17}
$$

where $\sigma^{ab}(x) = 2\epsilon_1\epsilon_2(u^{[a}v^{b]})$ is the element of area of the infinitesimal loop $\delta\gamma$. We will call this operator the loop derivative.

Notice that we have proved that for an arbitrary function of loop space, one does not have contributions from the terms Q, P, S in the expansion (1.16). If one considers functions of curves rather than of loops, these terms will in general be present. As an example, they are present if one considers the function given by the length of the curve. On the other hand, not every function of loop space is differentiable. For instance, we will see when we consider knot invariants — functionals of loops invariant under smooth deformations of the loops — that they are not strictly speaking loop differentiable. The reason for this is that sometimes appending an infinitesimal loop could enable us to change the topology of the knots and therefore to induce finite changes in the values of the functions.

Loop derivatives of various kinds were considered by several authors. The idea was introduced by Mandelstam [8]. Later generalizations can be found in the work of Chen [7], Makeenko and Migdal [10, 12], Polyakov [44], Gambini and Trias [13, 14, 15], Blencowe [16] and Brügmann and Pullin [26]. Other references can be found in Loll [17]. The various definitions are not equivalent, and many of them refer to objects that are in reality different from the loop derivative we are defining here. One of the main differences is that in many treatments the infinitesimal loop, instead of being appended at an arbitrary fixed point of the manifold defined by a path π_o^x as is our case, is appended to a point *that lies on the loop*. Since one is considering functions of arbitrary loops that means that the point where the derivative acts has to be redefined when considering its value on a new loop. In other words, the domain of the function that results when applying these kinds of derivatives is not the loop space defined in section 1.2, but the space of loops with a marked point. Makeenko and Migdal [12] noticed this fact and this drove them to call it "keyboard derivative."

Notice that this is not the case for the derivative we defined. The result of the application of the loop derivative to a function of a loop is also a function of a loop. For each arbitrary open path there is a different derivative. For these definitions to work it is crucial to have a basepoint, which provides a fixed point for any loop on which to attach the open path that defines the derivative. These considerations are of crucial importance. For instance, we will soon prove that our derivative satisfies Bianchi identities, a fact that cannot be proven for derivatives that act only on points of the loop. The relevance of the group of loops and the path dependence of the loop derivative were first recognized by Gambini and Trias [13, 15].

At the end of section 1.2 we noted that the elements of the group of loops have a natural action on open paths, giving as a result a deformation of the path. We can immediately find an example of this fact in terms of a differential operator defined by simply extending the definition of the loop derivative (1.17) to give for open paths

$$\Psi(\pi_o^x \circ \delta\gamma \circ \pi_x^o \circ \gamma_o^y) = (1 + \tfrac{1}{2}\sigma^{ab}(x)\Delta_{ab}(\pi_o^x))\Psi(\gamma_o^y). \qquad (1.18)$$

We will take some notational latitude to give the same name to the loop derivative acting on paths and on loops. In all cases the context will uniquely determine to which derivative we are referring. Notice that this extension to open paths is not at all clear for derivatives that depend on a point of the loop as is the case of the "keyboard derivative".

1.3.2　Properties of the loop derivative

• **Tensor character.** By its very definition, (1.17), it is immediate to see that the loop derivative has to behave as a tensor under local coordinate transformations containing the end point of the path π_o^x for loop differentiable functions. One need just require that the whole expression be invariant and notice that the loop derivative is contracted with the tensor σ^{ab}. Therefore by quotient law, it must be a tensor. Notice that the loop derivative is really associated with the surface spanned by du^a and dv^b rather than with the individual infinitesimal vectors, being invariant under vector transformations that preserve the element of area.

• **Commutation relations.** The loop derivatives are non-commutative operators. This, as we will see later, is naturally associated with the fact

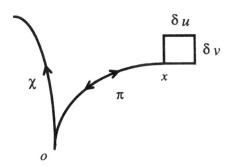

Fig. 1.3.   The two paths used to compute the commutation relation

that they correspond to the generators of a non-Abelian group. Their commutation relations can be computed directly from the geometric properties of the group of loops in the following way. Consider two infinitesimal loops $\delta\eta_1$, $\delta\eta_2$ given by

$$\delta\eta_1 = \pi_o^x \circ \delta u \delta v \delta \bar{u} \delta \bar{v} \circ \pi_x^o \quad \text{and} \quad \delta\eta_2 = \chi_o^y \circ \delta q \delta r \delta \bar{q} \delta \bar{r} \circ \chi_y^o \quad (1.19)$$

and with area elements

$$\sigma_1^{ab} = \epsilon_1 \epsilon_2 (u^a v^b - v^a u^b) \quad \text{and} \quad \sigma_2^{ab} = \epsilon_3 \epsilon_4 (q^a r^b - r^a q^b). \quad (1.20)$$

Then we can derive the following relation:

$$\Psi(\delta\eta_1 \circ \delta\eta_2 \circ (\delta\eta_1)^{-1} \circ (\delta\eta_2)^{-1} \circ \gamma) = (1 + \tfrac{1}{2}\sigma_1^{ab}\Delta_{ab}(\pi_o^x))$$
$$\times (1 + \tfrac{1}{2}\sigma_2^{cd}\Delta_{cd}(\chi_o^y))(1 - \tfrac{1}{2}\sigma_1^{ef}\Delta_{ef}(\pi_o^x))(1 - \tfrac{1}{2}\sigma_2^{gh}\Delta_{gh}(\chi_o^y))\Psi(\gamma) =$$
$$(1 + \tfrac{1}{4}\sigma_1^{ab}\sigma_2^{cd}[\Delta_{ab}(\pi_o^x), \Delta_{cd}(\chi_o^y)])\Psi(\gamma). \quad (1.21)$$

The first equality follows from the definition of the loop derivative and of the loops $\delta\eta_i$. To prove the second, one expands keeping only terms of first order in each ϵ_i and neglecting those of order ϵ_i^2.

We will now define an open path by composing the two paths we have been using

$$\chi_o'^y = \delta\eta_1 \circ \chi_o^y. \quad (1.22)$$

This allows us to rewrite the loop composed by the first three loops in the argument of Ψ in the left-hand side of equation (1.21) as,

$$\delta\eta_1 \circ \delta\eta_2 \circ (\delta\eta_1)^{-1} = \chi_o'^y \circ \delta q \delta r \delta \bar{q} \delta \bar{r} \circ \chi_y'^o. \quad (1.23)$$

Therefore,

$$\Psi(\delta\eta_1 \circ \delta\eta_2 \circ (\delta\eta_1)^{-1} \circ (\delta\eta_2)^{-1} \circ \gamma) =$$
$$(1 + \tfrac{1}{2}\sigma_2^{ab}\Delta_{ab}(\chi_o^{\prime y}))(1 - \tfrac{1}{2}\sigma_2^{cd}\Delta_{cd}(\chi_o^y))\Psi(\gamma). \qquad (1.24)$$

And again expanding in ϵs and keeping only the first order in each ϵ_i we get

$$(1 + \tfrac{1}{2}\sigma_2^{ab}\Delta_{ab}(\chi_o^{\prime y}))(1 - \tfrac{1}{2}\sigma_2^{cd}\Delta_{cd}(\chi_o^y))\Psi(\gamma) =$$
$$(1 + \tfrac{1}{4}\sigma_1^{ab}\sigma_2^{cd}\Delta_{ab}(\pi_o^x)[\Delta_{cd}(\chi_o^y)])\Psi(\gamma), \qquad (1.25)$$

where in the last expression $\Delta_{ab}(\pi_o^x)[\Delta_{cd}(\chi_o^y)]$ represents the action of the first loop derivative only on the path dependence of the second derivative.

All this implies

$$[\Delta_{ab}(\pi_o^x), \Delta_{cd}(\chi_o^y)] = \Delta_{cd}(\chi_o^y)[\Delta_{ab}(\pi_o^x)], \qquad (1.26)$$

from which it is immediate to show that

$$\Delta_{ab}(\pi_o^x)[\Delta_{cd}(\chi_o^y)] = -\Delta_{cd}(\chi_o^y)[\Delta_{ab}(\pi_o^x)]. \qquad (1.27)$$

These expressions highlight the path dependence of the loop derivative, in the sense that they express the variation of the derivative when the path is varied. We will see at the end of this subsection how these expressions can be naturally interpreted as a group commutator when we prove that the loop derivative is a generator of the group of loops.

This commutation relation can be viewed in a different light by considering its integral expression. In order to do this, we will introduce a loop dependent operator $U(\alpha)$ on the space of functions of loops which has the effect of introducing a finite deformation in the argument of the function,

$$U(\alpha)\Psi(\gamma) \equiv \Psi(\alpha \circ \gamma). \qquad (1.28)$$

The operator has a naturally defined inverse,

$$U(\alpha)^{-1} = U(\alpha^{-1}), \qquad (1.29)$$

and has a natural composition law,

$$U(\alpha)U(\beta)\Psi(\gamma) = U(\alpha \circ \beta)\Psi(\gamma). \qquad (1.30)$$

We now consider the action of the loop derivative evaluated along a deformed path, shown in figure 1.4, on a function of loop, and applying the definition of loop derivative (1.17) we get

$$(1 + \tfrac{1}{2}\sigma^{ab}\Delta_{ab}(\alpha \circ \pi_o^x))\Psi(\gamma) = \Psi(\alpha \circ \pi_o^x \circ \delta\gamma \circ \pi_x^o \circ \alpha^{-1} \circ \gamma), \qquad (1.31)$$

where $\delta\gamma$ is the infinitesimal loop associated with the area element σ^{ab}. We then use the definition of the operator U (1.28) to get

$$\Psi(\alpha \circ \pi_o^x \circ \delta\gamma \circ \pi_x^o \circ \alpha^{-1} \circ \gamma) = U(\alpha)(1 + \tfrac{1}{2}\sigma^{ab}\Delta_{ab}(\pi_o^x))U(\alpha)^{-1}\Psi(\gamma), \qquad (1.32)$$

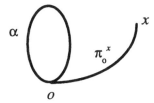

Fig. 1.4. The deformed path used to derive the integral expression of the commutation relations

from which we can read off the identity,

$$\Delta_{ab}(\alpha \circ \pi_o^x) = U(\alpha)\Delta_{ab}(\pi_o^x)U(\alpha)^{-1}, \qquad (1.33)$$

which expresses the transformation property of the loop derivative under finite deformations of its path dependence. We will see at the end of this chapter that this expression is the reflection in the language of loops of the gauge covariance of the field tensor in a gauge theory.

• **Bianchi identities.** There is a second set of relations that again can be directly obtained from the geometric properties of the group of loops. One can readily see that they are a reflection of the usual Bianchi identities of Yang–Mills theories. In order to describe them we need to introduce a new differential operator, which we will call the end point derivative or Mandelstam covariant derivative [8], that acts on functions of open paths.

Given a function of an open path $\Psi(\pi_o^x)$, a local coordinate chart at the point x and a vector in that chart u^a, we define the Mandelstam derivative by considering the change in the function when the path is extended from x to $x + \epsilon u$ by the infinitesimal path δu shown in figure 1.5 as

$$\Psi(\pi_o^x \circ \delta u) = (1 + \epsilon u^a D_a)\Psi(\pi_o^x). \qquad (1.34)$$

We denote the new path as $\pi_o^{x+\epsilon u}$. If one performs a coordinate transformation, noting that ϵu^a is a vector and applying the quotient law, it is immediate to see that D_a transforms as a one-form.

Having introduced this operator, we are now ready to derive the Bianchi identity. As usual, the fundamental idea is that "the boundary of a boundary vanishes". In the group of loops language, this can be expressed by considering a thin loop ι. A representative curve of this loop has the shape of a box with sides δu, δv and δw, connected to the origin by the path π_o^x as shown in figure 1.6.

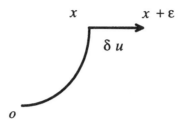

Fig. 1.5.　The extended path defining the Mandelstam derivative

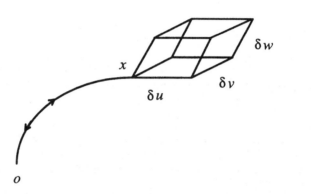

Fig. 1.6.　The loop used to derive the Bianchi identity for the loop derivative.

The curve δu represents paths that go from a generic point x to $x + \epsilon_1 u$ and similarly for δv and δw with increments $\epsilon_2 v$ and $\epsilon_3 w$ respectively. Explicitly,

$$
\begin{aligned}
\iota = {}& \pi_o^x \circ \delta u \delta v \delta w \delta \bar{v} \delta \bar{w} \delta \bar{u} \circ \pi_x^o \circ \pi_o^x \circ \delta u \delta w \delta \bar{u} \delta \bar{w} \circ \pi_x^o \\
& \circ \pi_o^x \circ \delta w \delta u \delta v \delta \bar{u} \delta \bar{v} \delta \bar{w} \circ \pi_x^o \circ \pi_o^x \circ \delta w \delta v \delta \bar{w} \delta \bar{v} \circ \pi_x^o \\
& \circ \pi_o^x \circ \delta v \delta w \delta u \delta \bar{w} \delta \bar{u} \delta \bar{v} \circ \pi_x^o \circ \pi_o^x \circ \delta v \delta u \delta \bar{v} \delta \bar{u} \circ \pi_x^o.
\end{aligned} \tag{1.35}
$$

Now, since ι is a tree and therefore indistinguishable from the identity loop, we have

$$
\Psi(\gamma) = \Psi(\iota \circ \gamma) \quad \forall \gamma. \tag{1.36}
$$

Noting that the tree is built by six infinitesimal loops (the "faces" of the box shown in figure 1.6), each connected to the origin via the path π and the "sides" of the box, we can rewrite this identity in terms of loop derivatives as

$$\Psi(\gamma) = (1 + \epsilon_2 \epsilon_3 v^a w^b \Delta_{ab}(\pi_o^{x+\epsilon_1 u}))(1 + \epsilon_1 \epsilon_3 u^c w^d \Delta_{cd}(\pi_o^x))$$
$$\times (1 + \epsilon_1 \epsilon_2 u^e v^f \Delta_{ef}(\pi_o^{x+\epsilon_3 w}))(1 + \epsilon_3 \epsilon_2 w^g v^h \Delta_{gh}(\pi_o^x))$$
$$\times (1 + \epsilon_3 \epsilon_1 w^i u^j \Delta_{ij}(\pi_o^{x+\epsilon_2 v}))(1 + \epsilon_2 \epsilon_1 v^k u^l \Delta_{kl}(\pi_o^x))\Psi(\gamma). \quad (1.37)$$

Collecting the terms of first order in each ϵ_i, applying the definition of the Mandelstam derivative and noting that u, v, w are arbitrary we get the final form of the Bianchi identities for the loop derivatives,

$$D_a \Delta_{bc}(\pi_o^x) + D_b \Delta_{ca}(\pi_o^x) + D_c \Delta_{ab}(\pi_o^x) = 0. \quad (1.38)$$

We will soon see applications of the above derived identities and their crucial role in the formulation of gauge theories in terms of loops.

• **The Ricci identity.** Consider the action of four Mandelstam covariant derivatives along the vectors u, v on a function of an open path $\Psi(\pi_o^x)$. Keeping the terms of first order in $\epsilon_1 \epsilon_2$ in the left-hand side of the next expression we get

$$(1 + \epsilon_1 u^a D_a)(1 + \epsilon_2 v^b D_b)(1 - \epsilon_1 u^c D_c)(1 - \epsilon_2 v^d D_d)\Psi(\pi_o^x) =$$
$$(1 + \epsilon_1 \epsilon_2 u^a v^b [D_a, D_b])\Psi(\pi_o^x). \quad (1.39)$$

The action of the four covariant derivatives is equivalent to appending an infinitesimal loop at the end of the path π_o^x and therefore can be written in terms of the loop derivative,

$$[D_a, D_b]\Psi(\pi_o^x) = \Delta_{ab}(\pi_o^x)\Psi(\pi_o^x). \quad (1.40)$$

This expression is the loop analogue of the usual expression of the commutator of covariant derivatives in terms of the curvature and we will again see its implications for gauge theories at the end of this chapter.

• **The loop derivative as a generator of the group of loops.** Let us now show that we can, by superposition of loop derivatives, generate any finite loop homotopic to the identity. We need to introduce a parametrization for this proof. Let $\gamma(s)$ be a parametrized curve belonging to the equivalence class defining the finite loop γ with $s \in [0, 1]$. Consider a one-parameter family of parametrized loops $\eta(s, t)$ interpolating smoothly between $\gamma(s)$ and the identity loop, such that $\eta(s, 0)$ is in the equivalence class of the identity loop and $\eta(s, 1) = \gamma(s)$. Consider the curves $\eta(1, s)$ $(= \gamma(s))$ and $\eta(1 - \epsilon, s)$. The two curves are drawn in figure 1.7 and differ by an infinitesimal element of area. The whole purpose of our proof will be to cover the infinitesimal area separating the two mentioned curves with a "checkerboard" of infinitesimal closed curves

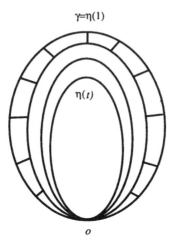

Fig. 1.7.   The construction of a finite loop from the loop derivative. The curves $\delta\eta_i$ are determined by two elements in the family $\eta(t)$.

such that along each of them one can define a loop derivative. One can therefore express the curve $\gamma(s)$ as

$$\gamma(s) = \lim_{n\to\infty} \eta(s, 1 - \epsilon) \circ \delta\eta_1 \circ \cdots \circ \delta\eta_n, \qquad (1.41)$$

where the $\delta\eta_i$ are shown in figure 1.7. Analytically, in terms of differential operators on functions of loops we can write[||]

$$
\begin{aligned}
\Psi(\eta(1)) = {}& \Psi(\eta(1 - \epsilon)) \\
& + \epsilon \oint_0^1 ds\dot\eta^a(1 - \epsilon, s)\eta'^b(1 - \epsilon, s)\Delta_{ab}(\eta(1 - \epsilon)_o^s)\Psi(\eta(1 - \epsilon)),
\end{aligned}
$$

$$(1.42)$$

where $\dot\eta(t, s) \equiv d\,\eta(t, s)/ds$ and $\eta'(t, s) \equiv d\,\eta(t, s)/dt$ It is immediate to proceed from $\eta(1 - \epsilon, s)$ inwards just by repeating the same construction, and so continuing until the final curve is the identity. The end result is

$$\Psi(\eta(1)) = \mathrm{T}\exp\left(\int_0^1 dt \oint_0^1 ds\dot\eta^a(t, s)\eta'^b(t, s)\Delta_{ab}(\eta(t)_o^s)\Psi(\eta(0))\right) \quad (1.43)$$

where the outer integral is ordered in t (T-ordered). This result is the loop version of the non-Abelian Stokes theorem of gauge theories [18] and it shows that the loop derivative is a generator of loop space, i.e., it allows us to generate any finite loop homotopic to the identity.

Notice that the expression for the finite element of the group involves a superposition of an infinite number of generators associated with different

---

[||] We drop the s dependence of η where it is not relevant.

paths. This is not the usual situation that one encounters in Lie groups, where it is enough to exponentiate one generator to obtain any element of the group. This is another indication of the non-Lie character of the group of loops. It is a direct consequence of the impossibility of defining a non-integer number of powers of $\delta\eta$.

Finally, identifying the loop derivative as a generator of the group of loops allows us to rewrite in a revealing form the commutation relations of the loop derivatives. Applying the definition of the loop derivative, equation (1.26) can be cast in the following way

$$[\Delta_{ab}(\pi_o^x), \Delta_{cd}(\chi_o^y)] = \lim_{\epsilon_i \to 0} \frac{1}{\sigma^{ab}}(\Delta_{cd}(\delta\eta_1 \circ \chi_o^y) - \Delta_{cd}(\chi_o^y)), \qquad (1.44)$$

which is the usual expression of the commutator in terms of a linear combination of elements of the algebra. So we see that the group of loops formally obeys commutation relations similar to those of a Lie group.

1.3.3   Connection derivative

In section 1.3.1 we introduced the loop derivative. We saw in the previous section that this operator has several properties resembling those of the curvature or field tensor of a gauge theory. We will now introduce a differential operator with properties similar to those of the connection or vector potential of a gauge theory [14, 15, 4]. This operator appears naturally as an intermediate step in the construction of gauge theories from the group of loops. Although one could formulate a gauge theory completely in terms of the path dependent loop derivative alone, the treatment that we will follow will lead us to a more familiar formulation of gauge theories.

Let us consider a covering of the manifold with overlapping coordinate patches. We attach to each coordinate patch $\mathcal{P}^i$ a path $\pi_o^{y_0^i}$ going from the origin of the loop to a point y_0^i in $\mathcal{P}^i$. We also introduce a continuous function with support on the points of the chart $\mathcal{P}^i$ such that it associates to each point x on the patch a path $\pi_{y_0^i}^x$. Given a vector u at x, the connection derivative of a continuous function of a loop $\Psi(\gamma)$ will be obtained by considering the deformation of the loop given by the path $\pi_o^{y_0^i} \circ \pi_{y_0^i}^x \circ \delta u \circ \pi_{x+\epsilon u}^{y_0^i} \circ \pi_{y_0^i}^o$ shown in figure 1.8. The path δu goes from x to $x + \epsilon u$. We will say that the connection derivative δ_a exists and is well defined if the loop dependent function of the deformed loop admits an expansion in terms of ϵu^a given by

$$\Psi(\pi_o^x \circ \delta u \circ \pi_{x+\epsilon u}^o \circ \gamma) = (1 + \epsilon u^a \delta_a(x))\Psi(\gamma), \qquad (1.45)$$

where we have written π_o^x to denote the path $\pi_o^{y_0^i} \circ \pi_{y_0^i}^x$ and similarly for its inverse.

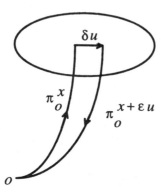

Fig. 1.8.   The path that defines the connection derivative

The definition of the connection derivative can be immediately extended to act on functions of open paths, in a similar way to that used for the loop derivative. We will take some notational latitude to give the same name to the connection derivative acting on functions of paths or loops. In all cases the context will uniquely determine to which derivative we are referring.

Notice that the deformation introduced in order to define the connection derivative could have been generated by application of successive loop derivatives, as seen in the non-Abelian Stokes theorem. This implies that any function that is loop differentiable should be connection differentiable, and that there is a natural relation between the two derivatives. We will now prove the converse relation between the connection and loop derivatives. It will be quite reminiscent of the well known relation between the connection and the curvature (or vector potential and field) in a gauge theory. In order to do this, let us start by considering the following identity in loop space,

$$\delta\gamma \equiv \pi_o^x \circ \delta u \delta v \delta \bar{u} \delta \bar{v} \circ \pi_x^o$$
$$= \pi_o^x \circ \delta u \circ \pi_{x+\epsilon u}^o \circ \pi_o^{x+\epsilon u} \circ \delta v$$
$$\circ \pi_{x+\epsilon_1 u+\epsilon_2 v}^o \circ \pi_o^{x+\epsilon 1 u+\epsilon_2 v} \circ \delta\bar{u} \circ \pi_{x+\epsilon_2 v}^o \pi_o^{x+\epsilon_2 v} \circ \delta\bar{v} \circ \pi_x^o, \quad (1.46)$$

corresponding to the path shown in figure 1.9. Notice that the first definition is written in such a way that it has the structure of the paths we have used to define loop derivatives, whereas the second has the form of the paths used to define connection derivatives. Specifically, it implies the following identity between differential operators,

$$(1 + \epsilon_1\epsilon_2 u^a v^b \Delta_{ab}(\pi_o^x))\Psi(\gamma) = (1 + \epsilon_1 u^a \delta_a(x))(1 + \epsilon_2 v^b \delta_b(x + \epsilon_1 u))$$
$$\times (1 - \epsilon_1 u^c \delta_c(x + \epsilon_1 u + \epsilon_2 v))(1 - \epsilon_2 v^d \delta_d(x + \epsilon_2 v))\Psi(\gamma). \quad (1.47)$$

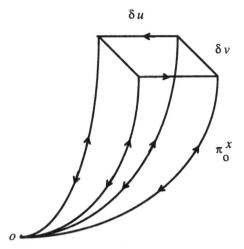

Fig. 1.9. The path that defines the relation between connection derivatives and loop derivatives.

We now expand to first order in $\epsilon_1 \epsilon_2$ and get

$$\Delta_{ab}(\pi_o^x) = \partial_a \delta_b(x) - \partial_b \delta_a(x) + [\delta_a(x), \delta_b(x)]. \tag{1.48}$$

Notice that we have obtained the loop derivative for the path π_o^x given by $\pi_o^{y_0^i} \circ \pi_{y_0^i}^x$. This path is uniquely prescribed by the function defining the connection derivative. The loop derivative defined by (1.48) automatically satisfies the Bianchi identities due to the fact that it is obtained by a construction in loop space that is totally similar to that used to derive the identities themselves.

The idea of introducing the connection derivative was to provide us in the language of loops with a notion of connection or vector potential similar to that of gauge theories. However, the connection in a gauge theory is gauge dependent. How does this dependence manifest itself in the language of loops? It does so through the choice of prescription of path used to compute the connection derivative. We will now study this in some detail.

Let us consider a connection derivative at a point x and two prescriptions for choosing the path from the origin to x, given by the continuous functions $\pi_o^x = f(x)$ and $\chi_o^x = g(x)$, as shown in figure 1.10. We again consider two equivalent paths in the group of loops,

$$\chi_o^x \circ \delta u \circ \chi_{x+\epsilon u}^o = \chi_o^x \circ \pi_x^o \circ \pi_o^x \circ \delta u \circ \pi_{x+\epsilon u}^o \circ \pi_o^{x+\epsilon u} \circ \chi_{x+\epsilon u}^o. \tag{1.49}$$

We also introduce a point dependent operator $U(x)$ constructed from the loop dependent deformation operator $U(\gamma)$ defined in (1.28) and the loop

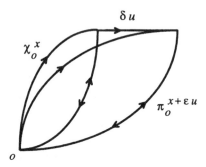

Fig. 1.10.   The path dependence of the connection derivative.

associated with the point x, $\chi_o^x \circ \pi_x^o$, by $U(x) \equiv U(\chi_o^x \circ \pi_x^o)$. This gives
the identity between operators:

$$(1 + \epsilon u^a \delta_a^{(\chi)}(x))\Psi(\gamma) = U(x)(1 + \epsilon u^a \delta_a^{(\pi)}(x))U^{-1}(x + \epsilon u)\Psi(\gamma). \quad (1.50)$$

From which we can immediately compute the change in the connection
derivative due to a change in the prescription of the path,

$$\delta_a^{(\chi)}(x) = U(x)\delta_a^{(\pi)}(x)U(x)^{-1} + U(x)\partial_a U(x)^{-1}, \quad (1.51)$$

and we see that it is totally analogous to the transformation law for a
gauge connection under changes of gauge.

The usual relation between connections and holonomies in a local chart
in a gauge theory can also be written in this language. This relation is just
an expression of the fact that the infinitesimal generators associated with
connections allow us to construct finite loops. As shown in figure 1.11
one uses infinitesimal increments generated by the connection derivative
to build a loop. The expression of this fact in loop space is

$$\gamma = \lim_{n \to \infty} \delta\gamma_1 \circ \delta\gamma_2 \circ \cdots \circ \delta\gamma_n, \quad (1.52)$$

where $\delta\gamma_i = \pi_o^{x_i} \delta u_i \pi_{x_i + \epsilon u_i}^0$, and this relation connects the deformation
operator with the connection derivative,

$$U(\gamma) = \lim_{k \to \infty} (1 + (x_2 - x_1)^a \delta_a(x_1))$$
$$\times (1 + (x_3 - x_2)^a \delta_a(x_2)) \cdots (1 + (x_1 - x_k)^a \delta_a(x)). \quad (1.53)$$

We can rewrite the above expression as the following path ordered
exponential,

$$U(\gamma) = \mathrm{P} \exp \left(\int_\gamma dy^a \delta_a(y) \right). \quad (1.54)$$

This again is reminiscent of the familiar expression for gauge theories,
which yields the holonomy in terms of the path ordered exponential of a

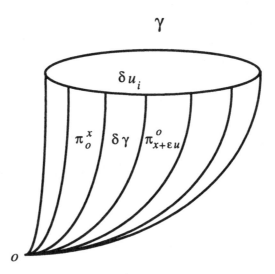

Fig. 1.11. How to generate a finite loop using the infinitesimal generators.

connection.

1.3.4 Contact and functional derivatives

The differential operators that we have introduced up to now are characteristic of the group structure of the group of loops and find no direct analogues in the space of parametrized curves. It is worthwhile analyzing whether there is any relation between these operators and the usual functional derivative $\delta/\delta p(s)$ acting on functionals of parametrized curves $\Psi[p(s)]$. The operator in loop space that allows us to make this connection is the contact derivative $\mathcal{C}_a(x)$, which plays an important role in diffeomorphism invariant theories, as we will see later.

We define the contact derivative of a function of loops in terms of the expression

$$\mathcal{C}_a(x)\Psi(\gamma) = \oint_\gamma dy^b \delta(x-y)\Delta_{ab}(\gamma_o^y)\Psi(\gamma), \qquad (1.55)$$

where γ_o^y is the portion of the loop γ going from the basepoint to y. This operator was first introduced in the chiral formulation of Yang–Mills theories in the loop representation by Gambini and Trias [13]. It can be considered as the projection of the loop derivative on the tangent to the loop γ,

$$X^a(x,\gamma) = \oint_\gamma dy^a \delta(x-y). \qquad (1.56)$$

This expression involving the tangent to the loop will play a role in the ideas of chapter 2.

An important property of the contact derivative is that it is the generator of diffeomorphisms on functions of loops. Given the infinitesimal diffeomorphism

$$x \longrightarrow x'^a = x^a + \epsilon u^a, \tag{1.57}$$

the expression for $\Psi(\gamma'_\epsilon)$, where γ'_ϵ is the loop obtained by "dragging along" γ with the diffeomorphism (1.57), is given by

$$\Psi(\gamma'_\epsilon) = \left(1 + \epsilon \int d^3x\, u^a(x) \mathcal{C}_a(x)\right) \Psi(\gamma)$$

$$= \left(1 + \epsilon \oint_\gamma dy^b u^a(y) \Delta_{ab}(\gamma^y_o)\right) \Psi(\gamma). \tag{1.58}$$

To prove that $\mathcal{C}_a$ actually is a generator of diffeomorphisms, we will show that it satisfies the corresponding algebra,

$$\left[\int d^3x\, N^a(x) \mathcal{C}_a(x), \int d^3y\, M^a(y) \mathcal{C}_a(y)\right] = \int d^3x\, \mathcal{L}_{\vec{M}} N^a(x) \mathcal{C}_a(x), \tag{1.59}$$

where N^a, M^a are arbitrary vector fields on the three-manifold.

We will start with an auxiliary calculation that will prove useful in what follows. We will evaluate the action of the loop derivative on a contact derivative. To this end we construct the following expression, which holds due to the very definition of loop derivative,

$$(1 + \tfrac{1}{2}\sigma^{cd}\Delta_{cd}(\gamma^z_o)) \int_\gamma dy^a \delta(y - x) \Delta_{ab}(\gamma^y_o) \Psi(\gamma) :=$$

$$\int_{\delta\gamma_z \circ \gamma} dy^a \delta(y - x) \Delta_{ab}((\delta\gamma_z \circ \gamma)^y_o) \Psi(\delta\gamma_z \circ \gamma). \tag{1.60}$$

In this expression, $\delta\gamma_z$ is the infinitesimal loop added to γ through a path from the origin up to the point z. That is, we evaluate the action of an infinitesimal deformation of area σ^{cd} acting on the contact derivative.

We now expand the right-hand side of (1.60), partitioning the domain of integration into the portions after and before the action of the deformation and use the definition of the loop derivative to expand $\Psi(\delta\gamma \circ \gamma)$,

$$\int_o^z dy^a \Delta_{ab}(\gamma^y_o) \delta(y - x)(1 + \tfrac{1}{2}\sigma^{cd}\Delta_{cd}(\gamma^z_o)) \Psi(\gamma)$$

$$+ \{u^a \delta(z - x) \Delta_{ab}(\gamma^z_o) + v^a \delta(z + u - x) \Delta_{ab}(\gamma^{z+u}_o)$$

$$- u^a \delta(z + u + v - x) \Delta_{ab}(\gamma^{z+u+v}_o) - v^a \delta(z + u - x) \Delta_{ab}(\gamma^{z+u+v+\bar{u}}_o)\}$$

$$\times (1 + \tfrac{1}{2}\sigma^{cd}\Delta_{cd}(\gamma^z_o)) \Psi(\gamma)$$

$$+ \int_z^1 dy^a \delta(y - x) \Delta_{ab}((\delta\gamma_z \circ \gamma)^y_o)(1 + \tfrac{1}{2}\sigma^{cd}\Delta_{cd}(\gamma^z_o)) \Psi(\gamma). \tag{1.61}$$

The last term in this expression can be rewritten as

$$\oint_\gamma dy^a \delta(x-y)\Theta(y-z)(1+\tfrac{1}{2}\sigma^{cd}\Delta_{cd}(\gamma_o^z))[\Delta_{ab}(\gamma_o^y)](1+\tfrac{1}{2}\sigma^{ef}\Delta_{ef}(\gamma_o^z))\Psi(\gamma) \tag{1.62}$$

where $\Theta(y-z)$ is a Heaviside function that orders points along the loop, i.e., it is 1 if z precedes y and zero otherwise. We will be able to combine the zeroth order contribution of this term with the first term in (1.61). It should be noticed that the first loop derivative does not act on everything to its right but only on the path inside the second loop derivative γ_o^y, a fact that as before we denote by enclosing it in brackets. We now consider the expansion of the terms containing the infinitesimally shifted loop. They can be expressed with the use of the Mandelstam derivative,

$$\Delta_{ab}(\gamma_o^{z+u}) = (1+u^c D_c)\Delta_{ab}(\gamma_o^z), \tag{1.63}$$

$$\Delta_{ab}(\gamma_o^{z+u+v}) = (1+v^d D_d)(1+u^c D_c)\Delta_{ab}(\gamma_o^z), \tag{1.64}$$

$$\delta(z+u-x) = (1+u^a \partial_a)\delta(z-x). \tag{1.65}$$

We now expand (1.61) again

$$\oint_\gamma dy^a \delta(y-x)\Delta_{ab}(\gamma_o^y)(1+\tfrac{1}{2}\sigma^{cd}\Delta_{cd}(\gamma_o^z))\Psi(\gamma)$$

$$+\{u^a\delta(z-x)\Delta_{ab}(\gamma_o^z) + v^a(1+u^e\partial_e)\delta(z-x)(1+u^d D_d)\Delta_{ab}(\gamma_o^z)$$

$$-u^a(1+v^c D_c)\delta(z-x)(1+v^d D_d)\Delta_{ab}(\gamma_o^z) - v^a\delta(z-x)\Delta_{ab}(\gamma_o^z)\}$$

$$\times(1+\tfrac{1}{2}\sigma^{cd}\Delta_{cd}(\gamma_o^z))\Psi(\gamma) + \tfrac{1}{2}\oint_\gamma dy^a\delta(y-x)\Theta(y-z)\sigma^{cd}$$

$$\times\Delta_{cd}(\gamma_o^z)[\Delta_{ab}(\gamma_o^y)](1+\tfrac{1}{2}\sigma^{cd}\Delta_{cd}(\gamma_o^z))\Psi(\gamma). \tag{1.66}$$

Of the terms in braces, it can be readily seen that only contributions proportional to $u^a v^b$ are present, neglecting terms of higher order. The other terms combine to give the original expression. We can finally read off the contribution of the loop derivative,

$$\Delta_{cd}(\gamma_z)\oint_\gamma dy^a\delta(x-y)\Delta_{ab}(\gamma_o^y)\Psi(\gamma) =$$

$$2(\partial_{[c}\delta(z-x)\Delta_{d]b}(\gamma_o^z) + \delta(z-x)D_{[c}\Delta_{d]b}(\gamma_o^z))\Psi(\gamma)$$

$$+\oint_\gamma dy^a\Theta(y-z)\delta(y-x)\Delta_{cd}(\gamma_o^z)[\Delta_{ab}(\gamma_o^y)]\Psi(\gamma)$$

$$+\oint_\gamma dy^a\delta(y-x)\Delta_{ab}(\gamma_o^y)\Delta_{cd}(\gamma_o^z)\Psi(\gamma). \tag{1.67}$$

With this calculation in hand, it is straightforward to compute the successive action of two diffeomorphisms,

$$C(\vec{N})\,C(\vec{M})\,\Psi(\gamma) = \int d^3w N^d(w) \oint_\gamma dz^c \delta(w-z) \Delta_{cd}(\gamma_o^z)$$

$$\times \int d^3x M^b(x) \oint_\gamma dy^a \delta(y-x) \Delta_{ab}(\gamma_o^y) \Psi(\gamma). \quad (1.68)$$

Expanding this expression we get six terms

$$\int d^3w \int d^3x N^d(w) M^b(x)$$

$$\times \{ \oint_\gamma dz^c \delta(w-z) \partial_c \delta(z-x) \Delta_{db}(\gamma_o^z) \Psi(\gamma)$$

$$- \oint_\gamma dz^c \delta(w-z) \partial_d \delta(z-x) \Delta_{cb}(\gamma_o^z) \Psi(\gamma)$$

$$+ \oint_\gamma dz^c \delta(w-z) \delta(z-x) D_c \Delta_{db}(\gamma_o^z) \Psi(\gamma)$$

$$- \oint_\gamma dz^c \delta(w-z) \delta(z-x) D_d \Delta_{cb}(\gamma_o^z) \Psi(\gamma)$$

$$+ \oint_\gamma dz^c \oint_\gamma dy^a \delta(w-z) \delta(y-x) \Theta(y-z) \Delta_{cd}(\gamma_o^z)[\Delta_{ab}(\gamma_o^y)]\Psi(\gamma)$$

$$+ \oint_\gamma dz^c \oint_\gamma dy^a \delta(w-z) \delta(y-x) \Delta_{ab}(\gamma_o^y) \Delta_{cd}(\gamma_o^z) \Psi(\gamma) \}. \quad (1.69)$$

We should now subtract the same terms with the replacement $\vec{N} \leftrightarrow \vec{M}$. Since the calculation is tedious but straightforward we describe in words how the terms combine. The fifth and sixth terms, when combined with the similar terms coming from the substitution $\vec{N} \leftrightarrow \vec{M}$ cancel taking into account the commutation relations for the loop derivatives (1.26). The first and third terms combined with the first of the substitution $\vec{N} \leftrightarrow \vec{M}$ form a total derivative. The fourth term, combined with the third and fourth of the substitution $\vec{N} \leftrightarrow \vec{M}$ cancel due to the Bianchi identities of the loop derivatives. Finally, the second terms combine to produce exactly $C(\mathcal{L}_{\vec{N}}\vec{M})$, which is the correct result of the calculation.

We end by pointing out the relation between the contact derivative and the usual functional derivative. This can be immediately recognized by noticing that we can write $\int d^3x u^a(x) \mathcal{C}_a(x)$ in terms of parametrized curves by

$$\int d^3x u^a(x) \mathcal{C}_a(x) = \oint_0^1 ds u^a(p(s)) \frac{\delta}{\delta p^a(s)}, \quad (1.70)$$

where $p(s)$ is one of the parametrized curves in the equivalence class of the loop γ. The way to see this is to notice that if γ is the equivalence class of curves $[p^a(s)]$, then $\Psi(\gamma') = \Psi[p^a(s) + \epsilon u^a(p(s))]$. Then,

$$\mathcal{C}_a(x) = \oint ds \delta(x - p(s)) \frac{\delta}{\delta p(s)^a}, \qquad (1.71)$$

which relates the two derivatives. It should be noticed that these two derivatives only agree when acting on functions of loops. The expression on the right of (1.70) can act on functions of parametrized curves, on which the contact derivative is not defined.

1.4 Representations of the group of loops

In previous sections we derived several relations between generators of the group of loops. These relations were independent of any particular representation. We will now study their form in the context of a particular representation in terms of a given gauge group. We will see that from them emerges the kinematical structure of gauge theories.

Gauge theories arise as representations (homomorphisms $\mathcal{H}$) of the group of loops onto some gauge group G,

$$\mathcal{H} : \mathcal{L}_0 \to G, \qquad (1.72)$$

i.e.,

$$\gamma \longrightarrow H(\gamma), \qquad (1.73)$$

such that $H(\gamma_1)H(\gamma_2) = H(\gamma_1 \circ \gamma_2)$.

Let us assume we are considering a specific Lie group, for instance $SU(N)$, with $N^2 - 1$ generators X^i such that $\mathrm{Tr} X^i = 0$ and

$$[X^i, X^j] = C_k^{ij} X^k, \qquad (1.74)$$

where C_k^{ij} are the structure constants of the group in question. We will assume that the representation is loop differentiable[**]. This will enable us to obtain the usual local objects associated with the gauge theory (curvature and connection) from the loop language.

Let us compute the action of the connection derivative in this representation. We use the same prescriptions as in the previous section

$$(1 + \epsilon u^a \delta_a(x)) H(\gamma) = H(\pi_o^x \circ \delta u \circ \pi_{x+\epsilon u}^o \circ \gamma) = H(\pi_o^x \circ \delta u \circ \pi_{x+\epsilon u}^o) H(\gamma). \qquad (1.75)$$

Since the loop $\pi_o^x \circ \delta u \circ \pi_{x+\epsilon u}^o$ is close to the identity loop (with the topology of loop space) and since H is a continuous, differentiable representation,

$$H(\pi_o^x \circ \delta u \circ \pi_{x+\epsilon u}^o) = 1 + i\epsilon u^a A_a(x), \qquad (1.76)$$

[**] If it is not loop differentiable, instead of dealing with holonomies we will have "generalized" holonomies, which are not derived from a smooth connection.

where $A_a(x)$ is an element of the algebra of the group, in our example of $SU(N)$. That is, $A_a(x) = A_a^i X^i$. Therefore, we see that through the action of the connection derivative,

$$\delta_a(x)H(\gamma) = iA_a(x)H(\gamma). \qquad (1.77)$$

Following similar steps one obtains the action of the loop derivative,

$$\Delta_{ab}(\pi_o^x)H(\gamma) = iF_{ab}(x)H(\gamma), \qquad (1.78)$$

where F_{ab} is an algebra-valued antisymmetric tensor field.

Remember that π_o^x is only fixed by the prescription. Changing the prescription for π_o^x is the way to change the gauge. Suppose we change the prescription by $\pi_o^x \to \pi_o'^x = \pi_o'^x \circ \pi_x^o \circ \pi_o^x$. We then simply use equation (1.33) that in terms of the field reads

$$F'_{ab}(x) = H(x)F_{ab}(x)H(x)^{-1}, \qquad (1.79)$$

where $H(x)$ is a shorthand for $H(\pi_o'^x \circ \pi_x^o)$.

From equation (1.48) we immediately get the usual relation defining the curvature in terms of the potential,

$$F_{ab}(x) = \partial_a A_b(x) - \partial_b A_a(x) + i[A_a, A_b]. \qquad (1.80)$$

Gauge transformations in terms of the connection are immediate from equation (1.51),

$$A_a(x)' = H(x)A_a(x)H(x)^{-1} - iH(x)\partial_a H(x)^{-1}. \qquad (1.81)$$

Let us act with the deformation operator $U(\eta)$ introduced in section 1.3.3 on the representation $H(\gamma)$,

$$U(\eta)H(\gamma) = H(\eta \circ \gamma) = H(\eta)H(\gamma). \qquad (1.82)$$

Now, applying formula (1.54),

$$U(\eta)H(\gamma) = \mathrm{P}\exp\left(\oint_\eta dy^a \delta_a(y)\right) H(\gamma), \qquad (1.83)$$

and substituting equation (1.77) and comparing terms we get

$$H(\eta) = \mathrm{P}\exp\left(i \oint_\eta dy^a A_a(y)\right), \qquad (1.84)$$

which shows that the representation we are considering is given in terms of the usual expression for the holonomy of the connection A_a.

Up to now, open paths have not played any relevant physical role. We will now show that open paths are naturally related to the inclusion of material fields coupled to gauge theories. We will consider matter fields that transform under the fundamental representation of the gauge group considered in our example, $SU(N)$.

We will describe the matter field at the point x through a path dependent object $\Psi(\pi_o^x)$. The natural extension of the representation introduced at the beginning of this section to the case of open paths is to consider the composition of an open path and a loop, defined by

$$\Psi(\gamma \circ \pi_o^x) \equiv H(\gamma)\Psi(\pi_o^x). \qquad (1.85)$$

As in other cases, the role of the path choice will be to fix a gauge choice. A local description in a fixed gauge is obtained by fixing a family of paths, each of which are associated with each point in the manifold. The functions Ψ will now become functions of points labeled by the fixed prescription π used to determine the paths $\Psi^{(\pi)}(x)$. The prescription is given through a continuous function from the points of the manifold into the paths $f(x) = \pi_o^x$. Notice that if we change the prescription for the path $\pi_o^x \to \pi'^x_o = \pi'^x_o \circ \pi_x^o \circ \pi_o^x$ we get

$$\Psi^{(\pi')}(x) = H(\pi'^x_o \circ \pi_x^o)\Psi^{(\pi)}(x). \qquad (1.86)$$

The Mandelstam derivative D_a behaves as the usual covariant derivative of a gauge theory. Consider its action on a function of an open path,

$$(1 + \epsilon u^a D_a)\Psi(\pi_o^x) = \Psi(\pi_o^x \circ \delta u) = \Psi(\pi_o^x \circ \delta u \circ \pi_{x+\epsilon u}^o \circ \pi_o^{x+\epsilon u})$$
$$= (1 + \epsilon u^a \delta_a(x))(1 + \epsilon u^b \partial_b)\Psi(\pi_o^x), \qquad (1.87)$$

expanding in ϵ and keeping terms of first order, we get

$$D_a\Psi(\pi_o^x) = \partial_a\Psi^{(\pi)}(x) + \delta_a(x)\Psi(\pi_o^x). \qquad (1.88)$$

Using the relation between the function of a deformed path and the holonomy (1.85), we get

$$D_a\Psi(\pi_o^x) = \partial_a\Psi^{(\pi)}(x) + iA_a(x)\Psi^{(\pi)}(x). \qquad (1.89)$$

The usual form of the Ricci identity,

$$[D_a, D_a] = iF_{ab}, \qquad (1.90)$$

can be obtained directly from the above expression or by considering the representation of equation (1.40).

1.5 Conclusions

We have seen how gauge theories arise as representations of the group of loops. All the usual kinematical concepts of gauge theories are reflections of properties of the group of loops.

It is important to realize that the identities and properties that we proved in this chapter for the loop and connection derivative *do not* depend on any choice of gauge group to represent the group of loops. In this sense one can think of the corresponding generators of the group as associated with an "abstract" curvature and connection. It is only when one considers a particular representation of the group of loops in terms of a gauge group that these quantities adopt the usual meaning of connections and curvatures in gauge theories.

In the next chapter we will introduce more techniques that will put us in a better position to deal with loops in the context of quantum theories in the loop representation.

2

Loop coordinates and the extended group of loops

2.1 Introduction

Continuing with the idea of describing gauge theories in terms of loops, we will now introduce a set of techniques that will aid us in the description of loops themselves. The idea is to represent loops with a set of objects that are more amenable to the development of analytical techniques. The advantages of this are many: whereas there is limited experience in dealing with functions of loops, there is a significant machinery to deal with analytic functions. They even present advantages for treatment with computer algebra.

Surprisingly, we will see that the end result goes quite beyond our expectations. The quantities we originally introduced to describe loops immediately reveal themselves as having great potential to replace loops altogether from the formulation and go beyond, allowing the development of a reformulation of gauge theories that is entirely new. This formulation introduces new perspectives with respect to the loop formulation that have not been fully developed yet, though we will see in later chapters some applications to gauge theories and gravitation.

The plan for the chapter is as follows: in section 2.2 we will start by introducing a set of tensorial objects that embody all the information that is needed from a loop to construct the holonomy and therefore to reconstruct any quantity of physical relevance for a gauge theory. In section 2.3 we will show how the group of loops is a subgroup of a Lie group with an associated Lie algebra, the extended loop group. The generators of this Lie group will turn out to be coordinates in the extended loop space, which we discuss in section 2.4. In section 2.5 we will study how the differential operators introduced in the previous chapter act on the loop coordinates. In particular we will study the action of the generator of diffeomorphisms. In section 2.6 we will discuss how to construct

29

diffeomorphism invariant quantities in terms of loop coordinates and, in particular, knot invariants. In the conclusion we will discuss the differences and similarities between the group structures we have introduced and the usual Lie groups. The subject of this chapter has been discussed in detail in reference [20], the reader is referred to it for a more technical approach.

2.2   Multitangent fields as description of loops

As we discussed in the previous chapter, all the gauge invariant information present in a gauge field can be retrieved from the holonomy. Therefore the only information we really need to know from loops is that used in the definition of the holonomy,

$$H_A(\gamma) = P \exp\left(i \oint_\gamma A_a \, dy^a \right). \tag{2.1}$$

We can write this definition more explicitly as

$$H_A(\gamma) = 1 + \sum_{n=1}^\infty i^n \int dx_1^3 \dots dx_n^3 A_{a_1}(x_1) \dots A_{a_n}(x_n) X^{a_1 \dots a_n}(x_1, \dots, x_n, \gamma), \tag{2.2}$$

where the loop dependent objects X are given by

$$X^{a_1 \dots a_n}(x_1, \dots, x_n, \gamma) =$$
$$\oint_\gamma dy_n^{a_n} \int_o^{y_n} dy_{n-1}^{a_{n-1}} \dots \int_o^{y_2} dy_1^{a_1} \delta(x_n - y_n) \dots \delta(x_1 - y_1) =$$
$$\oint_\gamma dy_n^{a_n} \dots \oint_\gamma dy_1^{a_1} \delta(x_n - y_n) \dots \delta(x_1 - y_1) \Theta_\gamma(o, y_1, \dots, y_n) \tag{2.3}$$

and $\Theta_\gamma(o, y_1, \dots, y_n)$ is a generalized Heaviside function that orders the points along the contour starting at the origin of the loop, i.e.,

$$\Theta_\gamma(o, y_1, \dots, y_n) = \begin{cases} 1 \text{ if } o < y_1 < y_2 < \dots y_n \text{ along the loop} \\ \quad\quad\quad 0 \text{ otherwise.} \end{cases} \tag{2.4}$$

These relations define the X objects of "rank" n. We shall call them the multitangents of the loop γ.

By writing the holonomy in the non-standard form (2.2) we have been able to isolate all the loop dependent information in the multitangents of the loop. No more information from the loop is needed in order to compute the holonomy than that present in the multitangents of all orders.

In what follows, it will be convenient to introduce the notation

$$X^{\mu_1 \dots \mu_n}(\gamma) \equiv X^{a_1 \, x_1 \dots a_n \, x_n}(\gamma) \equiv X^{a_1 \dots a_n}(x_1, \dots, x_n, \gamma), \tag{2.5}$$

with $\mu_i \equiv (a_i x_i)$, which is more suggestive of the role played by the x variables under diffeomorphisms. The X objects transform as multivector densities (they behave as a vector density at the point x_i on the index a_i) under the subgroup of coordinate transformations that leaves the base point o fixed. In other words if

$$x^a \longrightarrow x'^a = D^a(x) \tag{2.6}$$

then

$$X^{a_1 x'_1 \ldots a_n x'_n}(D\gamma) = \frac{\partial x'^{a_1}_1}{\partial x^{b_1}_1} \cdots \frac{\partial x'^{a_n}_n}{\partial x^{b_n}_n} \frac{1}{J(x_1)} \cdots \frac{1}{J(x_n)} X^{b_1 x_1 \ldots b_n x_n}(\gamma), \tag{2.7}$$

where J is the Jacobian of the transformation.

The Xs are not really "coordinates" in the sense that they are not independent. They are constrained by algebraic and differential relations.

The algebraic constraints stem from relations satisfied by the generalized Heaviside function,

$$\Theta_\gamma(o, y_1, y_2, y_3) + \Theta_\gamma(o, y_2, y_1, y_3) + \Theta_\gamma(o, y_2, y_3, y_1) = \Theta_\gamma(o, y_2, y_3),$$
$$\Theta_\gamma(o, y_1) = 1, \qquad \Theta_\gamma(o, y_1, y_2) + \Theta_\gamma(o, y_2, y_1) = 1, \tag{2.8}$$

which imply the following kind of relations among the Xs,

$$X^{\mu_1 \mu_2} + X^{\mu_2 \mu_1} = X^{\mu_1} X^{\mu_2},$$
$$X^{\mu_1 \mu_2 \mu_3} + X^{\mu_2 \mu_1 \mu_3} + X^{\mu_2 \mu_3 \mu_1} = X^{\mu_1} X^{\mu_2 \mu_3}. \tag{2.9}$$

And in general,

$$X^{\underline{\mu_1 \ldots \mu_k} \mu_{k+1} \ldots \mu_n} \equiv \sum_{P_k} X^{P_k(\mu_1 \mu_n)} = X^{\mu_1 \ldots \mu_k} X^{\mu_{k+1} \ldots \mu_n}, \tag{2.10}$$

where the sum goes over all the permutations of the μ variables which preserve the ordering of the $\mu_1, \ldots, \mu_k$ and the $\mu_{k+1}, \ldots, \mu_n$ among themselves. We have introduced the notation of underlined indices to symbolize the permutation for future use.

The differential constraint ensures that the holonomy has the correct transformation properties under gauge transformations, and can be readily derived from equation (2.2). It is given by

$$\frac{\partial}{\partial x_i^{a_i}} X^{a_1 x_1 \ldots a_i x_i \ldots a_n x_n} =$$
$$\left(\delta(x_i - x_{i-1}) - \delta(x_i - x_{i+1}) \right) X^{a_1 x_1 \ldots a_{i-1} x_{i-1} a_{i+1} x_{i+1} \ldots a_n x_n}. \tag{2.11}$$

In this expression, both x_0 and x_{n+1} represent the base point of the loop.

An important property of the differential constraint is that *any* multitensor density $D^{a_1 x_1 \ldots a_n x_n}$ that satisfies it can be put into equation (2.2) and the resulting object is a *gauge covariant* quantity. When restricted to the multitangents of a loop, the resulting object is the holonomy. It is this property that exhibits the relevance of this formulation. In it, loops are only a particular case. One can, in general, deal with arbitrary multitensor densities and construct gauge invariant objects, for instance by taking the trace. The multitensor densities need not have the same distributional character as the multitangents associated with a loop. Their divergence structure is dictated by the differential constraint, which requires its solutions to be distributional. This will have important consequences later. We will call the space of all multitensors that satisfy the differential constraints $\mathcal{D}_o$.

With this construction in hand, one could go further and forget loops and holonomies altogether. Since one can represent any gauge covariant object using the Ds, one could represent a gauge theory *entirely* in terms of Ds. This has not been done up to present for non-Abelian theories in a complete fashion (nor for gravity), but it can be easily worked out for an Abelian theory, as we will do in chapter 4.

When one allows arbitrary multitensors in (2.2) the convergence of the series is not guaranteed. There is no easy way to prescribe multitensors such that the series converges, so we will assume from now on that we work only with multitensors such that the series converges. Even this requirement is not enough to produce an object with a gauge invariant trace. The differential constraint (2.11) only ensures that if one performs a gauge transformation on the trace of the holonomy of a multitensor the resulting series has terms that cancel in pairs. For this to imply gauge invariance, it has to happen that [222]

$$\sum_{k=1}^{N} A_{\mu_1} \ldots A_{\mu_k} [A, \Lambda]_{\mu_k} A_{\mu_{k+1}} \ldots A_{\mu_n} X^{\mu_1 \cdots \mu_n} \qquad (2.12)$$

goes to zero as $N \to \infty$. Λ is the parameter of the gauge transformation and is therefore an arbitrary function. Notice that the vanishing of (2.12) is not guaranteed by the convergence of the holonomy alone. The question of selecting an appropriate set of multitensors in a precise way in order to ensure convergence of these expressions is at present not settled, see reference [21].

2.3   The extended group of loops

When we introduced the group of loops in the previous chapter, we noticed that no one-parameter subgroup existed (since one could only define integer powers of the generators) and therefore it did not form a Lie group. In this section we will introduce a Lie group, the "extended loop group". The group of loops will be a subgroup of it. This construction is of interest in itself, since it is clear that it is a great advantage to have at our disposal all the machinery of Lie groups to analyze loops. Among other results, by identifying the free parameters of the algebra associated with the extended loop group we will be able to solve automatically the homogeneous part of the differential and algebraic constraints (2.10), (2.11) of section 2.2. With some additional construction, we will have a definition for the portion of the multitensor density fields that is unconstrained, i.e., *that we can freely specify*. They can therefore genuinely be called "coordinates" and contain as a subspace the "loop coordinates" or coordinates on loop space. We will elaborate more on this concept in section 2.4. Now we will proceed to construct the extended loop group.

2.3.1 The special extended group of loops

Let us start by considering arbitrary* multitensor densities similar to those introduced in section 2.2 and define a quantity $\mathbf{E}$ by

$$\mathbf{E} = (E, E^{\mu_1}, \ldots, E^{\mu_1 \cdots \mu_n}, \ldots) \equiv (E, \vec{E}), \qquad (2.13)$$

where E is a real number and $E^{\mu_1, \ldots, \mu_n}$ (for any $n \neq 0$) is an arbitrary multivector density field. It can be readily checked that the set of these quantities has the structure of a vector space (denoted as $\mathcal{E}$) with the usual composition laws of addition and multiplication.

We will now introduce a product law in $\mathcal{E}$ in the following way: given two vectors $\mathbf{E}_1$ and $\mathbf{E}_2$, we define $\mathbf{E}_1 \times \mathbf{E}_2$ as the vector with components

$$\mathbf{E}_1 \times \mathbf{E}_2 = (E_1 E_2, E_1 \vec{E}_2 + \vec{E}_1 E_2 + \vec{E}_1 \times \vec{E}_2), \qquad (2.14)$$

where $\vec{E}_1 \times \vec{E}_2$ is given by

$$(\vec{E}_1 \times \vec{E}_2)^{\mu_1 \cdots \mu_n} = \sum_{i=1}^{n-1} E_1^{\mu_1 \cdots \mu_i} E_2^{\mu_{i+1} \cdots \mu_n} \quad . \qquad (2.15)$$

For any value of n, the rank n component of the $\times$-product of elements of $\mathcal{E}$ can be expressed as

$$(\mathbf{E}_1 \times \mathbf{E}_2)^{\mu_1 \cdots \mu_n} = \sum_{i=0}^{n} E_1^{\mu_1 \cdots \mu_i} E_2^{\mu_{i+1} \cdots \mu_n} \qquad (2.16)$$

with the convention

$$E^{\mu_1 \cdots \mu_0} = E^{\mu_{n+1} \cdots \mu_n} = E \quad . \qquad (2.17)$$

---

* In this chapter we will always discuss *real* multitensor fields. It is obvious that the formalism is unchanged if one allows complex fields. In some applications they seem to play an important role, as we will see in section 3.4.2 (see also [19]).

The product law is associative and distributive with respect to the addition of vectors. It has a null element (the null vector) and an identity element, given by

$$\mathbf{I} = (1, 0, \ldots, 0, \ldots) . \tag{2.18}$$

An inverse element exists for all vectors with non-vanishing zeroth rank component. It is given by

$$\mathbf{E}^{-1} = E^{-1}\mathbf{I} + \sum_{i=1}^{\infty}(-1)^i E^{-i-1}(\mathbf{E} - E\mathbf{I})^i, \tag{2.19}$$

such that

$$\mathbf{E} \times \mathbf{E}^{-1} = \mathbf{E}^{-1} \times \mathbf{E} = \mathbf{I} . \tag{2.20}$$

When evaluating the components of $\mathbf{E}^{-1}$ it should be noticed that the sum involved in (2.19) is actually finite due to the fact that $(\mathbf{E} - E\mathbf{I})$ is a vector with its zeroth rank component equal to zero. Therefore,

$$\left[\,\vec{E}\times \overset{i}{\ldots} \times\vec{E}\,\right]^{\mu_1\ldots\mu_n} = \left[\,(\mathbf{E} - E\mathbf{I})^i\,\right]^{\mu_1\ldots\mu_n} = 0 \text{ if } n < i . \tag{2.21}$$

The set of all vectors with non-vanishing zeroth rank component (notice the role of E^{-1} in equation (2.20)) forms a group with the $\times$-product as composition law.

The $\times$-product law has an interesting property when restricted to multitangents. In this case it just corresponds to the composition law of loops,

$$\mathbf{X}(\gamma_1) \times \mathbf{X}(\gamma_2) = \mathbf{X}(\gamma_1 \circ \gamma_2), \tag{2.22}$$

where $\mathbf{X}(\gamma) = (1, X^{\mu_1}(\gamma), \ldots, X^{\mu_1\ldots\mu_n}(\gamma), \ldots)$. Therefore we see that the product law that gave rise to the group of loops is the same product law we are generalizing to the case of arbitrary multitensor fields. The $\times$-product law can also represent more general compositions than those of two loops sharing a common basepoint, such as the composition of an open path with a loop at its end, assuming a generalization of the definition of multitangents to open paths.

After all this construction, let us now make contact with the group of loops. First, let us restrict attention to multitensors (not necessarily associated with a loop) that satisfy the constraints (2.10), (2.11). Consider the set of vectors $\mathcal{X} \in \mathcal{E}$ that have their zeroth rank component equal to one, $\mathbf{X} = (1, \vec{X})$.

The set $\mathcal{X}$ is closed under the $\times$-product law. If $\mathbf{X}_1 \in \mathcal{X}$ and $\mathbf{X}_2 \in \mathcal{X}$, it is clear from the definition of the group product that $\mathbf{X}_1 \times \mathbf{X}_2$ satisfies the differential constraint. One can also demonstrate that $\mathbf{X}_1 \times \mathbf{X}_2$ satisfies the algebraic constraint. In a similar way one can show that the inverse $\mathbf{X}^{-1}$ given by (2.19) satisfies the constraints if $\mathbf{X}$ does. A detailed proof of these properties can be seen in the appendices of reference [20]. These results show that the group structure under the ×-composition law is preserved by the imposition of the algebraic and differential constraints. We call $\mathcal{X}$ the Special-extended Loop group (SeL group)[†]. Note that the zeroth rank component of $\mathbf{E}$ plays a role analogous to the determinant in a group of matrices. For this reason we introduce the name Special when selecting $E = 1$.

The group of loops is a subgroup of the SeL group since $\mathbf{X}(\gamma) \in \mathcal{X}$ and the composition law of the group of loops ∘ is mapped via (2.22) to the ×-product.

An important question at this point is: is the group SeL just a fancy rewriting of the group of loops, or is it actually a more general structure? We will show that SeL is actually larger than the group of loops by direct construction. Consider the group element $\mathbf{X}^m \equiv \mathbf{X} \times \overset{m}{\ldots} \times \mathbf{X}$. Note that if $\mathbf{X}$ gives the multitangent field of certain loop γ, $\mathbf{X}^m$ would be the multitangent field of the loop γ swept itself m times. Applying the binomial expansion we get,

$$\mathbf{X}^m \equiv [\mathbf{I} + (\mathbf{X} - \mathbf{I})]^m = \mathbf{I} + \sum_{i=1}^{m} \binom{m}{i}(\mathbf{X} - \mathbf{I})^i \ . \tag{2.23}$$

The extension of (2.23) to real values of m is straightforward, being defined as

$$\mathbf{X}^\lambda = \mathbf{I} + \sum_{i=1}^{\infty} \binom{\lambda}{i}(\mathbf{X} - \mathbf{I})^i \tag{2.24}$$

with λ real. We usually call this object the analytic extension of $\mathbf{X}$. Note that for $\lambda = -1$ we recover the expression of the inverse of $\mathbf{X}$. Also in this case, due to (2.21) the analytic extension is well defined for all elements of $\mathcal{X}$. One can prove that if $\mathbf{X}$ is constrained by the differential and algebraic identities, its analytic extension also satisfies the constraints (again see the appendices of [20]). So, the analytic extension of any $\mathbf{X}$ is in $\mathcal{X}$. Moreover, we have

$$\mathbf{X}^\lambda \times \mathbf{X}^\mu = \mathbf{X}^{\lambda+\mu} \ . \tag{2.25}$$

We conclude that the set $\{\mathbf{X}^\lambda / \lambda \in \mathrm{R} \text{ and } \mathbf{X} \text{ a given element of } \mathcal{X}\}$ defines an Abelian one-parameter subgroup of the $\mathcal{X}$ group.

For non-integer values of λ, the λth power of a multitangent is not a multitangent. This fact explicitly shows that there exist in $\mathcal{X}$ other elements besides the loop coordinates.

[†] Tavares [43] has also considered this group. His "shuffle product" is associated with the algebraic constraint in our terminology.

Matrix representations of the SeL group can be generated through a natural extension of the holonomy. The extended holonomy associated with a non-Abelian connection $A_{ax} \equiv A_a(x)$ is defined as $H_A(\mathbf{X}) = \mathbf{A} \cdot \mathbf{X}$, where

$$\mathbf{A} \equiv (1, i A_{a_1 x_1}, \ldots, i^n A_{a_n x_n}, \ldots), \tag{2.26}$$

$$\mathbf{X} \equiv (1, X^{a_1 x_1}, \ldots, X^{a_1 x_1 \ldots a_n x_n}, \ldots), \tag{2.27}$$

and the dot acts like a generalized Einstein convention including contractions of the discrete indices a_i and integrals over the three-manifold in the continuous indices x_i. We have

$$H_A(\mathbf{X}_1) H_A(\mathbf{X}_2) = \sum_{k=0}^{\infty} \sum_{j=k}^{\infty} i^j A_{\mu_1 \ldots \mu_k} A_{\mu_{k+1} \ldots \mu_j} X_1^{\mu_1 \ldots \mu_k} X_2^{\mu_{k+1} \ldots \mu_j}$$

$$= \sum_{j=0}^{\infty} i^j A_{\mu_1 \ldots \mu_j} \left(\sum_{k=0}^{j} X_1^{\mu_1 \ldots \mu_k} X_2^{\mu_{k+1} \ldots \mu_j} \right) = H_A(\mathbf{X}_1 \times \mathbf{X}_2), \tag{2.28}$$

where convention (2.17) has been applied over all the indices. The correspondence $\mathbf{X} \to H_A(\mathbf{X})$ gives a representation of the SeL group into a particular gauge group. In the case of the $\mathcal{X}$ group and the connections A belonging to the algebra of a unitary group, $H_A(\mathbf{X})$ is an element of the given unitary group. If one considers multitensors that do not satisfy the algebraic constraint, one still has a group and can construct a representation by considering As that belong to a unitary gauge algebra. However, the corresponding representation will give a holonomy that is not an element of the gauge group. It will, in general, be an element of the general linear group of the same dimension as the gauge group. This highlights the role of the algebraic constraint in this formalism. The differential constraint imposed on $\mathbf{X}$ ensures that $H_A(\mathbf{X})$ is a gauge covariant quantity provided that the expressions involved in the proofs converge (see chapter 12 for some subtleties on this issue).

We have shown that the analytic extension of any element of the SeL group defines a one-parameter subgroup. By studying its properties one can find the algebra associated with the SeL group.

2.3.2   Generators of the SeL group

Consider the one-parameter subgroup $\{\mathbf{X}^\lambda\}$ and suppose that we increase λ by an infinitesimal amount. We can write

$$\mathbf{X}^{\lambda + d\lambda} = \mathbf{X}^\lambda \times \mathbf{X}^{d\lambda} = \mathbf{X}^\lambda + \frac{d\mathbf{X}^\lambda}{d\lambda} d\lambda , \tag{2.29}$$

and taking $\lambda = 0$ we get

$$\mathbf{X}^{d\lambda} = \mathbf{I} + \mathbf{F} d\lambda, \tag{2.30}$$

where

$$\mathbf{F} \equiv \frac{d\mathbf{X}^\lambda}{d\lambda}\Big|_{\lambda=0} = \left(0, \sum_{i=1}^{\infty} \frac{(-1)^{i-1}}{i} \vec{X}^i\right) = (0, \vec{F}) . \qquad (2.31)$$

Introducing (2.30) in (2.29) we obtain the following differential equation for the elements of $\{\mathbf{X}^\lambda\}$

$$\frac{d\mathbf{X}^\lambda}{d\lambda} = \mathbf{X}^\lambda \times \mathbf{F} = \mathbf{F} \times \mathbf{X}^\lambda . \qquad (2.32)$$

This equation can be iteratively integrated to give

$$\mathbf{X}^\lambda = \mathbf{I} + \sum_{k=1}^{n} \frac{\lambda^k}{k!} \mathbf{F}^k + \mathbf{F}^{n+1} \times \int_0^\lambda d\lambda_1 \int_0^{\lambda_1} d\lambda_2 \ldots \int_0^{\lambda_n} d\lambda_{n+1} \mathbf{X}^{\lambda_{n+1}} . \qquad (2.33)$$

The process actually stops for any finite rank n component ($\mathbf{F}^{n+1} = \mathbf{F} \times {}^{n+1} \times \mathbf{F} = 0$ in this case). Therefore

$$\mathbf{X}^\lambda = \mathbf{I} + \sum_{k=1}^{\infty} \frac{\lambda^k}{k!} \mathbf{F}^k = \exp(\lambda\mathbf{F}) . \qquad (2.34)$$

We conclude that the vector $\mathbf{F}$ given by (2.31) is the generator of the one-parameter subgroup $\{\mathbf{X}^\lambda\}$. It is evident that the generator satisfies the differential constraint. We shall now prove the following fundamental property: $\mathbf{F}$ satisfies the homogeneous algebraic constraint (i.e., the sum over permutations defined in equation (2.10) vanishes). In other words, the generator of the one-parameter subgroup $\{\mathbf{X}^\lambda\}$ is the algebraic free part of $\mathbf{X}$.

We know that

$$(\mathbf{X}^\lambda)^{\underline{\mu_1 \ldots \mu_k}\mu_{k+1}\ldots\mu_n} = (\mathbf{X}^\lambda)^{\mu_1\ldots\mu_k}(\mathbf{X}^\lambda)^{\mu_{k+1}\ldots\mu_n} . \qquad (2.35)$$

Differentiating with respect to λ and evaluating for $\lambda = 0$ we get

$$\frac{d}{d\lambda}\left(\mathbf{X}^\lambda\right)^{\mu_1\ldots\mu_k\mu_{k+1}\ldots\mu_n}_{\lambda=0} = \left(\frac{d\mathbf{X}^\lambda}{d\lambda}\right)^{\mu_1\ldots\mu_k}_{\lambda=0} I^{\mu_{k+1}\ldots\mu_n}$$

$$+ I^{\mu_1\ldots\mu_k}\left(\frac{d\mathbf{X}^\lambda}{d\lambda}\right)^{\mu_{k+1}\ldots\mu_n}_{\lambda=0} . \qquad (2.36)$$

As $1 \le k < n$, we conclude

$$F^{\underline{\mu_1\ldots\mu_k}\mu_{k+1}\ldots\mu_n} = 0 , \quad 1 \le k < n . \qquad (2.37)$$

Reciprocally, one can demonstrate that the exponential of any algebraically free quantity produces an object that satisfies the algebraic constraint. It is important to stress that these results allow us to obtain the general solution for the algebraic constraint (equation (2.34) with $\lambda = 1$

and its inverse (2.31) give the relationship between an object that satisfies the algebraic constraint and its algebraic-free part).

The set of all $\mathbf{F}$s that satisfy the differential constraint and the homogeneous algebraic constraint forms a vector space $\mathcal{F}$. One can define a bilinear operation on $\mathcal{F}$ in the following way,

$$[\mathbf{F_1}, \mathbf{F_2}] = \mathbf{F_1} \times \mathbf{F_2} - \mathbf{F_2} \times \mathbf{F_1} \quad \text{for any } \mathbf{F_1}, \mathbf{F_2} \in \mathcal{F}. \qquad (2.38)$$

This operation is closed on $\mathcal{F}$. The vector space $\mathcal{F}$ together with the bracket operation (2.38) defines the Lie algebra associated with the SeL group.

2.4 Loop coordinates

The quantities $\mathbf{X}$ that we introduced in section 2.2 are not freely specifiable. That is, in order to be able to construct a gauge covariant object via equation (2.1), the $\mathbf{X}$s had to satisfy the differential and algebraic constraints (2.10), (2.11). That they are not freely specifiable is a natural thing, since they are elements of a group. That is why it was important to find the associated algebra, since its free parameters give us a chance to separate the part of the multitangents that we can freely specify. In the previous section we saw how to construct the set of objects $\mathcal{F}$. These objects had the advantage of being constrained not by the algebraic constraint, but by the homogeneous algebraic constraint. This latter constraint is very easily solvable, simply by requiring some symmetries on the $\mathbf{F}$s, given by equation (2.37). In terms of the $\mathbf{F}$s one immediately is able to compute a solution to both the differential and algebraic constraints making use of equation (2.34),

$$\mathbf{X} = \exp(\mathbf{F}). \qquad (2.39)$$

However, the $\mathbf{F}$s are far from freely prescribable since they are constrained by the differential constraint. The main intention of this section is to give a prescription for generating the $\mathbf{F}$s (and through them the $\mathbf{X}$s) from freely specifiable quantities. In order to do this we will need to introduce some technology to deal with transverse tensors. This technology will also be useful for dealing with knot invariants.

2.4.1 Transverse tensor calculus

First of all notice that the notion of transversality (divergence equal to zero) is well defined for vector densities, since their divergence can be computed without introducing an external metric. For instance, statements such as

$$\partial_{ax} E^{ax\,by} = 0 \qquad (2.40)$$

are well defined for an object like E which is a vector density on the index a at the point x.

Let us introduce the notion of transverse and longitudinal projectors in the multivector density space. In order to do this, it is convenient to endow the space of transverse vector densities of rank one with a natural metric structure. Given two transverse fields V^{ax} and W^{ax} one can define their inner product [22],

$$g(V,W) = \int d^3x \ V^a A_a^W,$$
$$\partial_a V^a = \partial_a W^a = 0, \qquad (2.41)$$

where A_a^W is a "potential" defined in the following way. Construct a two-form $W_{ab} = \epsilon_{abc} W^c$. This two-form is curl-free, $\partial_{[c} W_{ab]}$, due to the transversality of W^a. Then one can define the one-form ("potential") A_a^W by $\partial_{[b} A_{a]}^W = W_{ab}$. This one-form is defined up to the addition of a gradient. This will force us to give *ad-hoc* prescriptions when dealing with expressions in terms of A_a^W. However, the inner product (2.41) is well defined in a prescription independent way since the addition of a gradient to A_a^W only contributes a total divergence term.

The inner product introduced by (2.41) gives rise to a covariant metric on the space of transverse vectors,

$$g(V,W) = g_{0\ axby}\, V^{ax} W^{by}, \qquad (2.42)$$

which can be explicitly written, for instance, in the transverse (noncovariant) prescription,

$$\partial^a A_a^W = 0 \qquad (2.43)$$

as

$$g_{0\ axby} = -\frac{1}{4\pi}\, \epsilon_{abc}\, \frac{x^c - y^c}{\mid x - y \mid^3} \quad . \qquad (2.44)$$

Notice that due to the use of a non-covariant prescription the final object has both coordinate and background metric dependence. g_0 is a well known object in knot theory, where it plays the role of the kernel of the Gauss knot invariant, as we will see in section 2.6. It is the expression in a particular prescription of the covariant metric in the space of transverse vector densities defined by (2.41). Notice that in what follows we will not need to specify a background metric unless we want to give a specific prescription. In general, the covariant metric is defined up to gradients that change according to the prescription chosen,

$$g_{axby} = g_{0\ axby} + \rho_{ax\,y,b} + \rho_{by\,x,a} \ . \tag{2.45}$$

Transverse and longitudinal projectors may easily be written without the use of a background metric in terms of g and its inverse in the transverse space,

$$g^{axby} = \epsilon^{abc}\,\partial_c\,\delta(x - y) \ . \tag{2.46}$$

We define the quantities δ_T and δ_L (the transverse and longitudinal Dirac deltas) as

$$\delta_T{}^{ax}{}_{by} \equiv g^{ax\,cz}g_{cz\,by} \tag{2.47}$$

and

$$\delta_L{}^{ax}{}_{by} \equiv \delta^{ax}{}_{by} - \delta_T{}^{ax}{}_{by}, \tag{2.48}$$

where $\delta^{ax}{}_{by} = \delta^a{}_b\delta(x - y)$. It is straightforward to check that they have the desired projection properties,

$$\delta_T{}^{\mu}{}_{\rho}\ \delta_T{}^{\rho}{}_{\nu} = \delta_T{}^{\mu}{}_{\nu} \ ,$$
$$\delta_L{}^{\mu}{}_{\rho}\ \delta_L{}^{\rho}{}_{\nu} = \delta_L{}^{\mu}{}_{\nu} \ ,$$
$$\delta_L{}^{\mu}{}_{\rho}\ \delta_T{}^{\rho}{}_{\nu} = \delta_T{}^{\mu}{}_{\rho}\ \delta_L{}^{\rho}{}_{\nu} = 0 \ .$$

By using the explicit form of the covariant metric one can prove that

$$\delta_L{}^{ax}{}_{by} = \phi^{ax}{}_{y,b} \, , \tag{2.49}$$

where

$$\frac{\partial}{\partial x^a}\,\phi^{ax}{}_y = -\delta(x - y) \ . \tag{2.50}$$

The ambiguity in the definition of the metric induces an ambiguity in the decomposition into transverse and longitudinal parts. Each function ϕ that satisfies (2.50) determines a particular prescription of the decomposition. It is important to note that the transverse density fields and in particular the contravariant metric (2.46) are prescription independent. In the particular case in which we choose the transverse metric to be g_0 we have

$$\phi_0{}^{ax}{}_y = \frac{1}{4\pi}\,\frac{\partial}{\partial x^a}\frac{1}{|\,x - y\,|}, \tag{2.51}$$

$$\delta_{0T}{}^{ax}{}_{by} = \delta^{ax}{}_{by} + \frac{\partial^a\partial_b}{4\pi}\frac{1}{|\,x - y\,|} \ . \tag{2.52}$$

A transverse projector acting on the vector space $\mathcal{E}$ of multitensor densities can be immediately introduced through the matrix δ_T, defined in components as

$$\delta_T{}^{\mu_1\cdots\mu_n}{}_{\nu_1\ldots\nu_m} \equiv \delta_{n,m}\ \delta_T{}^{\mu_1}{}_{\nu_1}\ldots\delta_T{}^{\mu_n}{}_{\nu_n}\,. \qquad (2.53)$$

Given any multivector density $\mathbf{E}$ one can construct a multivector density $\mathbf{E}_T$ that is transverse or in other words that satisfies the homogeneous part of the differential constraint (2.11) by,

$$\mathbf{E}_T = \delta_T \cdot \mathbf{E}\,. \qquad (2.54)$$

The set of all $\mathbf{E}_T$s forms a linear vector space $\mathcal{E}_T$. The definition of $\mathbf{E}_T$ is not unique, it depends on the prescription used in the definition of the projector.

Since δ_T a projector, relation (2.54) is obviously not invertible in general. However, it turns out that it can be inverted on a subspace of $\mathcal{E}$ given by $\mathcal{E}_D$, the multitensor densities that satisfy the differential constraint (2.11). In order to do this, let us start by evaluating

$$E_D^{\mu_1\cdots\mu_n} = \delta^{\mu_1}{}_{\nu_1}\ldots\delta^{\mu_n}{}_{\nu_n}E_D^{\nu_1\cdots\nu_n}, \qquad (2.55)$$

making use of identity (2.48) and the differential constraint and recalling that the first rank component of $\mathbf{E}$ is transverse, we then get

$$\mathbf{E}_D = \sigma \cdot \mathbf{E}_T. \qquad (2.56)$$

The soldering quantities σ only depend on the function ϕ which characterizes the choice of decomposition in transverse and longitudinal parts,

$$\sigma^{\mu_1\cdots\mu_n}{}_{\nu_1\ldots\nu_m} = \begin{cases} \delta_T{}^{\mu_1\cdots\mu_n}{}_{\nu_1\ldots\nu_n}\,, & \text{if } m = n \\ Q^{\mu_1\cdots\mu_n}{}_{\rho_1\ldots\rho_{n-1}}\,\sigma^{\rho_1\cdots\rho_{n-1}}{}_{\nu_1\ldots\nu_m}\,, & \text{if } m < n \\ 0\,, & \text{if } m > n \end{cases} \qquad (2.57)$$

with

$$Q^{a_1x_1\ldots a_nx_n}_{c_1y_1\ldots c_{n-1}y_{n-1}} \equiv \sum_{j=1}^{n}\delta^{a_1x_1\ldots a_{j-1}x_{j-1}}_{c_1y_1\ldots c_{j-1}y_{j-1}}\left(\phi^{a_jx_j}_{y_j} - \phi^{a_jx_j}_{y_{j-1}}\right)\delta_T{}^{a_{j+1}x_{j+1}\ldots a_nx_n}_{c_jy_j\ldots c_{n-1}y_{n-1}}\,. $$

$$(2.58)$$

Again, this definition is not unique and will be prescription dependent. However, starting from a given $\mathbf{E}_D$ one can construct an $\mathbf{E}_T$ and then uniquely reconstruct the original $\mathbf{E}_D$ by applying σ.

A crucial property is that the quantities σ *satisfy the differential constraint in their upper indices*, as can be checked from their definition. That is, given an arbitrary transverse multitensor density $\mathbf{E}_T$, one can construct a solution of the differential constraint by applying equation (2.56).

The quantities σ have definite transversality properties

$$\delta_T \cdot \sigma = \delta_T \,, \tag{2.59}$$

$$\sigma \cdot \delta_T = \sigma \,. \tag{2.60}$$

Notice that due to these properties we can relax the requirement to construct a solution to the differential constraint, i.e., given an *arbitrary* multitensor $\mathbf{E}$, the quantity $\sigma \cdot \mathbf{E}$ is a solution of the differential constraint.

Under a change of the prescription $\phi_{1y}^{ax} \to \phi_{2y}^{ax}$ we get a $\sigma[\phi_2]$ satisfying

$$\sigma[\phi_1] = \sigma[\phi_2] \cdot \sigma[\phi_1]. \tag{2.61}$$

The operations δ_T and σ define an isomorphism between vector spaces, $\mathcal{E}_D$ the space of multitensors that solve the differential constraint and $\mathcal{E}_T$ via,

$$\mathbf{E}_T = \delta_T \cdot \mathbf{E}_D, \tag{2.62}$$

$$\mathbf{E}_D = \sigma \cdot \mathbf{E}_T. \tag{2.63}$$

The vector product can be introduced in the vector space $\mathcal{E}_D$ and, due to the isomorphism, it is simply given by

$$\mathbf{E}_{D1} \times \mathbf{E}_{D2} = \sigma \, \cdot \, (\mathbf{E}_{T1} \times \mathbf{E}_{T2}). \tag{2.64}$$

This last property will have useful applications in section 2.6 where we construct diffeomorphism invariants.

We are now ready to combine this construction with the ideas of the last section to define the loop coordinates.

2.4.2   Freely specifiable loop coordinates

We saw in section 2.3.2 that one could generate a solution to the differential and algebraic constraints $\mathbf{X}$ by considering

$$\mathbf{X} = \exp(\mathbf{F}) \tag{2.65}$$

but for this to hold $\mathbf{F}$ had to satisfy the differential constraint and the homogeneous algebraic constraint.

Let us now consider an arbitrary transverse multitensor $\mathbf{E}_T$. Applying the results of the last subsection, we notice that the quantity $\sigma \cdot \mathbf{E}_T$ satisfies the differential constraint. Unfortunately, it does not satisfy the homogeneous algebraic constraint (if it did, we would be done, since it would be an element of $\mathcal{F}$).

We will remedy this situation now. We define a new matrix, given by

$$\Omega^{\mu_1 \dots \mu_n}{}_{\nu_1 \dots \nu_m} \equiv \delta^{\mu_1 \dots \mu_n}{}_{\nu_1 \dots \nu_m} + \sum_{k=1}^{n-1} \frac{(n-k)}{n} (-1)^k \, \delta^{\underline{\mu_1 \dots \mu_k} \mu_{k+1} \dots \mu_n}{}_{\nu_1 \dots \nu_m},$$

$$\tag{2.66}$$

where

$$\delta^{\mu_1\ldots\mu_n}{}_{\nu_1\ldots\nu_m} \equiv \delta_{n,m}\ \delta^{\mu_1}{}_{\nu_1}\ldots\delta^{\mu_n}{}_{\nu_n}\ . \tag{2.67}$$

The matrix Ω has the following important property: it satisfies the homogeneous algebraic constraint in the upper indices. This fact immediately shows that Ω is a projector. Given an arbitrary vector $\mathbf{E}$, $\Omega \cdot \mathbf{E}$ is an algebraic-free object. In particular we have $\mathbf{F} = \Omega \cdot \mathbf{F}$.

Let us now introduce the following set of vectors

$$\mathcal{S}_{\nu_1\ldots\nu_m} = (\, 0,\ \vec{\mathcal{S}}_{\nu_1\ldots\nu_m}) \tag{2.68}$$

with

$$\mathcal{S} = \big(\sigma \cdot \Omega \big), \tag{2.69}$$

which written explicitly in components is

$$(\mathcal{S}_{\nu_1\ldots\nu_m})^{\mu_1\ldots\mu_n} = \sigma^{\mu_1\ldots\mu_n}{}_{\alpha_1\ldots\alpha_l}\Omega^{\alpha_1\ldots\alpha_l}{}_{\nu_1\ldots\nu_m}. \tag{2.70}$$

These vectors combine the action of σ, which converted an arbitrary multitensor into a solution of the differential constraint, and Ω, which projects into the space of solutions of the homogeneous algebraic constraint. That is, given an arbitrary multivector density $\mathbf{E}$, projecting it with $\mathcal{S}$ one obtains an element of $\mathcal{F}$. Simply by exponentiating this element, as we saw in section 2.3.2, we obtain a solution of the differential and algebraic constraint. That is, we just consider,

$$\mathbf{X} = \exp(\mathcal{S} \cdot \mathbf{E}), \tag{2.71}$$

and the $\mathbf{E}$s are unconstrained! Notice that expression (2.71) is the usual relation between elements of a Lie group ($\mathbf{X}$), a basis of generators $\mathcal{S}$ and their free parameters ($\mathbf{E}$).

Expression (2.71) does not really depend on the portion of the $\mathbf{E}$s that does not satisfy the homogeneous algebraic and differential constraints since the contraction with the $\mathcal{S}$s is independent of that portion. Therefore, one will usually concentrate on the set of transverse vectors $\mathbf{Y}$ that satisfy the homogeneous algebraic constraint, and we will call this set $\mathcal{Y}$,

$$\mathcal{Y}: \qquad \mathbf{Y} = \delta_T \cdot \mathbf{Y} \qquad \text{and} \qquad Y^{\underline{\mu_1\ldots\mu_k}\mu_{k+1}\cdots\mu_n} = 0\ ,\ \ 1 \le k < n\ . \tag{2.72}$$

The situation is totally analogous, for instance, to that of the Lorentz group. In that case the generators are antisymmetric matrices and therefore one usually works with free parameters that are antisymmetric matrices in spite of the fact that any kind of matrix would do. It is just that one can only code relevant information in its antisymmetric part. Similarly here, any arbitrary multitensor $\mathbf{E}$ would work as a free parameter, but only information coded in the portion that satisfies the homogeneous constraints will be relevant for constructing the $\mathbf{X}$s via equation (2.71),

$$\mathbf{X} = \exp(\mathcal{S} \cdot \mathbf{Y}). \qquad (2.73)$$

The elements of $\mathcal{Y}$ are immediately related to those of $\mathcal{F}$ by

$$\mathbf{Y} = \delta_T \cdot \mathbf{F}. \qquad (2.74)$$

When referring to multitangents rather than arbitrary multitensors we can therefore call the objects $\mathbf{Y}$ "loop coordinates" or coordinates in loop space. Abusing the terminology a bit we will also refer to them in this way when we talk about arbitrary multitensor densities not necessarily associated with loops.

Since they are solutions to the homogeneous algebraic and differential constraint, the $\mathcal{S}$s are elements of $\mathcal{F}$ and therefore they form a basis for the algebra as we suggested above. Details of their construction, the proof that they satisfy the algebra and the determination of the structure constants of the SeL can be seen in reference [20].

2.5   Action of the differential operators

In the previous chapter we introduced a series of differential operators that represented the infinitesimal generators of the group of loops. The loop coordinates provide us with an explicit representation in terms of which we can explore the action of the differential operators. We will not discuss in detail the action of all the differential operators, since as we saw, they are related to each other. We will only concentrate on the action of the loop derivative and of the contact derivative. The former can be used as the starting point to compute any other derivative. The latter is related to diffeomorphism invariance and therefore deserves a detailed treatment.

Let us therefore start by computing the action of the loop derivative on a multitangent field. By the definition of the loop derivative (1.17),

$$(1 + \tfrac{1}{2}\sigma^{ab}\Delta_{ab}(\pi_o^z))X^{a_1 x_1 \ldots a_n x_n}(\gamma) \equiv X^{a_1 x_1 \ldots a_n x_n}(\pi_o^z \circ \delta u \delta v \delta \bar{u} \delta \bar{v} \circ \pi_z^o \circ \gamma), \qquad (2.75)$$

and recalling the relation between the $\times$-product and the composition law (2.22), we can write

$$X^{a_1 x_1 \ldots a_n x_n}(\pi_o^z \circ \delta u \delta v \delta \bar{u} \delta \bar{v} \circ \pi_z^o \circ \gamma) =$$
$$(\mathbf{X}(\pi_o^z) \times \mathbf{X}_z(\delta u \delta v \delta \bar{u} \delta \bar{v}) \times \mathbf{X}(\pi_z^o) \times \mathbf{X}(\gamma))^{a_1 x_1 \ldots a_n x_n}. \qquad (2.76)$$

Notice that $\mathbf{X}_z(\delta u \delta v \delta \bar{u} \delta \bar{v})$ is a multitangent basepointed at z, which is in line with the fact that it is composed with an open path that ends at z.

We therefore need to evaluate $\mathbf{X}(\delta u \delta v \delta \bar{u} \delta \bar{v})$ applying the definition of the multitangents (2.3). We can do this order by order. We will only make explicit the calculation of the first order,

$$\mathbf{X}_z(\delta u \delta v \delta \bar{u} \delta \bar{v})^{a_1 x_1} = \epsilon_1 u^{a_1} \delta(x_1 - z) + \epsilon_2 v^{a_1} \delta(x_1 + \epsilon_1 u - z)$$
$$-\epsilon_1 u^{a_1} \delta(x_1 + \epsilon_1 u + \epsilon_2 v - z) - \epsilon_2 v^{a_1} \delta(x_1 + \epsilon_2 v - z), \quad (2.77)$$

we now expand the Dirac deltas

$$\delta(x_1 + \epsilon_1 u - z) = \delta(x_1 - z) + \epsilon_1 u^b \partial_b \delta(x_1 - z), \quad (2.78)$$

and noticing that all linear terms cancel, we collect terms of order $\epsilon_1 \epsilon_2$ to get

$$\mathbf{X}_z(\delta u \delta v \delta \bar{u} \delta \bar{v})^{a_1 x_1} = \tfrac{1}{2} \sigma^{ab} \delta^{a_1 c}_{ab} \delta_{,c}(x_1 - z). \quad (2.79)$$

In this last expression $\sigma^{ab} = 2\epsilon_1 \epsilon_2 u^{[a} v^{b]}$ as usual and we have introduced the antisymmetrized Kronecker delta $\delta^{cd}_{ab} = \tfrac{1}{2}(\delta^c_a \delta^d_b - \delta^c_b \delta^d_a)$ and the notation $\delta_{,c}(x - z) = \partial_c \delta(x - z)$.

With this in mind, similar calculations follow for higher order multitangents. The results are

$$\Delta_{ab}(\pi^z_o) X^{a_1 x_1}(\gamma) = \delta^{a_1 c}_{ab} \delta_{,c}(x_1 - z), \quad (2.80)$$
$$\Delta_{ab}(\pi^z_o) X^{a_1 x_1 a_2 x_2}(\gamma) = \delta^{a_1 a_2}_{ab} \delta(x_1 - z)\delta(x_2 - z)$$
$$+\delta^{a_2 c}_{ab} \delta_{,c}(x_2 - z) X^{a_1 x_1}(\pi^z_o) + \delta^{a_1 c}_{ab} \delta_{,c}(x_1 - z) X^{a_2 x_2}(\pi^o_z \circ \gamma), \quad (2.81)$$

and, in general,

$$\Delta_{ab}(\pi^z_o) X^{a_1 x_1 \ldots a_n x_n}(\gamma) =$$
$$\delta^{a_1 c}_{ab} \delta_{,c}(x_1 - z) X^{a_2 x_2 \ldots a_n x_n}(\pi^o_z \circ \gamma)$$
$$+\delta^{a_n c}_{ab} \delta_{,c}(x_n - z) X^{a_1 x_1 \ldots a_{n-1} x_{n-1}}(\pi^z_o)$$
$$+\delta^{a_1 a_2}_{ab} \delta(x_1 - z)\delta(x_2 - z) X^{a_3 x_3 \ldots a_n x_n}(\pi^o_z \circ \gamma)$$
$$+\delta^{a_{n-1} a_n}_{ab} \delta(x_{n-1} - z)\delta(x_n - z) X^{a_1 x_1 \ldots a_{n-2} x_{n-2}}(\pi^z_o)$$
$$+\sum_{j=1}^{n-2} \delta^{a_{j+1} c}_{ab} \delta_{,c}(x_{j+1} - z) X^{a_1 x_1 \ldots a_j x_j}(\pi^z_o) X^{a_{j+2} x_{j+2} \ldots a_n x_n}(\pi^o_z \circ \gamma)$$
$$+\sum_{j=1}^{n-3} \delta^{a_{j+1} a_{j+2}}_{ab} \delta(x_{j+1} - z)\delta(x_{j+2} - z)$$
$$\times X^{a_1 x_1 \ldots a_j x_j}(\pi^z_o) X^{a_{j+3} x_{j+3} \ldots a_n x_n}(\pi^o_z \circ \gamma). \quad (2.82)$$

In terms of these expressions for the loop derivative one can reconstruct the action of any other differential operator. We will consider as an example the expressions for the contact derivative.

The expression of the action of the contact derivative on a multitangent is,

$$\mathcal{C}_a(z)X^{a_1 x_1 \ldots a_n x_n}(\gamma) \equiv \oint_\gamma dy^b \delta(z-y)\Delta_{ab}(\gamma_o^y)X^{a_1 x_1 \ldots a_n x_n}(\gamma) =$$

$$\sum_{j=1}^{n} \delta_{ab}^{a_j c}\delta_{,c}(x_j - z)X^{a_1 x_1 \ldots a_{j-1}x_{j-1}bza_{j+1}x_{j+1}\ldots a_n x_n}(\gamma) +$$

$$\sum_{j=1}^{n-1} \delta_{ab}^{a_j a_{j+1}}\delta(x_j - z)\delta(x_{j+1} - z)X^{a_1 x_1 \ldots a_{j-1}x_{j-1}bza_{j+2}x_{j+2}\ldots a_n x_n}(\gamma).$$

$$(2.83)$$

This expression can be written as a linear transformation of the Xs. This is just an expression of the fact that a "passive" diffeomorphism where one deforms the loop is the same as an "active" diffeomorphism where one maintains the loop fixed but changes coordinates. Let us take a minute to explore this result in detail. We rewrite the expression for the contact derivative as

$$\mathcal{C}_a(z)X^{a_1 x_1 \ldots a_n x_n}(\gamma) = \sum_{j=1}^{n} A_{az}{}^{a_j x_j}{}_{by}X^{a_1 x_1 \ldots a_{j-1}x_{j-1}bya_{j+1}x_{j+1}\ldots a_n x_n}(\gamma)$$

$$+ \sum_{j=1}^{n-1} B_{az}{}^{a_j x_j a_{j+1} x_{j+1}}{}_{by}X^{a_1 x_1 \ldots a_{j-1}x_{j-1}bya_{j+2}x_{j+2}\ldots a_n x_n}, \qquad (2.84)$$

with

$$A_{az}{}^{a_1 x_1}{}_{by} = \delta_{ab}^{a_1 c}\delta_{,c}(x_1 - z)\delta(y - z), \qquad (2.85)$$

$$B_{az}{}^{a_1 x_1 a_2 x_2}{}_{by} = \delta_{ab}^{a_1 a_2}\delta(x_1 - z)\delta(x_2 - z)\delta(y - z), \qquad (2.86)$$

where we have used a generalized Einstein convention on the index y.

Sometimes it will be useful to compute the action of differential operators on cyclic multitangents, for instance, if one wants to evaluate the contact derivative of a Wilson loop, which only depends on the cyclic portion of the multitangents,

$$X_c^{\mu_1 \ldots \mu_n} = \frac{1}{n}\left(X^{\mu_1 \ldots \mu_n} + X^{\mu_2 \ldots \mu_n \mu_1} + \ldots X^{\mu_n \ldots \mu_1}\right). \qquad (2.87)$$

It is given by

$$\mathcal{C}_a(z)X_c^{a_1 x_1 \ldots a_n x_n}(\gamma) = \sum_{j=1}^{n} C_{az}{}^{a_j x_j}{}_{by}X_c^{a_1 x_1 \ldots a_{j-1}x_{j-1}bya_{j+1}x_{j+1}\ldots a_n x_n}(\gamma),$$

$$(2.88)$$

where

$$C_{az}{}^{a_1 x_1}{}_{by} = \delta_{ab}^{a_1 c}\delta_{,c}(x_1 - z)\delta(y - z) - \delta_a^{a_1}\delta(x_1 - z)\delta_{,b}(z - y). \qquad (2.89)$$

Equation (2.84) can also be rearranged in terms of the linear transformation matrix C making use of the differential constraint, which was also used to derive (2.88).

These expressions allow us to write the expression for the transformation law of the multitangents under an infinitesimal coordinate transformation $x^a \longrightarrow x'^a = D^a(x) \equiv x^a + N^a(x)$ simply by computing

$$\left(1 + \int d^3 x N^a(x) \mathcal{C}_a(x)\right) X^{\mu_1 \ldots \mu_n} = \Lambda_D{}^{\mu_1}_{\nu_1} \cdots \Lambda_D{}^{\mu_n}_{\nu_n} X^{\nu_1 \ldots \nu_n}, \qquad (2.90)$$

with the coordinate transformation matrices given by

$$\Lambda_D{}^{ay}{}_{bx} = \frac{1}{J(x)} \frac{\partial D^a(x)}{\partial x^b} \delta(x - D^{-1}(y)) = \frac{\partial D^a(x)}{\partial x^b} \delta(D(x) - y)), \quad (2.91)$$

where $J(x)$ is the Jacobian of the coordinate transformation.

2.6 Diffeomorphism invariants and knots

Any vector $\mathbf{F}$ belonging to the SeL algebra behaves as a multivector density under a diffeomorphism that leaves the basepoint fixed. In matrix form the transformation law corresponding to a coordinate transformation $x^a \longrightarrow x'^a = D^a(x)$ is

$$\mathbf{F}' = \Lambda_D \cdot \mathbf{F} \qquad (2.92)$$

where

$$\Lambda_D{}^{\mu_1 \ldots \mu_n}{}_{\nu_1 \ldots \nu_m} \equiv \delta_{n,m} \ \Lambda_D{}^{\mu_1}{}_{\nu_1} \cdots \Lambda_D{}^{\mu_n}{}_{\nu_n}. \qquad (2.93)$$

From here it is immediate just by inspecting equation (2.62) to derive the transformation law for the transverse algebraic-free vectors $\mathbf{Y}$,

$$\mathbf{Y}' = \delta_T \cdot \mathbf{F}' = \mathcal{L}_D \cdot \mathbf{Y}, \qquad (2.94)$$

where

$$\mathcal{L}_D \equiv \delta_T \cdot \Lambda_D \cdot \sigma. \qquad (2.95)$$

The diffeomorphism transformation given by (2.92) is just a particular example of a more general family of transformations: the automorphisms of the algebra. Other automorphisms can be considered, for instance, the conjugation $\mathbf{F}' = \mathbf{X} \times \mathbf{F} \times \mathbf{X}^{-1}$.

The isomorphism between the vector spaces $\mathcal{E}_D$ and $\mathcal{E}_T$ makes $\mathcal{L}_D$ a representation of the diffeomorphism group. This representation emerges as the push-forward of the natural action of diffeomorphisms on the space of solutions of the differential constraint through the isomorphism of that space with the space of transverse vectors $\mathcal{E}_T$.

The presence of the non-diagonal matrix σ in $\mathcal{L}_D$ makes this representation highly non-trivial. This is an important result, due to the possibility of introducing objects that transform under the adjoint representation of the diffeomorphism group. In fact, the isomorphism guarantees the following property of the σs

$$\sigma = \Lambda_D \cdot \sigma \cdot \mathcal{L}_{D^{-1}}. \qquad (2.96)$$

This relationship clearly shows the role played by the σs as the soldering quantities between the fundamental representation Λ_D and the adjoint representation $\mathcal{L}_D$. It is straightforward to see that the subspaces $\mathcal{F}$ and $\mathcal{Y}$ are invariant under diffeomorphisms.

Our task is to construct quantities invariant under automorphisms. To illustrate the procedure to follow, let us consider what is usually done to construct invariants of a group, say $SU(2)$. One takes elements of the group $\omega_i \sigma^i$, where σ^i are the usual Pauli matrices and ω_i free parameters, and computes their trace

$$\mathrm{Tr}(\omega_i \sigma^i \omega_j \sigma^j) = \mathrm{Tr}(\sigma^i \sigma^j) \omega_i \omega_j = G^{ij} \omega_i \omega_j. \qquad (2.97)$$

The result is obviously an invariant and it has the form of a metric G^{ij} (in this particular case equal to δ^{ij}), which is invariant under the action of the automorphisms of the group, contracted with the free parameters of the group. Analogously one can take traces of higher order products of elements and one would end up with invariants of the form $G^{i_1 \cdots i_n} \omega_{i_1} \ldots \omega_{i_n}$. We will generically call the Gs "invariant metrics".

We will now follow a similar procedure to find invariants under automorphisms of the SeL group. Since we showed that diffeomorphisms are just a particular case of automorphisms, the result will be diffeomorphism invariant. Consider a covector in the space $\mathcal{Y}$, $\mathbf{g} = (0, g_{\mu_1 \mu_2}, \cdots, g_{\mu_1 \ldots \mu_n}, \cdots)$ with the following properties:

$$\mathbf{g} = \mathbf{g} \cdot \mathcal{L}_D, \qquad (2.98)$$

$$g_{\mu_1 \ldots \mu_n} = g_{(\mu_1 \ldots \mu_n)_{cyclic}}. \qquad (2.99)$$

With it, we can define a multilinear form from $\mathcal{Y} \times \cdots \times \mathcal{Y}$ into the complex numbers,

$$I_n = \mathbf{g} \cdot (\mathbf{Y}_1 \times \cdots \times \mathbf{Y}_n) \qquad (2.100)$$

that is invariant with respect to all automorphisms described above. The invariance property (2.98) ensures that (2.100) is invariant under diffeomorphisms, (2.99) ensures invariance under conjugation. Why do we require the extra cyclicity property (2.99)? The reader should remember that all the multitangent formalism is basepointed, i.e., there is a preferred point in the manifold as was obvious, for instance, when writing the differential constraint (2.11). The diffeomorphisms under which the constructed quantity would end up being invariant would be those that leave the basepoint fixed. This is not what one is usually interested in, not even in the case of knot invariants, when the multitensors really are multitangents to loops. The cyclicity property ensures that the quantities constructed do not depend on any basepoint.

Unfortunately, we do not have a general technique for constructing the invariant tensors g. Taking traces as in the $SU(2)$ example does not work since we want objects not only invariant under conjugacy but also under other automorphisms, specifically the ones that represent diffeomorphisms and the traces are not invariant under these transformations. Some invariant tensors g are known and we will discuss them in some detail later.

This formalism appears to be a very powerful technique for constructing invariants associated with three-manifolds. Its implications have not been worked out in detail yet, so we will end the generic discussion here. However, it is quite clear that this construction can immediately be particularized to the case in which one is not dealing with arbitrary multitensor fields, but with multitangents associated with loops. The resulting invariants would be *knot invariants*. There is an abundant literature on the subject and therefore we will find it worthwhile to explore the implications of our formalism in some detail for this case in order to make contact with well known results.

Therefore, we will now consider the quantities

$$I_n(\gamma) = \mathbf{g} \cdot (\mathbf{Y}(\gamma) \times \cdots \times \mathbf{Y}(\gamma)), \tag{2.101}$$

and it is evident by construction that $I_n(\gamma) = I_n(\gamma')$ if γ and γ' are related by a diffeomorphism.

Let us consider some particular examples of these quantities. Take $n = 2$. In this case, the invariant metric has only one non-vanishing component,

$$g_{G\,\mu_1\ldots\mu_n} = \delta_{n,2}\, g_{\mu_1\mu_2}, \tag{2.102}$$

where $g_{\mu_1\mu_2}$ is the metric on the space of order one multitangents, already introduced in (2.45). It leads to the following invariant:

$$I_G(\gamma) = \mathbf{g}_G \cdot (\mathbf{Y}(\gamma) \times \mathbf{Y}(\gamma)) = g_{\mu_1\mu_2} Y^{\mu_1}(\gamma) Y^{\mu_2}(\gamma). \tag{2.103}$$

For a first order multitangent $Y^\mu(\gamma) = X^\mu(\gamma)$; replacing the definition of the Xs (2.3) and of g (2.44) and performing the integrals over the three-manifold explicitly we get

$$GL = -\frac{1}{4\pi} \oint_\gamma ds \oint_\gamma dt\, \dot{\gamma}^a(s) \dot{\gamma}^b(t) \epsilon_{abc} \frac{(\gamma(s)^c - \gamma(t)^c)}{|\gamma(s) - \gamma(t)|^3}. \tag{2.104}$$

The reader may recognize in this expression the Gauss linking number.

Since we computed it for only one curve, it is a "self-linking number", a quantity which is in general ill-defined and to which we will return in chapter 10.

Although there is not a systematic procedure for constructing the invariant metrics, an infinite family of them can be constructed applying results from Chern–Simons theories, a class of topological field theories that has recently attracted great attention [45]. Using these techniques other invariant metrics have been computed in explicit fashion [187, 47], but we will postpone their discussion until chapter 10 when we discuss Chern–Simons theory in some detail.

The metrics are prescription dependent objects, as can be readily seen from equation (2.98). The knot invariants, however, should be prescription independent. In order to see this let us fix some prescription for $\mathbf{g}$, $\mathbf{g}_1 = \mathbf{g}_1 \cdot \delta_{T1}$. Then

$$\mathbf{g}_1 \cdot \vec{Y}_1 = \mathbf{g}_1 \cdot \delta_{T1} \cdot \vec{F} = \mathbf{g}_1 \cdot \vec{F}. \tag{2.105}$$

But $\vec{F} = \sigma_2 \cdot \vec{Y}_2$, then

$$\mathbf{g}_1 \cdot \vec{Y}_1 = \mathbf{g}_1 \cdot \sigma_2 \cdot \vec{Y}_2 = \mathbf{g}_2 \cdot \vec{Y}_2, \tag{2.106}$$

where

$$\mathbf{g}_2 = \mathbf{g}_1 \cdot \sigma_2 \tag{2.107}$$

is the invariant tensor in the prescription 2. Using the algebraic-free coordinates we have

$$\mathbf{g}_1 \cdot \vec{F} = \mathbf{g}_1 \cdot \vec{Y}_1 = \mathbf{g}_2 \cdot \vec{Y}_2 = \mathbf{g}_2 \cdot \vec{F}. \tag{2.108}$$

If one is considering a specific representation of the group of loops in terms of a gauge group, as we will start to do in the next chapter, functionals of a loop and of multiloops will be related by a series of identities called the Mandelstam identities. With these identities one can build and relate invariants of links of more than one component. We will return to this subject in chapter 10.

2.7　Conclusions

In this chapter we introduced a series of analytic techniques for describing loops. We exhibited the important role of multitensor densities as representations of loops. In fact we noticed that multitensor density fields can play a more fundamental role than loops in physics altogether. We showed how to represent the group of loops and how to extend it to form a Lie group in terms of multitensor fields. We found, by constructing the associated Lie algebra and its free parameters, a set of freely prescribable multitensors that can be used as fundamental objects to describe loops or to build a more general framework. We showed how the diffeomorphisms are represented in terms of these objects and how to use them to construct invariants of three-manifolds and of knots. All these techniques will play a fundamental role in chapters 10 and 11 in the applications to quantum gravity. They will be especially useful for revealing the relations between quantum gravity and topological field theories and will possibly become the calculational bridge between the beautiful notions of knot theory and the Einstein equations. Of all the mathematical technology that we will introduce in chapters 1-3, the extended loop calculus is the most recently discovered and its implications are least explored. A great degree of improvement in the understanding of these issues is likely to appear in the years to come.

3

The loop representation

3.1 Introduction

At the beginning of the 1970s gauge theories and in particular Yang–Mills theories appeared as the fundamental theories that described particle interactions. Two main perturbative results were established: the unification of electromagnetic and weak interactions and the proof of the renormalizability of Yang–Mills theory. However, the advent of proposals to describe strong interactions in terms of gauge theories — and in particular the establishment of QCD and the quark model for the hadrons — required the development of new non-perturbative techniques. Problems such as that of confinement, chiral symmetry breaking and the $U(1)$ problem spawned interest in various non-perturbative alternatives to the usual treatment of quantum phenomena in gauge theories. Both at the continuum and lattice levels various attempts were made [44, 48, 12, 49, 50] to describe gauge theories in terms of extended objects as Wilson loops and holonomies. Some of these treatments started at a classical level [44], with the intention of completely reformulating and solving classical gauge theories in terms of loops. Other proposals were at the quantum mechanical level; for instance, trying to find a Schwinger–Dyson formulation in order to obtain a generating functional for the Green functions of gauge theories using the Wilson loop. Among these latter proposals we find the loop representation [5, 34], based on constructing a quantum representation of Hamiltonian gauge theories in terms of loops. In this context, the main advantage of the loop representation was to do away with the first class constraint of the theories (the Gauss law), and therefore with the redundancy introduced by gauge symmetries. It allowed researchers to work directly in the space of physical states.

The idea that a non-perturbative quantization is possibly the only viable solution to the problems presented by the quantization of general relativity is not new. However, the failure of various attempts based on perturbation theory over the last two decades has increased the belief that non-perturbative methods may be the only alternative to approach the quantization of gravity. In particular, the striking example of 2+1 gravity, which for many years was considered perturbatively as pathological as 3+1 gravity until it was proven by Witten [46] that it can be exactly quantized, has contributed to the belief that perturbative methods in general relativity can be quite misleading. Simultaneously, the introduction of a new set of variables by Ashtekar [51] that cast general relativity in the same language as gauge theories provided the natural framework for the introduction of loop techniques as a natural non-perturbative avenue for the quantization of Einstein's theory. As the Hamiltonian was the most promising scenario for the new variables, the loop representation appeared to be the most natural application of loop techniques to the problem [38, 39]. Moreover it was apparent from the beginning that the use of the loop representation allowed various new insights, in particular it revealed a new connection between general relativity and geometry, but now at a quantum level. Wavefunctions in the loop representation appeared in the pioneering work of Rovelli and Smolin as intimately related with various notions of mathematics, in particular those of the newly flourishing branch of knot theory. This connection was highlighted when the Jones polynomial was found to play the role of a possible state of quantum general relativity [52].

In this chapter we will briefly discuss various physical results that we will need, in combination with the loop techniques introduced in the first two chapters, to introduce the idea of a loop representation. These ideas will be used extensively to discuss the applications in subsequent chapters of the book. The level of rigor and depth that we will maintain in this chapter is only the one needed to discuss the applications. Many of the topics covered in this chapter would, in general, require a book by themselves if they were to be discussed in detail. The idea of this chapter is therefore to fix notation for the advanced reader and to introduce the beginner to these topics in order to allow a first reading of the rest of the book.

The organization of this chapter is as follows. We will start in section 3.2 with a discussion of the canonical formulation and quantization of field theories. The idea is to lay down the formalism that we will use to treat both Yang–Mills theories and gravity. In both cases we will be dealing with systems with constraints and we will briefly discuss their treatment. In section 3.3 we discuss Yang–Mills theories in the canonical formulation both at a classical and a quantum mechanical level, highlighting the role of the Gauss law. We will then discuss the role of Wilson loops as a basis of solutions of the Gauss law and their properties in section 3.4. In section

3.5 we will discuss, in general, the formulation of the loop representation and its implications. We analyze in some detail two possible definitions of the loop representation that we will use throughout the book. We will end with a summary and discussion in section 3.6

3.2 Hamiltonian formulation of systems with constraints

3.2.1 Classical theory

The subject of constrained Hamiltonian systems was pioneered by Dirac [27] in the 1950s and is well established by now. Abundant literature exists on the subject and treatments vary from elementary to very sophisticated, since the subject is endowed with a rich geometrical structure. The intention of this section is just to fix notation and to remind the reader briefly of the ideas involved. More extensive treatments can be found in [27, 28, 29, 30] and those who want to explore the geometrical framework are referred to [31, 32, 33].

Physical theories are not usually described in terms of the minimum possible number of variables. In general, descriptions are made in terms of quantities that present a certain degree of redundancy which results in the fact that the system is invariant under certain symmetries. For instance, one does not usually describe the free electromagnetic field in terms of the two helicity components of the electric field, but rather in terms of the vector potential. The resulting formulation is invariant under gauge transformations. What will happen in general is that given a set of initial data the end result of the evolution will not be unique but will lie on a set of equivalent physical configurations related by the symmetries of the theory. Systems as simple as the free relativistic particle are usually formulated with redundant variables due to the Lorentz symmetry which does not specify a unique choice of time.

We will assume that one has a Hamiltonian system (possibly with an infinite number of degrees of freedom), described by a set of canonical variables q_i and canonical momenta p_i with Poisson bracket relations,

$$\{q_i, p_j\} = \delta_{ij}. \tag{3.1}$$

When one formulates canonically a system with redundant variables and symmetries, the resulting canonical formulation has constraints. The constraints are a set of relations $\phi_m(p_i, q_i) = 0, \quad i = 1, \ldots, m$ among the canonical variables. Some constraints become manifest when one performs the Legendre transform from the Lagrangian formulation. These are called "primary" constraints. When one requires that these constraints be preserved by evolution, new constraints may appear, called "secondary", which in turn have to be preserved by evolution and so on.

There is a further distinction between constraints. A constraint ϕ_k will be said to be of first class if its Poisson bracket with all the other constraints is a linear combination of the constraints,

$$\{\phi_k, \phi_i\} = C_{ki}^j \phi_j \quad \forall i. \tag{3.2}$$

Other constraints are called second class. In this book we will only discuss first class constraints. This is due to three reasons. First, there is a procedure ("Dirac brackets") [27] to convert a set of second class constraints to first class ones by redefining the Poisson bracket structure of phase space. Second, most formulations of the theories of interest in this book such as Yang–Mills theories and general relativity only present first class constraints. Third, although certain gauge fixed formulations of gauge theories involve second class constraints, the loop formulation does not require any gauge fixing, since it is automatically gauge invariant.

The effect of having constraints in the theory is to restrict the dynamics to taking place on a surface $\hat{\Gamma}$ in the phase space Γ called the "constraint surface". The dynamical trajectories on $\hat{\Gamma}$ are not well defined. Each dynamical evolution is represented by an infinite family of trajectories that are physically equivalent. This is the representation in this picture of what is usually called "gauge". The family of trajectories are "gauge equivalent". This is due to the fact that there is an ambiguity in extending quantities from $\hat{\Gamma}$ to Γ since two quantities that differ by a combination of constraints are equal on the constraint surface. In particular, the Hamiltonian is not well defined and two Hamiltonians differing by linear combinations of the constraints will generate two physically equivalent gauge related trajectories

$$H \sim H' = H + \lambda^m \phi_m, \tag{3.3}$$

where λ^m do not depend on the canonical variables.

After an infinitesimal amount of time, two equivalent dynamical evolutions which started from the same initial conditions differ by terms proportional to the commutators of the dynamical variables with the constraints. That is, one can view the commutator of any function of phase space with a constraint as a representation on phase space of the infinitesimal generator of the symmetry associated with the constraint,

$$\Delta f(p, q) = \{\lambda^m \phi_m(p, q), f(p, q)\} \Delta t. \tag{3.4}$$

Strictly speaking, these symmetries generated by the first class constraints of a theory should be called "gauge" symmetries of a theory. For the case of usual Yang–Mills theories on trivial fiber bundles the symmetries generated by the constraints coincide with the usual idea of gauge symmetries. In general, however, this equivalence is only local and global inequivalences may give rise to observable physical phenomena.

Non-vanishing quantities whose Poisson brackets with the constraints vanish on the constraint surface are called "observables" of the system, since they are the quantities that are invariant under the symmetries generated by the constraints.

3.2.2   Quantum theory

A procedure for quantizing Hamiltonian systems with first class constraints was first proposed by Dirac [27]. Although the original formulation was presented for systems with a finite number of degrees of freedom, it is readily generalizable to the case of field theories. The method consists basically of five steps.

• Select an algebra of quantities in the classical theory general enough to be able to express any quantity of physical interest in terms of the selected quantities. In the simplest examples, one usually takes the canonical coordinates with their Poisson relations as such an algebra.

• Represent this algebra as a set of operators acting on a functional space $\mathcal{V}$ and promote the Poisson bracket relations to relations between commutators of operators. No particular restriction is imposed on the functional space at this level. Again, as an example one can choose representation on functionals of the configuration variables $\Psi[q]$, and represent the fundamental operators as $\hat{q}\Psi[q] = q\Psi[q]$, $\hat{p}\Psi[q] = -i\hbar(\delta\,\Psi[q]/\delta q)$ and their commutation relation as $[\hat{q},\hat{p}] = i\hbar$. (From now on we will choose units such that $\hbar = 1$.)

• Promote the constraint equations to wave equations acting on the space of functions $\mathcal{V}$. This process is, in general, not unique, depending on regularizations and factor orderings. Moreover, it should be performed in such a way as to promote the classical Poisson brackets of the constraint to consistent commutation relations of the wave equations. The space of solutions to the wave equations will, in general, be a restriction of $\mathcal{V}$ and will contain the wavefunctions of physical relevance: we call it $\hat{\mathcal{V}}$.

• Determine the evolution as a function of the parameter of evolution of the associated classical theory of the states (Schödinger picture) or observables (Heisenberg picture) with the use of either the Schrödinger equation for the states,

$$i\frac{\partial\Psi}{\partial t} = \hat{H}\Psi, \tag{3.5}$$

where $\hat{H}$ is the Hamiltonian operator, or the Heisemberg equations for the observables. Notice that the evolution is unambiguous since in the previous point we imposed the constraints on the wavefunctionals. That is, adding a combination of constraints to the Hamiltonian does not change the evolution, since they annihilate the wavefunctionals.

• Introduce an inner product on $\hat{\mathcal{V}}$ such that it becomes a Hilbert space, the observables become self-adjoint operators and the wavefunctions of physical interest become normalizable.

With these steps completed one is in position to do physics by taking expectation values of physical observables using the inner product on the Hilbert space of wavefunctions. Notice that apart from some subtleties, this is what most physicists would recognize as the "usual" procedure of canonical quantization. However, several points need further comment.

First of all notice that in the first step we are allowing the use of a non-canonical algebra to perform the quantization. This is not, strictly speaking, what Dirac originally proposed, since he only considered the use of the algebra of canonical quantities. Allowing a non-canonical (and possibly overcomplete) algebra is more flexible in the sense that it can accommodate dynamical systems which do not naturally have a canonical algebra or situations where to use a canonical algebra is not convenient. It will be important in the formulation of the loop representation.

It could happen that when one performs the Legendre transform to determine the Hamiltonian the end result is a quantity that vanishes on the constraint surface. That is, the Hamiltonian of the theory is a combination of constraints. In this case the Schrödinger (or Heisenberg) equations simply say that the states (or observables) do not evolve with the classical parameter of evolution. In this case the notion of "time" in the system has to be retrieved in a different way. One possibility is to isolate one of the canonical variables as a "time" T and "deparametrize" the theory in such a way that the Hamiltonian constraint can be written as $H \equiv \pi_T - \bar{H}$, where π_T is the variable canonically conjugate to T ("energy"). Then one considers $\bar{H}$ as a true non-vanishing Hamiltonian and T as an evolution variable. The evolution in the "time" T is generated with the Hamiltonian $\bar{H}$ and its corresponding Schrödinger (or Heisenberg) equation. This procedure is generically by no means trivial and in many systems it is not known how to perform it in a consistent way. Many systems have vanishing Hamiltonians and almost any system can be written in such a way that the Hamiltonian vanishes (these are usually called "parametrized theories"). Other systems, however, come naturally "already parametrized". An example of this behavior, as we shall see, is general relativity. Other examples are the relativistic free particle and string. A comprehensive discussion of these and other related issues is the review article by Kuchař [57].

Another thing that can happen is that the theory could have symmetries that are not reflected in the appearance of constraints. This is usually the case with global symmetries, like "large" gauge transformations or diffeomorphisms. We will largely ignore these in this book. In principle, one should require that wavefunctions transform as unitary representations of the respective symmetry. This can put constraints on the inner product one selects, as has been emphasized by Peldán [58].

In the method presented above the first three steps contained specific proposals. Although in each of them one is faced with many inequivalent choices, one can always pick one of them and proceed. A different situation arises in the last step, where no prescription for the introduction of an inner product is made. The Dirac quantization procedure does not specify how to introduce an inner product and in this sense it is incomplete. This situation is particularly complex in systems where one does not have extra auxiliary structures that in some sense determine a preferred inner product. For instance, in usual field theories on a flat background the Poincaré invariance uniquely fixes the inner product. However, in the gravitational case, for instance, one does not have at hand such a guiding principle. There are proposals to extend Dirac's method of quantization in such a way as to have a program that chooses an inner product without resorting to any additional symmetries or structures. Among these proposals is that of Ashtekar [2] who suggests endowing the phase space with a star algebra structure which may be sufficient to fix the inner product. The issue of the inner product in non-linear field theories is by no means completely understood at present and in our book we will discuss it only tangentially.

3.3 Yang–Mills theories

Yang–Mills theories have proven to be very useful as descriptions of the physics of the elementary particles. An extensive literature has dealt with them from various viewpoints and at present there is a good understanding of many of their features. It is therefore reasonable to introduce Yang–Mills theories at this point to illustrate various concepts we will need in the rest of the book, especially in the applications to gravity. In particular Yang–Mills theories have proven an adequate ground to develop techniques related to loops. Many techniques and results that are only conjectured to hold for the gravitational case have actually been proved and exhaustively studied for the Yang–Mills case.

In this section we will introduce the canonical formulation of classical and quantum Yang–Mills theories in terms of the traditional variables. In particular we will study the meaning of the Gauss law as a constraint and generator of the gauge symmetries of the theory. In subsequent sections we will review these results in the language of loops.

3.3.1 Canonical formulation

Yang–Mills theories are based on an algebra valued connection $\mathbf{A}_\mu$ on a flat manifold $\eta_{\mu\nu}$ with an action

$$S \equiv \int d^4x \tilde{\mathcal{L}} = -\tfrac{1}{2} \int d^4x \sqrt{\eta} \eta^{\mu\lambda} \eta^{\nu\rho} \mathrm{Tr}(\mathbf{F}_{\mu\nu} \mathbf{F}_{\lambda\rho}), \qquad (3.6)$$

where $\mathbf{F}_{\mu\nu} = \partial_\mu \mathbf{A}_\nu - \partial_\nu \mathbf{A}_\mu + i[\mathbf{A}_\mu, \mathbf{A}_\nu]$ and $[\,,\,]$ is the commutator in the Lie algebra associated with the gauge group. We denote by $\sqrt{\eta}$ the square root of the absolute value of the determinant of the metric. Elements of the Lie algebra will be denoted with boldface characters. Sometimes it will also be convenient to introduce the notation in components in terms of the basis of generators of the Lie algebra, for instance,

$$\mathbf{A}_\mu = A_\mu^i \mathbf{X}^i, \qquad (3.7)$$

where $\mathbf{X}^i$ are the generators of the Lie algebra satisfying

$$[\mathbf{X}^i, \mathbf{X}^j] = C_k^{ij} \mathbf{X}^k, \qquad (3.8)$$

where C_k^{ij} are the structure constants of the group in question.

We take as background metric $\eta = \mathrm{diag}(-1, 1, 1, 1)$ and consider as configuration variables $\mathbf{A}_0$ and $\mathbf{A}_a$ and compute their canonical momenta[*], $\tilde{\pi}^0$ and $\tilde{\pi}^a$

$$\tilde{\pi}^0 \equiv \frac{\delta S}{\delta \dot{\mathbf{A}}_0} = 0, \qquad (3.9)$$

$$\tilde{\pi}^a \equiv \frac{\delta S}{\delta \dot{\mathbf{A}}_a} = \sqrt{\eta} \eta^{ab} (\dot{\mathbf{A}}_b - \partial_b \mathbf{A}_0 + i[\mathbf{A}_0, \mathbf{A}_a]) = \tilde{\mathbf{E}}^a. \qquad (3.10)$$

The momentum conjugate to $\mathbf{A}_0$ vanishes. This will be a primary constraint. We now perform the Legendre transform to define the Hamiltonian density $\tilde{\mathcal{H}} = 2\mathrm{Tr}(\tilde{\mathbf{E}}^a \dot{\mathbf{A}}_a) - \tilde{\mathcal{L}}$,

$$\tilde{\tilde{\mathcal{H}}} = \mathrm{Tr}\left((\tilde{\mathbf{E}}^a \tilde{\mathbf{E}}^b \eta_{ab} + \tilde{\mathbf{B}}^a \tilde{\mathbf{B}}^b \eta_{ab}) - \sqrt{\eta} \mathbf{A}_0 (\mathbf{D}_a \tilde{\mathbf{E}}^a) \right) \qquad (3.11)$$

where $\mathbf{D}_a \equiv \partial_a + i[\mathbf{A}_a,]$ is the gauge covariant derivative. We can now extend the Hamiltonian including the primary constraint $\tilde{\mathcal{H}}' = \tilde{\mathcal{H}} + \mathrm{Tr}(\lambda_0 \tilde{\pi}^0)$, where λ_0 is a group-valued Lagrange multiplier.

---

[*] In this book we will use an overtilde to denote tensor densities of weight +1, and an undertilde for weight -1, a notation that is becoming standard. The only exceptions will be the Dirac delta function, which is a +1 density, but which we will denote as $\delta(x - y)$ to adhere to usual conventions, the square root of the determinant of metrics, since it is obvious, and — in an effort to try not to clutter the notation — the multitensor densities $X^{a_1 x_1 \cdots a_n x_n}$, since their tensor density character has been abundantly emphasized.

To compute the equations of motion of this theory we take the Poisson brackets of the phase space variables with the Hamiltonian. In particular, one observes that the time evolution of $\tilde{\pi}_0$ is given by

$$\dot{\tilde{\pi}}^0 = -\frac{\delta\tilde{\mathcal{H}}}{\delta\mathbf{A}_0} = \mathbf{D}_a\tilde{\mathbf{E}}^a = \partial_a\tilde{\mathbf{E}}^a + i[\mathbf{A}_a, \tilde{\mathbf{E}}^a] = 0. \qquad (3.12)$$

This equation guarantees the preservation in time of the primary constraint. It is in itself a new (secondary) constraint. It can be checked that this constraint is automatically conserved. Moreover, the primary and secondary constraints are first class, i.e.,

$$\{\mathcal{P}(\mu), \mathcal{P}(\lambda)\} = 0, \qquad (3.13)$$

$$\{\mathcal{P}(\mu), \mathcal{G}(\lambda)\} = 0, \qquad (3.14)$$

$$\{\mathcal{G}(\lambda), \mathcal{G}(\mu)\} = \mathcal{G}([\lambda, \mu]), \qquad (3.15)$$

where we have introduced the notation of "smeared out constraints",

$$\mathcal{G}(\lambda) \equiv \int d^3x \lambda D_a\tilde{\mathbf{E}}^a, \qquad (3.16)$$

$$\mathcal{P}(\mu) \equiv \int d^3x \mu \tilde{\pi}^0. \qquad (3.17)$$

From now on, every time we refer to a constraint as dependent on a parameter we will assume that the parameter has the needed index structure to be contracted with the constraint and an integration over the manifold has been performed. This enables us to avoid dealing with distributional expressions. Notice the geometric interpretation of the Gauss law as a generator of infinitesimal gauge transformations associated with the arbitrary group valued function λ,

$$\{\mathcal{G}(\lambda), \mathbf{A}_a\} = \mathbf{D}_a\lambda, \qquad (3.18)$$

$$\{\mathcal{G}(\lambda), \tilde{\mathbf{E}}^a\} = [\lambda, \tilde{\mathbf{E}}^a]. \qquad (3.19)$$

With this point of view of the Gauss law as a generator of gauge transformations notice that one can interpret the commutator (3.15) in the following way: the commutator of the infinitesimal gauge transformation parametrized by λ and that parametrized by μ is an infinitesimal gauge transformation parametrized by $[\lambda, \mu]$. The primary constraint simply states that the zeroth component of the vector potential can be arbitrarily rescaled,

$$\{\mathcal{P}(\mu), \mathbf{A}_0(x)\} = -\mu(x). \qquad (3.20)$$

This tells us that $\mathbf{A}_0$ and π^0 could be eliminated from the classical theory by appropriate rescalings. This fact will find a counterpart in quantum theory. This ends the discussion of the classical theory.

3.3.2 Quantization

We will now apply the program of quantization that we introduced in section 3.2.2 to the canonical formulation of Yang–Mills theories. We start by picking an algebra of classical quantities, in our case simply the canonical algebra in terms of the Poisson brackets,

$$\{A_a^i(x), \tilde{E}_j^b(y)\} = \delta_a^b \delta_j^i \delta(x - y), \tag{3.21}$$

$$\{A_0^i(x), \tilde{\pi}_j^0(y)\} = \delta_j^i \delta(x - y), \tag{3.22}$$

with all other brackets vanishing. We also pick a polarization for the wavefunctionals $\Psi[\mathbf{A}, \mathbf{A}_0]$ where by $\mathbf{A}$ we mean the one form on the spatial surface with components $\mathbf{A}_a$

We now find a representation of the canonical algebra by defining,

$$\hat{\mathbf{A}}_a \Psi[\mathbf{A}, \mathbf{A}_0] = \mathbf{A}_a \Psi[\mathbf{A}, \mathbf{A}_0], \tag{3.23}$$

$$\hat{\tilde{\mathbf{E}}}^a \Psi[\mathbf{A}, \mathbf{A}_0] = -i \frac{\delta}{\delta \mathbf{A}_a} \Psi[\mathbf{A}, \mathbf{A}_0], \tag{3.24}$$

$$\hat{\mathbf{A}}_0 \Psi[\mathbf{A}, \mathbf{A}_0] = \mathbf{A}_0 \Psi[\mathbf{A}, \mathbf{A}_0], \tag{3.25}$$

$$\hat{\tilde{\pi}}^0 \Psi[\mathbf{A}, \mathbf{A}_0] = -i \frac{\delta}{\delta \mathbf{A}_0} \Psi[\mathbf{A}, \mathbf{A}_0]. \tag{3.26}$$

Notice that up to now we have performed several arbitrary choices, which in general would yield inequivalent quantum theories if performed in a different way. For example, we could have added the functional gradient of an arbitrary function of $\mathbf{A}$ to the definition of the conjugate momenta $\hat{\tilde{\mathbf{E}}}^a$ and this would yield the same quantum commutator algebra.

We now promote the constraints to quantum equations and impose them on the wavefunctions. The primary constraint can be satisfied immediately, simply by noticing that it requires the wavefunctionals not to depend on $\mathbf{A}_0$,

$$\mathcal{P}(\mu)\Psi[\mathbf{A}, \mathbf{A}_0] = -i \int d^3x \mu \frac{\delta}{\delta \mathbf{A}_0} \Psi[\mathbf{A}, \mathbf{A}_0]. \tag{3.27}$$

First of all, notice that we have imposed the "smeared out" form of the constraint, and we will usually do this. This is equivalent to imposing the constraint point by point in the manifold since the equation should hold for an arbitrary smearing function μ. Moreover, it is instructive to view the action of the constraint in the following way. Consider the action of $(1 + i\epsilon\mathcal{P}(\mu))$ on a wavefunction in the limit $\epsilon \to 0$,

$$(1 + i\epsilon\mathcal{P}(\mu))\Psi[\mathbf{A}, \mathbf{A}_0] = \Psi[\mathbf{A}, \mathbf{A}_0 + \epsilon\mu]. \tag{3.28}$$

We see that the quantum constraint acts as the infinitesimal generator on the wavefunctions of the symmetry that we mentioned in the classical theory: that the component $\mathbf{A}_0$ of the vector potential could be rescaled arbitrarily. From now on we will therefore concentrate on functionals that only depend on the spatial part of the connection, $\Psi[\mathbf{A}]$.

Let us now focus on the Gauss law. We can promote it to a quantum operator in the following way,

$$\hat{\mathcal{G}}(\lambda)\Psi[\mathbf{A}] = -i \int d^3x \lambda^k \left(\partial_a \frac{\delta}{\delta A_a^k} + C_{lm}^k A_a^l \frac{\delta}{\delta A_a^m} \right) \Psi[\mathbf{A}]. \qquad (3.29)$$

Notice that, in principle, there is a factor ordering ambiguity when representing the non-Abelian term of the covariant derivative. The reader may check that due to the symmetries of the structure constants of any compact group (in particular $SU(N)$) it is immaterial which ordering is picked for the non-Abelian term.

Let us again study the infinitesimal action of the Gauss law on wavefunctionals,

$$(1 + i\epsilon\mathcal{G}(\lambda))\Psi[\mathbf{A}] = \Psi[\mathbf{A} + \epsilon\mathbf{D}\lambda]. \qquad (3.30)$$

We see that it acts as an infinitesimal generator of gauge transformations on the wavefunctionals. It is therefore immediate to solve the constraint. One just has to consider wavefunctionals which are gauge invariant functions of the connection and they will automatically be annihilated by the Gauss law.

Notice that the Gauss law, both at a classical and quantum mechanical level, only generates gauge transformations connected to the identity. "Large" gauge transformations are not included and their presence can give rise to observable physical effects. This is a generic feature of constrained systems. Constraints usually only generate local gauge symmetries. In the case of Yang-Mills theories the presence of large gauge transformations gives rise to the Θ-vacua, connected with the instanton structure of the theory [59]. Similar effects arise for gravity [61]. For both Yang–Mills and gravity "large" gauge transformations are responsible for the presence of fractional spin states [60].

One should now study the evolution of the wavefunctionals (let us adopt for the sake of argument the Schrödinger picture). For that we have to promote the Hamiltonian of the theory to an operator. This can be accomplished with a straightforward factor ordering (though a regularization is needed). One can then study the eigenstates and spectra of eigenvalues of the theory. In Yang-Mills theories the interpretation of the eigenvalues would be the masses of the particle spectra of the theory. This formulation would lead to a non-perturbative solution of Yang-Mills theories if one could implement the evolution equation and introduce an inner product. The treatment of this problem is involved and there is not a closed solution for it in the continuum, although lattice techniques have been applied to it. We will return to these issues in chapter 6.

3.4 Wilson loops

Observable quantities in gauge theories need to be gauge invariant. Wavefunctions in a quantum representation also need to be gauge invariant. It will therefore be very useful to introduce a set of quantities involving the connection $\mathbf{A}_a$ in terms of which any gauge invariant quantity can be written. These objects are called Wilson loops, are gauge invariant under both small and large gauge transforations and are constructed taking traces of the holonomy,

$$W_{\mathbf{A}}(\gamma) = \mathrm{Tr}\left[\mathrm{P}\exp\left(i\oint_{\gamma} dy^a \mathbf{A}_a\right)\right]. \tag{3.31}$$

The gauge invariance of these quantities follows immediately from the properties of the connection and the holonomy that were introduced in chapter 1 and the cyclicity of the traces. Because of this, they are observables in the canonical sense (they have vanishing Poisson brackets with all the constraints of the theory).

The objects are dependent on a loop and have a non-local dependence on the gauge connection. In general they are complex numbers. We can write the Wilson loop using the notation of chapter 2 as

$$W_{\mathbf{A}}(\gamma) = \mathrm{Tr}(I) + \sum_{n=1}^{\infty} i^n \mathrm{Tr}(\mathbf{A}_{a_1}(x_1)\dots\mathbf{A}_{a_n}(x_n))X^{a_1 x_1\dots a_n x_n}(\gamma). \tag{3.32}$$

Observe that the trace $\mathrm{Tr}(A_{a_1}(x_1)\dots A_{a_n}(x_n))$ is cyclic in the indices $a_1 x_1\dots a_n x_n$, and therefore the Wilson loop only depends on the cyclic portion of the multitangents. As we mentioned in chapter 2, this removes all information about the basepoint of the loop. That is, Wilson loops are functions of non-basepointed loops.

Wilson loops have two fundamental properties, the discussion of which will occupy the rest of this section:

- the Mandelstam identities;

- the reconstruction property.

The Mandelstam identities are a set of relations between Wilson loops which reflect the structure of the particular gauge group considered. The reconstruction property will tell us that given the Wilson loop functions evaluated for all possible loops we can reconstruct all the gauge invariant information present in the gauge connection. Both properties together will imply that Wilson loops constitute an overcomplete basis of solutions of the Gauss law constraint.

3.4.1   The Mandelstam identities

The Mandelstam identities are the reflection in the language of Wilson loops of the particular properties of the gauge group used to define the holonomies and of generic properties of traces. In terms of them we will see reflected group properties such as unitarity, the dimension of the representation and value of the determinant of matricial representations. They will allow us to express products of Wilson loops in terms of sums of products involving a smaller number of Wilson loops.

These identities were first introduced by Mandelstam [9] for the $O(3)$ group. Giles [35] extended them for groups $GL(N)$ and Gambini and Trias [34] extended them to the case of special and unitary groups. Loll [23] discussed the case of certain non-compact groups.

Let us consider gauge groups that admit fundamental representations in terms of $N \times N$ matrices, for instance, $GL(N)$, $SL(N)$, $U(N)$, $SU(N)$. The Mandelstam identities arise as a consequence of the properties of the traces of $N \times N$ matrices. There are two kinds of identities, called identities of the first and second kinds.

The Mandelstam identities of the first kind are a simple consequence of the cyclic property of the traces, which we mentioned in the previous section[†],

$$W(\gamma_1 \circ \gamma_2) = W(\gamma_2 \circ \gamma_1). \qquad (3.33)$$

These identities hold for *any* gauge group of any dimension.

There are various identities of the second kind. The first family are a set of non-linear constraints that ensure that $W_{\mathbf{A}}(\gamma)$ is a trace of an $N \times N$ matrix. They can be obtained in the following way.

Observe first that in N dimensions any object with $N + 1$ totally antisymmetric indices vanishes,

$$\delta_{[B_1}^{A_1} \delta_{B_2}^{A_2} \cdots \delta_{B_{N+1}]}^{A_{N+1}} = 0. \qquad (3.34)$$

Then contract this with $N + 1$ holonomies,

$$H(\gamma_1)_{A_1}^{B_1} \ldots H(\gamma_{N+1})_{A_{N+1}}^{B_{N+1}} \qquad (3.35)$$

where $A_1, B_1, \ldots, A_{N+1}, B_{N+1}$ are matrix indices in the matricial representation of the group. The result is an identically vanishing sum of products of traces of products of holonomies. From here one can work out explicitly the identities for any order. For example, if $N = 1$, as in a

---

[†] In this section and the following we will omit writing the dependence of the Wilson loop on the connection $\mathbf{A}$ since the results proven will not depend on the choice of a particular connection

$U(1)$ group, the identity reads

$$W(\gamma_1)W(\gamma_2) - W(\gamma_1 \circ \gamma_2) = 0. \tag{3.36}$$

There is a compact way of writing this identity for an arbitrary order in terms of the quantities M_K, depending on K loops and defined by the following recurrence relations

$$(K+1)M_{K+1}(\gamma_1,\ldots,\gamma_{K+1}) \equiv W(\gamma_{K+1})M_K(\gamma_1,\ldots,\gamma_K)$$
$$-M_K(\gamma_1 \circ \gamma_{K+1}, \gamma_2,\ldots,\gamma_K) - \ldots - M_K(\gamma_1,\gamma_2,\ldots\gamma_K \circ \gamma_{K+1}), \tag{3.37}$$
$$M_1(\gamma) \equiv W(\gamma). \tag{3.38}$$

In terms of the Ms, the identity for an $N \times N$ matrix group can be written as

$$M_{N+1}(\gamma_1,\ldots,\gamma_{N+1}) = 0. \tag{3.39}$$

Notice that for the case of $N \times N$ matrices it is also true that

$$M_L(\gamma_1,\ldots,\gamma_L) = 0 \; \forall L > N+1. \tag{3.40}$$

An immediate consequence of the recurrence relation (3.37), obtained identifying the loop $N+1$ with ι (the identity loop), is

$$(N+1)M_{N+1}(\gamma_1,\ldots,\gamma_N,\iota) = (W(\iota) - N)M_N((\gamma_1,\ldots,\gamma_N) = 0, \tag{3.41}$$

from which we see that

$$W(\iota) = N. \tag{3.42}$$

Let us examine another example, for 2×2 matrices. One can expand the product of three traces in terms of two,

$$W(\gamma_1)W(\gamma_2)W(\gamma_3) = W(\gamma_1 \circ \gamma_2)W(\gamma_3) + W(\gamma_2 \circ \gamma_3)W(\gamma_1)$$
$$+W(\gamma_3 \circ \gamma_1)W(\gamma_2) - W(\gamma_1 \circ \gamma_2 \circ \gamma_3) - W(\gamma_1 \circ \gamma_3 \circ \gamma_2). \tag{3.43}$$

For instance $SU(2)$, $SU(1,1)$ and other groups that admit fundamental representations in terms of 2×2 matrices give rise to Wilson loops that satisfy the identity (3.43). These groups also admit other identities that reflect other properties apart from the 2×2 matricial nature of their representation.

Notice that because we are working with non-basepointed loops, the composition of two loops $\gamma_1 \circ \gamma_2$ in general is not well defined. For the remainder of this section whenever a composition of two loops appears, we will assume an arbitrary basepoint has been chosen to perform the composition. One simply links both loops to the basepoint through arbitrary retraced paths. The Mandelstam identities are independent of the basepoint chosen to define the composition of the loops.

Another identity appears for special groups, i.e., groups that admit fundamental representations in terms of matrices of unit determinant. As was proved in reference [34], for a group with fundamental representation in terms of $N \times N$ matrices of unit determinant the following identity in terms of the Ms holds:

$$M_N(\gamma_1 \circ \gamma, \gamma_2 \circ \gamma, \ldots, \gamma_N \circ \gamma) = M_N(\gamma_1, \gamma_2, \ldots, \gamma_N) \tag{3.44}$$

from which it trivially follows that $M_N(\gamma, \gamma, \ldots, \gamma) = 1$. These identities allow us, for a special group, to express the product of N Wilson loops in terms of that of $N - 1$ by taking $\gamma = \gamma_i$ for some i in equation (3.44). For example, for any special 2×2 matrix group (such as $SU(2)$, $SL(2, C)$, etc),

$$M_2(\gamma_1, \gamma_2) = M_2(\gamma_1 \circ \gamma_2^{-1}, \iota) \tag{3.45}$$

and

$$M_2(\gamma_1, \gamma_2) = \tfrac{1}{2}(W(\gamma_1)W(\gamma_2) - W(\gamma_1 \circ \gamma_2)), \tag{3.46}$$

$$M_2(\gamma_1 \circ \gamma_2^{-1}, \iota) = \tfrac{1}{2}(W(\gamma_1 \circ \gamma_2^{-1})W(\iota) - W(\gamma_1 \circ \gamma_2^{-1})), \tag{3.47}$$

therefore,

$$W(\gamma_1)W(\gamma_2) = W(\gamma_1 \circ \gamma_2^{-1}) + W(\gamma_1 \circ \gamma_2). \tag{3.48}$$

Finally, we will discuss the Mandelstam identities of the second kind that reflect the fact that a group is unitary. That is, if the group admits a fundamental representation in terms of unitary $N \times N$ matrices, the Wilson loops satisfy

$$W(\gamma) = W^*(\gamma^{-1}), \tag{3.49}$$

where * indicates the complex conjugate.

In general, apart from the Mandelstam identities, Wilson loops satisfy a series of inequalities. For instance, for unitary groups, the following inequality holds trivially

$$|W(\gamma)| \leq |W(\iota)| = N. \tag{3.50}$$

These inequalities contain additional information that is not present in the identities we discussed previously. For instance, all the identities we have discussed so far are the same for the groups $SU(2)$ and $SU(1,1)$. It is by considering inequalities in terms of the Wilson loops that one can determine which of these two groups is being considered. A discussion of inequalities and their consequences can be found in reference [23].

Let us end by summarizing the Mandelstam identities for the group $SU(2)$, which we will use extensively in this book:

Identity of the first kind,

$$W(\gamma_1 \circ \gamma_2) = W(\gamma_2 \circ \gamma_1). \tag{3.51}$$

Identity of the second kind,

$$W(\gamma_1)W(\gamma_2) = W(\gamma_1 \circ \gamma_2^{-1}) + W(\gamma_1 \circ \gamma_2). \qquad (3.52)$$

From here it is immediate to prove, choosing $\gamma_1 = \iota$,

$$W(\gamma) = W(\gamma^{-1}), \qquad (3.53)$$

and from this and the unitarity property it follows that $W(\gamma)$ is real and less than or equal to 2 in absolute value.

In spite of their simple appearance, successive combinations of the Mandelstam identities can lead to very non-trivial relations among Wilson loops. In any formulation in which one wants to use the Wilson loops as basic variables, these relations imply an additional complication in the formulation of the theory, since there is no definite way to determine which are the freely specifiable functions [24]. In particular there is no systematic way of generating the set of all possible relations between products of Wilson loops that are derived from the Mandelstam identities [24]. An important development in this area is the recognition by Rovelli and Smolin that spin networks might be used to characterize a complete set of independent products of Wilson loops [146]. We will return to these issues when we discuss the loop representation.

3.4.2 Reconstruction property

In the previous section we introduced a set of identities satisfied by Wilson loops. In this section we will study the opposite question: to what extent does a prescribed function of loops, satisfying the Mandelstam identities, qualify to become a Wilson loop? In particular, can we reconstruct the holonomy given such a function?

This question is of great importance. From the results of chapter 1 we have seen that one could use holonomies to describe gauge theories since they embody all the gauge invariant information of the connection. What we are about to do is to show that all the information present in a holonomy can be reconstructed from the Wilson loops. That is, the Wilson loops will acquire a status of fundamental variables in themselves since we will be able to reconstruct all the gauge invariant information of a theory from them. This step will be of fundamental importance in following sections where we will formulate a quantum representation purely in terms of loops.

The proof that this can actually be accomplished, i.e., that given a function of loops satisfying the Mandelstam constraints one can reconstruct the gauge invariant information encoded in it is the subject of the so called "reconstruction theorems". The idea is the following. Given a function $W(\gamma)$, satisfying the Mandelstam constraints (3.33), (3.39) it is possible to construct explicitly a set of $N \times N$ matrices $\mathbf{H}_\gamma$ defined modulo a similarity transformation, such that their traces are $W(\gamma)$. The first such theorem was proved by Giles [35] for the case of $U(N)$. Loll considered the cases of $SU(2)$ [25]. Ashtekar and Lewandowski [40] refined many of the proofs presented early in the literature and introduced several generalizations. Here we will discuss a simplified derivation assuming the Mandelstam identities for a 2×2 matrix group and we will follow the presentation of Giles. An elegant, short, alternative derivation of the reconstruction property has recently been presented in reference [40].

One starts by defining an algebra associated with the group of loops. It is constructed in the following way. Take the group of loops $\mathcal{L}_o$. Define a formal sum and product by a complex number law for elements of $\mathcal{L}_o$. Construct then an algebra $F\mathcal{L}_o$ by appending to the elements of $\mathcal{L}_o$ all their possible finite complex linear combinations. The product law of the algebra will be induced by the composition law of $\mathcal{L}_o$.

We now consider the extension of the notion of Wilson loop to this algebra. For those elements of $F\mathcal{L}_o$ belonging to $\mathcal{L}_o$ it is defined in the usual way. For linear combinations of them it is given by

$$W(a_1\gamma_1 + a_2\gamma_2) = a_1 W(\gamma_1) + a_2 W(\gamma_2). \qquad (3.54)$$

Notice that $\gamma_1, \gamma_2 \in \mathcal{L}_o$ and therefore the $W(\gamma_{1,2})$ are well defined. From now on we will use the same notation for elements of $F\mathcal{L}_o$ and elements of $\mathcal{L}_o$, it will be clear from the context to which we are referring.

This algebra is isomorphic to a complexification of the algebra $\mathcal{E}_D$ that we introduced in chapter 2, obtained by allowing the multitensor densities $E_D^{a_1 x_1 \dots a_n x_n}$ that satisfied the differential constraint to become complex-valued.

We want to see if these extended Wilson loops can be obtained as traces of "extended" holonomies $H(\gamma)$ in the sense introduced in chapter 2 (traces of linear combinations of holonomies are allowed). We would like to think of $H(\gamma)$ as representations of $F\mathcal{L}_o$. Notice that $F\mathcal{L}_o$ is associated with an infinite-dimensional group (in particular because $\mathcal{L}_o$ is) whereas the vector space of extended holonomies is finite-dimensional (they are 2×2 matrices in our simplified derivation). Therefore many elements of $F\mathcal{L}_o$ are represented by the same matrix. We now introduce an equivalence relation such that two elements of $F\mathcal{L}_o$ are equivalent if they lead to the same matrix. We are then able to establish a correspondence between equivalence classes of elements of $F\mathcal{L}_o$ and the matrices.

We say that $\gamma_1 \sim \gamma_2$ if

$$W(\gamma_1 \circ \zeta) = W(\gamma_2 \circ \zeta) \quad \forall \ \zeta. \qquad (3.55)$$

By the definition (3.54) it is obvious that the equivalence relation defined is compatible with the sum and product times a complex number.

We now prove that it is compatible with the product law of the algebra. Suppose $\gamma_1 \sim \gamma_2$ and $\eta_1 \sim \eta_2$. Then

$$W(\gamma_1 \circ \eta_1 \circ \zeta) = W(\gamma_2 \circ \eta_1 \circ \zeta) = W(\eta_1 \circ \zeta \circ \gamma_2)$$
$$= W(\eta_2 \circ \zeta \circ \gamma_2) = W(\gamma_2 \circ \eta_2 \circ \zeta). \qquad (3.56)$$

We denote by $F\mathcal{L}_o/K$ the algebra of equivalence classes of extended loops, K being the kernel of the equivalence relation.

We now use the Mandelstam identity of the second kind (3.39), to derive an explicit form for the matrix associated with an equivalence class belonging to $F\mathcal{L}_o/K$. Let us explicitly consider the identity for the case of 2×2 matrices already introduced in equation (3.43) (notice that we do not, at this stage, know the dimension of the representation and we will *prove* that the representation is (2×2) dimensional based on this identity),

$$W(\gamma_1)W(\gamma_2)W(\zeta) = W(\gamma_1 \circ \gamma_2)W(\zeta) + W(\gamma_1)W(\gamma_2 \circ \zeta)$$
$$+ W(\gamma_2)W(\gamma_1 \circ \zeta) - W(\gamma_1 \circ \gamma_2 \circ \zeta) - W(\gamma_2 \circ \gamma_1 \circ \zeta). \quad (3.57)$$

We will interpret this identity in the following way. Consider two elements of the algebra γ_1 and γ_2. The identity should hold for arbitrary ζ. This means that the identity between elements of $F\mathcal{L}_o$ (3.57) induces an identity between equivalence classes given by

$$(W(\gamma_1)W(\gamma_2) - W(\gamma_1 \circ \gamma_2))\iota - W(\gamma_1)\gamma_2 - W(\gamma_2)\gamma_1$$
$$+ \gamma_1 \circ \gamma_2 + \gamma_2 \circ \gamma_1 = 0, \qquad (3.58)$$

where ι should be understood as the identity element of $F\mathcal{L}_o$.

We will now use this identity to determine the eigenvalues of a matrix associated with the loop γ. To this end, we put $\gamma_1 = \gamma_2 = \gamma$ in (3.58) and get

$$\tfrac{1}{2}(W(\gamma)^2 - W(\gamma^2))\iota - W(\gamma)\gamma + \gamma^2 = 0. \qquad (3.59)$$

This second order relation can be factorized as

$$(\gamma - \lambda_1\iota)(\gamma - \lambda_2\iota) = 0, \qquad (3.60)$$

where $\lambda_1 + \lambda_2 = W(\gamma)$ and $\lambda_1\lambda_2 = \tfrac{1}{2}(W(\gamma)^2 - W(\gamma^2))$. If we now want to represent γ by a matrix, we see that it has at most two different eigenvalues. Therefore, this proves that a 2×2 representation suffices.

Let us now assume[‡] that for at least one γ, which we will call γ_0, $\lambda_1 \neq \lambda_2$. We have therefore established the form of the matrix $H(\gamma_0)$ associated with a particular loop γ_0, and it is in diagonal form. Notice that because

---

[‡] This assumption is not really needed, see Giles [35] for the exceptional case in which no such element exists.

holonomies are defined modulo a single similarity transformation at the basepoint, it is impossible to diagonalize the holonomies simultaneously for all possible γs.

We will now determine the matrix element associated with an arbitrary element γ of the algebra. With this aim we define the elements of the algebra,

$$\phi_1 = \frac{\gamma_0 - \lambda_1 \iota}{\lambda_1 - \lambda_2}, \tag{3.61}$$

$$\phi_2 = \frac{\gamma_0 - \lambda_2 \iota}{\lambda_1 - \lambda_2}, \tag{3.62}$$

which behave as projectors, $\phi_1\phi_2 = \phi_2\phi_1 = 0$, $\phi_i^2 = \phi_i$, $\phi_1 + \phi_2 = \iota$. The reader can check by applying the definition (3.54) that $W(\phi_i) = 1$. The idea of introducing these elements is that in a matricial representation they will behave as projectors on the one-dimensional eigenspaces associated with each eigenvalue.

We now apply these projectors. Given an arbitrary element η of the algebra we define its "components" η_{ij} by

$$\eta_{ij} = \phi_i \eta \phi_j. \tag{3.63}$$

As can be readily seen from their definition and the definition of the projectors, these "components" satisfy

$$\eta = \sum_{i,j=1}^{2} \eta_{ij}, \tag{3.64}$$

$$(\eta_1\eta_2)_{ij} = \sum_{k=1}^{2} (\eta_1)_{ik}(\eta_2)_{kj}, \tag{3.65}$$

$$W(\eta_{ij}) = \delta_{ij} W(\eta_{jj}). \tag{3.66}$$

We will now use these definitions to compute the "diagonal" elements of the algebra η_{ii}. They are given by

$$\eta_{ii} = W(\eta_{ii})\phi_i \tag{3.67}$$

(no sum over i is assumed). Let us prove this for the "11 component", the proof being totally analogous for the other component. We apply the Mandelstam identity to the following elements of the algebra, ϕ_2, η_{11} and an arbitrary element ζ,

$$W(\phi_2)W(\eta_{11})W(\zeta) = W(\phi_2 \circ \eta_{11})W(\zeta) + W(\phi_2)W(\eta_{11} \circ \zeta)$$
$$+ W(\eta_{11})W(\phi_2 \circ \zeta) - W(\phi_2 \circ \eta_{11} \circ \zeta) - W(\eta_{11} \circ \phi_2 \circ \zeta), \tag{3.68}$$

and observing that $\phi_2 \circ \eta_{11} = 0$ and $W(\phi_2) = 1$,

$$W(\eta_{11})W(\zeta) - W(\eta_{11} \circ \zeta) - W(\eta_{11})W(\phi_2 \circ \zeta) = 0, \tag{3.69}$$

which implies the following relation among equivalence classes (since ζ is arbitrary),

$$\eta_{11} = W(\eta_{11})(\iota - \phi_2) = W(\eta_{11})\phi_1, \tag{3.70}$$

and therefore this is the expression of the "11 component" of the element η. All this language in terms of the elements of the algebra has a natural counterpart in terms of the actual components of the representation in terms of 2×2 matrices $H(\eta)$. In particular, the diagonal components of the matrix are therefore given by

$$H(\eta)_{ii} = W(\eta_{ii}). \tag{3.71}$$

The non-diagonal elements are not uniquely determined. Remember that to perform the construction we chose a particular γ_0 represented by a diagonal matrix. There exist similarity transformations that maintain the diagonality of $H(\gamma_0)$ but change the non-diagonal components of the representations of a generic element η.

To determine the non-diagonal components, Giles [35] introduces a procedure based on picking a second specific loop η_0 and fixing the value of some off-diagonal components of its matrix representation. In this way, the freedom to perform similarity transformations is frozen. In the 2×2 case, one needs to fix one component, say, $H(\eta_0)_{12} = 1$. The other component of this matrix is determined by

$$H(\eta_0)_{21} \equiv W(\phi_1 \circ \eta_0 \circ \phi_2 \circ \eta_0). \tag{3.72}$$

This completes the determination of all the matrix elements of the fixed element η_0. The matrix elements of an arbitrary element η are given by

$$H(\eta)_{21} = W(\eta_{21} \circ \eta_0), \tag{3.73}$$

$$H(\eta)_{12} = W(\eta_{12} \circ \eta_0)/H(\eta_0)_{21}. \tag{3.74}$$

With this construction one actually has a representation of the algebra,

$$H(\gamma \circ \eta)_{ij} = \sum_k H(\gamma)_{ik} H(\eta)_{kj}, \tag{3.75}$$

which can be verified by combining the following expression (which is a consequence of equation (3.67)),

$$W(\eta_{jj} \circ \gamma_{jj}) = W(\eta_{jj})W(\gamma_{jj}), \tag{3.76}$$

and equations (3.72), (3.74) and (3.73).

Let us review what has been accomplished so far. We have established a procedure to reconstruct a holonomy given a set of quantities that satisfies the Mandelstam identities. In particular, this proves that one can reconstruct a holonomy from Wilson loops. The holonomy so constructed constitutes a representation of the group of loops the traces of which satisfy the Mandelstam identities.

The representation found only reproduces the Mandelstam identities of a general (2×2) matrix, the ones we used explicitly in the reconstruction. One could extend the method to take into account more specific Mandelstam identities (or inequalities). For instance, if one applies the above construction to a set of $SU(2)$ Wilson loops satisfying identities (3.48), (3.49) one does not necessarily end up with an $SU(2)$ holonomy but with a holonomy that satisfies the said identities. This could be accomplished, for instance, by an $SU(1,1)$ holonomy as well.

An important point to notice is that the reconstructed holonomy from an arbitrary set of functions satisfying the Mandelstam identities will in general not correspond to a usual smooth connection, but rather to a generalized ("distributional") connection. Because of this, when we formulate gauge theories purely in terms of loops, as we will do in the following sections, the formulation will usually correspond to this kind of generalized connections. If one wished to work with genuine connections one could do so by requiring extra conditions on the Wilson loops or in the case of loop representations on the corresponding wavefunctions.

Another point is the relation between the formalism introduced for the reconstruction theorem and that of the extended loop group introduced in chapter 2. As we pointed out at the beginning, the starting algebra defined on loops is isomorphic to a complexification of $\mathcal{E}_D$, the algebra of multitensor densities that satisfy the differential constraint. This helps to elucidate the nature of the algebra introduced by Giles, in the sense that it includes objects that are more general than loops, as is obvious due to the isomorphism with $\mathcal{E}_D$. In fact, the reconstruction theorems naturally work on $\mathcal{E}_D$, allowing us in general to reconstruct the gauge covariant matrix associated with any multitensor density $E_D^{a_1 x_1 \ldots a_n x_n}$. In particular, one gets the generalized holonomies associated with the elements of the SeL group.

3.5   Loop representation

The results we introduced in the previous section show that Wilson loops are an overcomplete basis of solutions of the Gauss law. In other words, any gauge invariant function (and therefore any physically interesting quantity) can be expressed as a combination of products of Wilson loops. It is therefore natural to try to build a quantum representation purely in terms of loops. Two different constructions have been introduced that allow us to define a quantum representation for gauge theories purely in terms of loops. In the first one a transform is defined between the connection and loop representations. This procedure allows us to convert any gauge invariant operator or wavefunction into a corresponding object in the loop representation. The second procedure is to introduce at a classical level an algebra of quantities parametrized by loops and take this algebra as the one to be represented in the first step of the Dirac quantization procedure. The resulting quantum representation is the loop representation.

Let us consider an analogy with a finite-dimensional system which clarifies the ideas underlying the loop representation. Suppose one is quantizing the non-relativistic free particle in one spatial dimension. Classically, the system is defined in terms of the canonical coordinates x and p, with Poisson brackets $\{x, p\} = 1$. Quantum mechanically, we take as the space of wavefunctions the functions of x, $\Psi(x)$. We will now construct a new representation for the system applying the ideas we will use to construct the loop representation.

Let us start by considering a transform approach, We consider a basis of states $W_k(x) = \exp(ikx)$, parametrized by a continuous variable k. Any wavefunction can be expanded in terms of this basis. We introduce a k-representation with wavefunctions $\Psi(k)$ given by the integral

$$\Psi(k) \equiv \int dx W_k^*(x)\Psi(x). \tag{3.77}$$

This equation is just the Fourier transform, and the reader may immediately recognize the k-representation as the ordinary momentum representation. The basis of states is an improper basis in the sense that it is not normalizable. Any operator in the position representation $\hat{O}_x$ with a specific order in the canonical variables acting on functions $\Psi(x)$ can be translated into the k-representation by

$$\hat{O}_k \Psi(k) \equiv \int dx W_k^*(x)\hat{O}_x\Psi(x) = \int dx(\hat{O}_x^\dagger W_k(x))^*\Psi(x). \tag{3.78}$$

As an example of the use of the transform, let us consider the transform of a set of quantities that we will use in what follows. They are defined as $T^0(k) = \exp(ikx)$ and $T^1(k) = p\exp(ikx)$. It is immediate to see that one can express any classical quantity in terms of the Ts. They satisfy a non-canonical algebra,

$$\{T^0(k_1), T^0(k_2)\} = 0, \tag{3.79}$$

$$\{T^1(k_1), T^0(k_2)\} = -ik_2 T^0(k_1 + k_2), \tag{3.80}$$

$$\{T^1(k_1), T^1(k_2)\} = i(k_1 - k_2)T^1(k_1 + k_2). \tag{3.81}$$

We now introduce a quantum representation of the algebra via the Fourier transform

$$\hat{T}^0(k_1)\Psi(k) = \int dx \, \exp(-ikx)\exp(ik_1\hat{x})\Psi(x) = \Psi(k - k_1) \tag{3.82}$$

$$\hat{T}^1(k_2)\Psi(k) = -i \int dx \, \exp(-ikx) \frac{d}{dx} \left(\exp(ik_2\hat{x}) \Psi(x) \right) = k\Psi(k - k_2)$$

$$(3.83)$$

and it can be seen that the non-canonical classical Poisson algebra is reproduced by the quantum commutator algebra. Notice that k_1 and k_2 are arbitrary parameters.

It is important to notice that the action of these operators may be translated to the action on a space of kets $|k>$,

$$\hat{T}^0(k_1)\Psi(k) = < k|\hat{T}^0(k_1)|\Psi> = < k - k_1|\Psi>, \qquad (3.84)$$

thus,

$$\hat{T}^{0\dagger}(k_1)|k> = |k - k_1> = \exp(-ik_1\hat{x})|k>. \qquad (3.85)$$

Therefore,

$$\hat{T}^0(k_1)|k> = \exp(ik_1\hat{x})|k> = |k + k_1>, \qquad (3.86)$$

and analogously we find

$$\hat{T}^1(k_2)|k> = \exp(ik_2\hat{x})\hat{p}|k> = k|k + k_2>. \qquad (3.87)$$

Notice that there is a factor ordering involved in the quantum algebra. The resulting ordering in the ket space representation is the *opposite* than the one in the space of wavefunctions.

Now consider a gauge theory (for instance $SU(2)$) in three dimensions described by canonical coordinates $\mathbf{A}_a$ and $\mathbf{E}^a$ with the usual Poisson brackets. Quantum mechanically, we consider wavefunctions of the connection, $\Psi[A]$. An (overcomplete) basis of states is given by the Wilson loops $W_\gamma[A]$. Again, the basis is parametrized by a continuous parameter, in this case the loop γ. The loop representation is defined in terms of the transform,

$$\Psi(\gamma) \equiv \int dA \, W_{\mathbf{A}}^*(\gamma) \Psi[\mathbf{A}]. \qquad (3.88)$$

Again we can transform any operator by using the transform. Notice an important difference. In the case of the free particle we chose a basis of functions $\exp(ikx)$ whereas in the gauge theory case we chose a basis of *solutions* of the Gauss law $W_{\mathbf{A}}(\gamma)$. That is, by going to the loop representation one has automatically solved the Gauss law. Similar situations could arise in the case of the free particle (i.e., by choosing a basis of solutions of the Schrödinger equation) but we will not pursue these here, their meaning being quite transparent. Notice another crucial difference: while the transform used in the free particle case is a well known Fourier transform, the one used for the gauge theory case is only formal. Very little is known about integration in spaces of connections and the theory of measure in this case is not well developed in general (for further developments see references [36, 40, 66, 67]). We will return to these issues in the following sections. Notice that the introduction of the loop transform can be thought of as performing an inner product in the connection representation between a wavefunction $|\Psi>$ and elements of a basis $<\gamma|$. Therefore we can write

$$\Psi(\gamma) = <\gamma|\Psi> = \int DA <\gamma|A><A|\Psi> \qquad (3.89)$$

through the introduction of the identity $1 = \int DA|A><A|$ which means that having a correct definition of the transform is equivalent to having an inner product in the connection representation.

Finally, it is not strictly true that for an arbitrary gauge group *single* Wilson loops are a basis of gauge invariant functions, but rather one needs to consider products of Wilson loops. This can be readily done, and the resulting wavefunction is a function of multiloops. We will discuss this in section 3.5.3.

Let us now explore the second approach, i.e., quantizing a non-canonical algebra of quantities. Again we consider the free particle and on the classical phase space we define the quantities $T^0(k) = \exp(ikx)$ and $T^1(k) = p\exp(ikx)$ which satisfy the non-canonical algebra discussed above. It is evident that one can express any classical quantity of interest in terms of this algebra. If one has a well defined transform, as is the case for the Fourier transform, one could proceed as before and find a quantum realization of this non-canonical algebra using the transform. It is therefore evident that the quantization that one would achieve coincides with the one that was introduced before via the transform. If one does not have a well defined transform at hand one can propose a quantum realization of the algebra and check that one reproduces the classical algebra at the quantum commutator level.

In this particular case we would propose

$$\hat{T}^0(k_1)\Psi(k) = \Psi(k - k_1), \qquad (3.90)$$

$$\hat{T}^1(k_2)\Psi(k) = k\Psi(k - k_2) \qquad (3.91)$$

and check that this representation reproduces the non-canonical algebra through quantum commutators. Notice that a choice of factor ordering must be made in the process. One can find the quantum expression for any classical quantity simply by writing the classical expression in terms of the Ts and translating with due care for factor orderings.

Again a very similar construction (at least formally) can be performed for a gauge theory. Consider the set of classical quantities

$$T^0(\gamma) = W(\gamma), \qquad (3.92)$$

$$T^a(\gamma_x^x) = \text{Tr}(H_{\mathbf{A}}(\gamma_o^x)\tilde{E}^a(x)H_{\mathbf{A}}(\gamma_x^o)), \qquad (3.93)$$

where $W(\gamma)$ is the Wilson loop and $H_{\mathbf{A}}(\gamma_o^x)$ is the holonomy along the loop γ from the basepoint o to x. These quantities satisfy a closed non-canonical Poisson algebra and can be promoted to a quantum operator algebra. The result would be the loop representation. However, various detailed issues have to be discussed and we will postpone their treatment until section 3.5.2.

In spite of the appeal of these simplified analogies, the fact that gauge theories are infinite-dimensional systems of a non-Abelian nature implies that all the steps described above are considerably more involved. We will discuss these points in detail in the following chapters. Here we will discuss the definitions. In later chapters of this book the applications of the loop representation for gauge theories and general relativity will be explored in detail.

3.5.1   The loop transform

As we mentioned before, the loop transform involves a functional integral in the space of connections modulo gauge transformations. This makes it considerably more involved from a technical point of view than the transforms among representations of ordinary quantum mechanics which we discussed as an analogy. Little is known about integration theory in non-linear spaces both from a mathematical and a physical point of view.

The loop transform was introduced for the treatment of gauge theories in the early 1980s by Gambini and Trias [62]. At that time the approach was to assume that the transform existed and study *a posteriori* the physical implications of its existence. In a sense, a high degree of assurance of its existence was obtained through this approach, since it was proven in very concrete situations that results obtained via the transform coincided with those obtained via more traditional techniques. An important arena for this kind of test was the application of loop techniques in the lattice [95, 109]. In this case the loop transform is rigorously defined for any Yang–Mills theory in terms of the Haar measure of the group. For the case of general relativity, the loop transform in terms of Ashtekar variables was first introduced by Rovelli and Smolin [39] in again the same spirit. Only recently have studies of some mathematical rigor been performed on its existence. The main effort in this area is the result of the collaboration of Ashtekar, Isham, Lewandowski, Marolf, Mourão and Thiemann [36, 40, 203] and the work by Baez [66]. A particularly readable account from the point of view of physicists is given in reference [204], the pattern of which we follow in this section.

In this book we will use the transform as a heuristic tool to derive results. The correctness or otherwise of such results will be judged through their consistency both among themselves and with facts known through other means, not through the rigor of their derivation via the transform. We will discuss in each case in detail which are the arguments and consistency checks that support that result. In this section, however, we would like to give a brief glimpse of some of the mathematical developments that are taking place to put the loop transform and the results derived through it on a solid mathematical ground. It is yet to be seen if the particular results presented in this book will survive in the form presented when a rigorous operational definition of the transform is found.

The key idea that allows the definition of a measure of integration on the non-linear space of connections modulo gauge transformations is the use of the Wilson loop as a projection operator. This allows the definition of the so called "cylindrical measures", which reduce the infinite-dimensional integral to a finite set of integrals over the gauge group. By demanding consistency of the various projections one ends with a theory of integration in infinite-dimensional spaces. Let us discuss in detail how this is accomplished. To investigate the ideas in a simpler context we discuss the definition of a measure in an infinite-dimensional but linear space, that of a Klein–Gordon field.

Consider a scalar field ϕ in flat spacetime satisfying the Klein–Gordon equation. The classical configuration space of such a theory is the set of all smooth field configurations on a spatial manifold that fall off appropriately at infinity. Quantum states for the theory are functions on the space of classical configurations $\Psi(\phi)$. One would like to introduce an inner product through an expression of the type $\int D_\mu \phi \bar{\Psi}(\phi)\Phi(\phi)$ where the integral ranges over the configuration space and our task is to introduce a suitable measure μ to perform the integral.

In order to do so we need to consider some particular functions on the configuration space. Possibly the simplest kind of function we can introduce are the functionals F defined by test functions $f(x)$ (the set of which is called Schwarz space) of the spatial manifold which we convolute with the classical configurations,

$$F_f(\phi) \equiv \int d^3x f(x)\phi(x) \tag{3.94}$$

and we require that $f(x)$ have appropriate regularity and falloff conditions such that the integral is well defined. With the above definition of the functionals F we are now in a position to introduce the idea of cylindrical functions. Consider a finite-dimensional subspace V_n of the Schwarz space and a basis of functions in it $(e_1, \ldots, e_n)$. Given a classical configuration $\phi(x)$ we can define its "projection" on the finite-dimensional subspace which yields a set of n numbers,

$$(F_{e_1}(\phi), \ldots F_{e_n}(\phi)). \qquad (3.95)$$

A function of the classical configuration space is called cylindrical with respect to V_n if its dependence on the classical configurations is through the set of n numbers that we introduced above, for some set of e_is. That is, $g(\phi)$ is cylindrical iff

$$g(\phi) = G(F_{e_1}(\phi), \ldots, F_{e_n}(\phi)) \qquad (3.96)$$

for some function of n real variables G.

A cylindrical measure μ is a measure that allows us to integrate cylindrical functions. Each of these measures is defined by an infinite consistent family of measures $\{\mu_{e_1,\ldots,e_n}\}$ each defined on all finite-dimensional spaces $\mathbf{R}^n$ associated with each basis of vectors $(e_1, \ldots, e_n)$. With these measures the integral of a cylindrical function is simply defined as an integral on $\mathbf{R}^n$,

$$\int D_\mu(\phi)g(\phi) = \int G(x_1, \ldots, x_n) D_{\mu_{e_1,\ldots,e_n}}(x_1, \ldots, x_n). \qquad (3.97)$$

The key issue is that the above expression has to be well defined and consistent for any set V_n that one chooses. This restricts considerably the choice of the family of measures, imposing a set of consistency conditions. First consider the case of a function that is cylindrical with respect to two subspaces V_n and V'_m that are disjoint. Such functions are necessarily constants, so the integrals of such constants with $\mu_{e_1,\ldots,e_n}$ and $\mu_{e'_1,\ldots,e'_m}$ should be the same, which fixes a normalization condition for the measures. Next consider a function $g(\phi)$ that is cylindrical with respect to two subspaces $V_n \subset V'_m$. Such a function has associated with it two functions of n and m real variables $G(x_1, \ldots, x_n)$ and $G'(x_1, \ldots, x_m)$ that define it as a cylindrical function with respect to both spaces. Since the basis of V_n will be a linear combination of the basis of V'_m one can figure out the precise relationship between G and G'. Since the integral of G with the measures $\mu_{e_1,\ldots,e_n}$ has to be the same as the integral of G' with the measure $\mu_{e'_1,\ldots,e'_m}$ this imposes a consistency condition on the elements of the family $\{\mu_{e_1,\ldots,e_n}\}$.

An example of a family of measures that is compatible with the consistency conditions introduced above is given by appropriately chosen Gaussian measures on $\mathbf{R}^n$. The well known quantum field theory of free fields is based on such measures. One can obtain the Fock representation by taking the Cauchy completion with respect to the inner product defined by the measure of the space of cylindrical functions on the classical configuration space. The hope is that the quantum field theory of interacting fields will arise from non-Gaussian cylindrical measures, as has been shown in several particular cases [37]. It is important to notice that the above mentioned Cauchy completion leads to a quantum theory defined — in the case of field theories — by functions on an enlargement of the classical configuration space. This amounts to considering not only regular functions of the classical configurations but also distributions. We will see in chapter 11 that the consideration of distributional fields in the loop transform poses new challenges for the regularization of the theory in the loop (and extended) representation.

How do these constructions apply to gauge theories? For the case of Maxwell theory the construction is basically the same as above. The reason is that for an Abelian theory the space of connections modulo gauge transformations is again a linear space and one simply repeats the above construction considering functions of the classical configuration space given by the magnetic fields.

For the non-Abelian case the configuration space is a non-linear space. The way around this problem is to exploit the properties of holonomies to provide an analogue of the functionals introduced above. Given a fixed finite set of independent[§] loops $\beta_1, \ldots, \beta_n$ we now say a function $g(\mathbf{A})$ of the space of connections modulo gauge transformations is cylindrical with respect to this set of loops if and only if it depends on the connection through the value of the holonomies associated with the β_is,

$$g(\mathbf{A}) = G(H_{\mathbf{A}}(\beta_1), \ldots, H_{\mathbf{A}}(\beta_n)), \tag{3.98}$$

where G is a function defined on n copies of the gauge group.

A cylindrical measure is defined in a way analogous to that used before as a consistent family of measures $\mu_{\beta_1, \ldots, \beta_n}$ on the nth tensor power of the gauge group. Again, there are consistency conditions to be met, which are more involved than in the simple example described previously. The remarkable fact is that there exist consistent families which define measures. An example of this is given by n copies of the Haar measure defined on the gauge group. Since this measure is defined without the introduction of any background structure it is diffeomorphism invariant.

We therefore have not only succeeded in introducing in a rigorous way a measure on the space of connections modulo gauge transformations but the measure is diffeomorphism invariant. It is therefore the kind of measures one would expect to be useful for analyzing problems in diffeomorphism invariant theories such as quantum gravity.

As we will see in chapter 7, for quantum gravity there is an additional complication in the sense that the gauge group is a complexified version of

---

[§] By independent loops we mean loops that have at least a segment that is not shared by the other loops with at most a finite number of intersections with the other loops.

$SU(2)$. It is remarkable that in an unrelated development, Hall [211] introduced a generalization of the Gaussian measure for complexified gauge groups. His motivation (the Bargmann representation of the harmonic oscillator, see chapter 4) is basically the root of the complex nature of the gravitational variables. Therefore by replacing the Haar measure by the Hall measure in the discussion above one can have a measure that is appropriate for the gravitational case. Development in this area is very rapid at present and may allow us to put on a solid ground many results that in this book we can only present formally.

It is yet to be seen if these kinds of measures produce physical theories of interest or if they are just mathematical curiosities. However, one cannot overstress the fact that until recently there were almost no measures known in non-linear infinite dimensional spaces and with these developments one may be able to gain enough experience to define measures that yield physical theories of relevance.

There has been a rapid development of these ideas. In particular rigorous definitions of the constraints and states of quantum gravity for the Euclidean case (where the theory is real) are currently under study. Many of the rigorous results provide a formal setting for the ideas we will discuss in chapters 7, 8, 9 and 10 [203].

3.5.2   The non-canonical algebra

There is an alternative procedure for introducing a loop representation that avoids having to go through an intermediate representation. The way to proceed is to go back to step one of the canonical quantization procedure we introduced in section 3.2.2 and pick a different classical algebra to quantize. We introduce the following quantities on the classical phase space of any gauge theory (or general relativity written in terms of Ashtekar's variables),

$$T(\gamma) \equiv \mathrm{Tr}(H_{\mathbf{A}}(\gamma)) = W_{\mathbf{A}}(\gamma), \tag{3.99}$$

$$T^a(\gamma_x^x) \equiv \mathrm{Tr}(H_{\mathbf{A}}(\gamma_o^x)\tilde{\mathbf{E}}^a(x)H_{\mathbf{A}}(\gamma_x^o)), \tag{3.100}$$

$$T^{ab}(\gamma_x^y, \gamma_y^x) \equiv \mathrm{Tr}(H_{\mathbf{A}}(\gamma_o^x)\tilde{\mathbf{E}}^a(x)H_{\mathbf{A}}(\gamma_x^y)\tilde{\mathbf{E}}^b(y)H_{\mathbf{A}}(\gamma_y^o)), \tag{3.101}$$

where $\gamma_x^x \equiv \gamma_x^o \circ \gamma_o^x$ and generically,

$$T^{a_1 \ldots a_n}(\gamma_{x_1}^{x_2}, \ldots, \gamma_{x_n}^{x_1}) \equiv \mathrm{Tr}(H_{\mathbf{A}}(\gamma_o^{x_1})\tilde{\mathbf{E}}^{a_1}(x_1)H_{\mathbf{A}}(\gamma_{x_1}^{x_2})$$
$$\ldots H_{\mathbf{A}}(\gamma_{x_{n-1}}^{x_n})\tilde{\mathbf{E}}^{a_n}(x_n)H_{\mathbf{A}}(\gamma_{x_n}^o)), \tag{3.102}$$

where $\gamma \equiv \gamma_o^{x_1} \circ \ldots \circ \gamma_{x_n}^o$.

Notice that if the loop has multiple points the quantities depend on what sort of partition of the loop one performs and care should be taken to keep track of these dependences.

We see that the quantities $T(\gamma)$ are our well known Wilson loops and the other Ts consist of "breaking up" the holonomy at points x_i, inserting an electric field and continuing the holonomy until back at the basepoint. It can be checked that these quantities are gauge invariant, i.e., they commute with the Gauss law. Generically we will speak of the Ts with k electric fields inserted as "T^k". The T^ks behave as multivector densities on the indices $a_1, \ldots, a_k$ at the points of the manifold $x_1, \ldots, x_k$. They were first introduced by Gambini and Trias [34] (in their notation they were called Ws, naturally extending the notation of the Wilson loops, although we have favored here the notation that has become standard among relativists, introduced by Rovelli and Smolin [38]).

As we argued before, the Wilson loops contain enough information to construct any gauge invariant function of the connection. By introducing Ts of higher order the expectation is that one would be able to construct any quantity depending on the electric fields, and therefore have an algebra of classical quantities which is sufficiently general to express any quantity of physical interest in terms of them. We have already shown examples of how to construct quantities of physical interest in terms of the Wilson loops, for instance,

$$\Delta_{ab}(\gamma_o^x) W_{\mathbf{A}}(\gamma) = \mathrm{Tr}(\mathbf{F}_{ab}(x) H_{\mathbf{A}}(\gamma_x^x)). \qquad (3.103)$$

So we see we can retrieve information about the F_{ab}^i. One can also retrieve information about momentum dependent quantities from the Ts of higher order, for instance, a trace of two electric fields,

$$\mathrm{Tr}(\tilde{\mathbf{E}}^a(x)\tilde{\mathbf{E}}^b(x)) = \lim_{\gamma \to x} T^{ab}(\gamma)(x,y), \qquad (3.104)$$

where by $\lim_{\gamma \to x}$ we mean the limit in which we shrink the loop to a point at x (and consequently the point y tends to x). In terms of the Ashtekar new variables for general relativity this trace plays the role of the spatial metric.

We will not by any means prove here that one could reconstruct any quantity of physical relevance in terms of these quantities. It suffices to realize that most quantities that one is usually interested in can be written as limits of the Ts and that therefore they seem to span the classical phase space of the theory of interest.

An interesting point is that the T^0s with the T^1s close an algebra, the "small T algebra". Let us compute it in an explicit fashion for an $SU(N)$ Yang–Mills theory. Because the Poisson bracket of $\mathbf{A}_a$ with itself is zero it is immediate that

$$\{T(\gamma), T(\eta)\} = 0. \qquad (3.105)$$

In order to obtain the Poisson bracket of T with T^1 we compute, starting from the canonical brackets (3.21),

$$\{H_{\mathbf{A}}(\gamma)^A{}_B, \tilde{\mathbf{E}}^a_j(x)\} = i(X^j)^C{}_D H_{\mathbf{A}}(\gamma^x_o)^A{}_C H_{\mathbf{A}}(\gamma^o_x)^D{}_B \oint_\gamma dy^a \delta(x-y),$$

(3.106)

where the indices $A, B, \ldots$ refer to the fundamental representation of $SU(N)$ and run from 1 to N, and $(X^j)^A{}_B, \quad j = 1, \ldots, N^2 - 1$ are the generators of the algebra satisfying (3.8). Combining this equation with the following identity for the generators of $SU(N)$,

$$\sum_{i=1}^{N^2-1} (X^i)^A{}_B (X^i)^C{}_D = \delta^C{}_B \delta^A{}_D - \frac{1}{N} \delta^A{}_B \delta^C{}_D,$$

(3.107)

we get

$$\{T^a(\gamma^x_x), T(\eta)\} = -i \left(T(\gamma^x_x \circ \eta^x_x) - \frac{1}{N} T(\gamma) T(\eta) \right) X^{ax}(\eta),$$

(3.108)

where $X^{ax}(\eta)$ is the multitangent of order one. Notice that the Poisson bracket vanishes if η and γ do not have a common point.

Similarly, for two T^1's,

$$\{T^a(\gamma^x_x), T^b(\eta^y_y)\} = -i \left(T^b(\eta^x_y \circ \gamma^x_x \circ \eta^y_x) - \frac{1}{N} T(\gamma) T^b(\eta^y_y) \right) X^{ax}(\eta) +$$

$$+ i \left(T^a(\gamma^y_x \circ \eta^y_y \circ \gamma^x_y) - \frac{1}{N} T(\eta) T^a(\gamma^x_x) \right) X^{by}(\gamma).$$

(3.109)

In the general $SU(N)$ case, in the right-hand side of the Poisson brackets we have products of the elements of the non-canonical algebra. It is only for the case of $SU(2)$ that we can rearrange these terms as linear superpositions of elements of the non-canonical algebra. This means that if one wants to find a quantum representation, one needs to consider a non-canonical algebra incorporating products of the Ts. As a consequence, wavefunctions in the loop representation so constructed will have to depend on more than one loop. We will return to these issues in the next section.

For the $SU(2)$ case the algebra can be written in a very compact fashion. The Poisson brackets of the T's are a linear combination of T's evaluated on loops obtained from the original ones through very simple rules of fusion and rerouting through the intersection of the loops. The result is zero if the loops do not intersect. The action can be understood in a simple fashion through a graphical representation as shown in figure 3.1. The explicit form of the algebra is,

$$\{T^a(\gamma^x_x), T(\eta)\} = \frac{i}{2} \sum_{\epsilon=-1}^{1} \epsilon X^{ax}(\eta) T(\gamma \circ \eta^\epsilon)$$

(3.110)

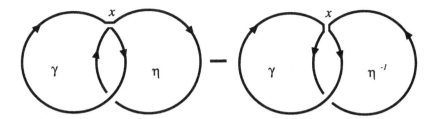

Fig. 3.1. The graphical representation of the commutator between T^1 and T^0. The commutator is zero if the "hand" (the point at which one inserted the triad) of the loop γ does not "grab" the loop η. The figure shows the two reroutings that arise in the two terms that result from the commutator.

$$\{T^a(\gamma_x^x), T^b(\eta_y^y)\} = -\frac{i}{2} \sum_{\epsilon=-1}^{1} \epsilon X^{ax}(\eta) T^b(\eta_y^x \circ (\gamma_x^x)^\epsilon \circ \eta_x^y)$$

$$+ \frac{i}{2} \sum_{\epsilon=-1}^{1} \epsilon X^{by}(\eta) T^a(\gamma_x^y \circ (\eta_y^y)^\epsilon \circ \gamma_y^x), \quad (3.111)$$

where η^ϵ represents either η or η^{-1}.

If one wants to consider higher order Ts, one needs Ts of arbitrarily large order in order to close the algebra, so strictly speaking it is not closed or only closes in a completion. For instance, for the $SU(2)$ case, the Poisson brackets are schematically

$$\{T^n, T^m\} \sim T^{n+m-1}. \quad (3.112)$$

The detailed commutation relations can be seen in reference [38].

The need to consider the infinite family of Ts to attain closure is just another manifestation of the overcompleteness of the loop basis. Although we know that we have "too many" loops, we are forced to include them all to span the classical phase space of the theory. It is tempting to try to construct a quantum theory by only representing the "small" algebra of T and T^1. Unfortunately it is not clear if these quantities are enough to span the classical phase space of gauge theories. There is a certain sense in which they do, though technicalities arise for the case of non-compact groups[68]. Even if they did in some particular cases, they are not very convenient for expressing some quantities of physical relevance, such as the Hamiltonian of Yang–Mills theories (and general relativity). Therefore from a practical point of view one resorts to the higher Ts to express quantities of interest.

Let us now sketch the quantization of this non-canonical algebra for the $SU(2)$ case. The idea is that we have now completed step one of the Dirac quantization procedure introduced in section 3.2.2: we have picked a (non-canonical) algebra of classical quantities that (modulo subtleties) spans the classical phase space of the theory. We now move on to the second step of the quantization program: to find a representation of this algebra in terms of operators acting on a space of wavefunctions. We pick wavefunctions of loops $\Psi(\gamma)$ and we represent the T operators by

$$\hat{T}(\eta)\Psi(\gamma) \equiv \Psi(\eta \circ \gamma) + \Psi(\eta \circ \gamma^{-1}), \qquad (3.113)$$

$$\hat{T}^a(\eta)(x)\Psi(\gamma) \equiv -\sum_{\epsilon=-1}^{1} \epsilon X^{ax}(\gamma)\Psi(\gamma \circ \eta^\epsilon). \qquad (3.114)$$

These kinds of expressions face regularization difficulties. They could be regularized by considering, for instance, "thickened out" loops and defining a regularized T^1 operator via a two parameter congruence of loops. A discussion of this can be found in references [2, 69].

Similar expressions for the quantum representation of the higher order Ts can be seen in reference [39]. One can check that these quantum operators satisfy quantum commutation relations that in the limit $\hbar \longrightarrow 0$ (the T^ns have a prefactor of $\hbar^n$ if one does not set $\hbar$ to one as we have been doing) reproduce the classical commutation relations mentioned above. All this is discussed in reference [39].

The resulting quantum theory is the loop representation that we introduced before. One can check all this — at least heuristically given the various ill-defined constructions that are involved — by formally using the transform. One can represent the T operators in the connection representation (using an appropriate factor ordering) and then transform them into operators in the loop representation. One immediately finds that the representation introduced above corresponds to ordering the electric fields to the left in the connection representation.

Is it preferable to introduce the loop representation via a quantization of a non-canonical algebra or via a transform? At this moment this is largely a matter of choice. Both definitions, as we have seen, face various points where ill-defined mathematical operations are rampant. In fact, it is not difficult to see that many of these difficulties are somewhat connected. The important point that we have shown in this section is that there is nothing "strange" about the loop representation. It is a quantum representation that can be obtained directly, applying the traditional Dirac quantization procedure. It is by no means "subordinated" to the connection representation and has an existence on the same footing as any other quantum representation. The main difference between the loop representation and other more traditional ones is the use of an overcomplete non-canonical set of operators.

3.5.3 Wavefunctions in the loop representation

Now that we have introduced the loop representation, what about the wavefunctions in such a representation. Will any function of a loop do as a wavefunction or are there other requirements to be met?

As we discussed in section 3.5.1 wavefunctions in the loop representation can be thought of as transforms of functionals of connections weighed by products of Wilson loops,

$$\Psi(\gamma_1, \ldots, \gamma_n) = \int d\mathbf{A}\, W_{\mathbf{A}}^*(\gamma_1) \cdots W_{\mathbf{A}}^*(\gamma_n) \Psi[\mathbf{A}]. \qquad (3.115)$$

An immediate property that follows from the fact that the Wilson loops appear in the transform as a product is that wavefunctions are symmetric under interchange of arguments,

$$\Psi(\gamma_1, \ldots, \gamma_i, \ldots, \gamma_j, \ldots, \gamma_n) = \Psi(\gamma_1, \ldots, \gamma_j, \ldots, \gamma_i, \ldots, \gamma_n). \qquad (3.116)$$

Wavefunctions in the loop representation will inherit a series of properties of Wilson loops. To begin with, they are functions with domain in the group of loops $\mathcal{L}_o$. Since the Wilson loops are traces of holonomies, they are actually functions of conjugacy classes of the group of loops; for example, for a function of a single loop,

$$\Psi(\gamma) = \Psi(\eta \circ \gamma \circ \eta^{-1}) \quad \forall \eta. \qquad (3.117)$$

For functions of multiloops a similar expression holds at each entry. It is immediate from the previous expression that

$$\Psi(\gamma \circ \eta) = \Psi(\eta \circ \gamma). \qquad (3.118)$$

It is here that the machinery introduced in chapter 1 and 2 will become useful, since we will all the time be operating on functions of the group of loops. In previous approaches wavefunctions in the loop representation were considered as functionals of parametrized curves with additional restrictions and functional derivatives played the role of differential operators. The consistency of this approach is delicate since one must ensure that the application of differential operators preserves the conditions imposed on the functional space. These issues are automatically taken care of by considering functions on the group of loops and the corresponding differential operators discussed in chapter 1.

Another important property is that wavefunctions inherit the Mandelstam identities among Wilson loops that we discussed in section 3.4.1.

To begin with, the Mandelstam identities relate products of Wilson loops of different orders. In particular for any group of $N \times N$ matrices, this allows us to express a product of Wilson loops in terms of expressions involving at most N factors and consequently to reduce any wavefunction to one depending on at most N loops.

Let us now discuss in detail the implications of the Mandelstam identities for the case of two-dimensional special groups (such as $SU(2)$, $SL(2,C)$, etc). In this case, the fundamental identity reads

$$\Psi(\gamma_1, \ldots, \gamma_i, \gamma_j, \ldots, \gamma_n) = \Psi(\gamma_1, \ldots, \gamma_i \circ \gamma_j, \ldots, \gamma_n)$$
$$+ \Psi(\gamma_1, \ldots, \gamma_i \circ \gamma_j^{-1}, \ldots, \gamma_n). \quad (3.119)$$

An important consequence of this identity is that it will make it possible to express *any functional of an arbitrary number of loops in terms of a functional of a single loop*. That is, in these cases one can construct a loop representation considering functions of a single loop.

That wavefunctions depend on a single loop does not imply that they are unconstrained, since many identities for wavefunctions of single loops can be derived from (3.119). Consider expression (3.119) for the case of two entries and put $\gamma_j = \iota$ and $\gamma_i = \eta$. Then

$$\Psi(\eta, \iota) = 2\Psi(\eta), \quad (3.120)$$

which implies, considering (3.119) with $\gamma_i = \iota$, $\gamma_j = \eta$, that

$$\Psi(\eta) = \Psi(\eta^{-1}). \quad (3.121)$$

Finally applying (3.119) to

$$\Psi(\gamma \circ \eta, \beta) = \Psi(\eta \circ \gamma, \beta), \quad (3.122)$$

we get

$$\Psi(\gamma \circ \eta \circ \beta) + \Psi(\gamma \circ \eta \circ \beta^{-1}) = \Psi(\eta \circ \gamma \circ \beta) + \Psi(\eta \circ \gamma \circ \beta^{-1}). \quad (3.123)$$

Given this set of identities one can reconstruct the identities for multiloops.

On a practical note, although these identities are fundamental in the sense that any other can be derived from them, they can imply very nontrivial relations between wavefunctions even at the single loop level.

Apart from these identities, as we mentioned in section 3.4.1 there are inequalities in terms of holonomies that reflect properties of the group (for instance that tell us if the group is $SU(2)$ rather than $SU(1,1)$). At the moment the treatment of these inequalities is unclear. For instance, it is not established if they imply any restrictions on the wavefunctions. They imply restrictions on the quantities that one quantizes. This would not be the first time that a quantization was attempted in terms of variables that satisfy inequalities. For instance, usual quantizations of gravity based on metric variables have to deal with the fact that the metric of space must have a Euclidean signature. Or in a more simplified situation, consider the quantization of the hydrogen atom in the position representation in spherical coordinates, where the radial variable has to be positive definite. Dealing with the detailed problems posed by the fact that one is quantizing in a representation where variables satisfy inequalities requires a degree of sophistication of the theory that has not yet been attained. For a deeper discussion of the problems of inequalities and quantization see reference [41].

3.6 Conclusions

In this chapter we introduced several physical techniques for the analysis of gauge theories. In terms of these, many of the notions of loops that we introduced in the first two chapters find a natural application. We introduced the loop representation and have shown that wavefunctions in the loop representation are simply functions of the group of loops. To develop in some detail the relationships known at present between the loop techniques of chapters one and two and the physical theories of chapter three is the subject of the rest of this book. In chapter 4, 5 and 6 of the book we will apply these techniques to gauge theories. In chapters 7-11 we will apply them to general relativity in terms of Ashtekar's variables.

4

Maxwell theory

In this chapter we will study the quantization of the free Maxwell theory. Admittedly, this is a simple problem that certainly could be tackled with more economical techniques, and this was historically the case. However, it will prove to be a very convenient testing ground to gain intuitive feelings for results in the language of loops. It will also highlight the fact that the loop techniques actually produce the usual results of more familiar quantization techniques and guide us in the interpretation of the loop results.

We will perform the loop quantization in terms of real and Bargmann [70] coordinates. The reason for considering the complex Bargmann coordinatization is that it shares many features with the Ashtekar one for general relativity. It also provides a concrete realization of the introduction of an inner product purely as a consequence of reality conditions, a feature that is expected to be useful in the gravitational case.

The Maxwell field was first formulated in the language of loops by Gambini and Trias [62]. The vacuum and other properties are discussed in reference [63] and multiphoton states are discussed in referece [64]. The loop representation in terms of Bargmann coordinates was first discussed by Ashtekar and Rovelli [65].

The organization of this chapter is as follows: in section 4.1 we will first detail some convenient results of Abelian loop theory, which will simplify the discussion of Maxwell theory and will highlight the role that Abelian theories play in the language of loops. In section 4.2 we will discuss the classical theory. We will discuss the Fock representation in section 4.3. We will then discuss in section 4.4 the quantization of the Maxwell theory in terms of real loop variables. We will recover the usual Fock space and the photon states in terms of loops, and study the interpretation of loop observables in terms of familiar notions of field theory. We will introduce an inner product and an interpretation of the wavefunctions in terms of loops. In section 4.5 we will summarize the loop quantization of Maxwell theory in terms of the Bargmann representation and see how this quantization leads, perhaps more naturally, to the same results as the previous section. This will also serve as the motivation and background for the discussion of the gravitational case. Finally, we will discuss in section 4.6 the quantization in the extended loop representation in terms of loop coordinates. We will show how one can reconstruct a classical canonical theory in terms of loops, the quantization of which leads to the loop representation. We will see that the loop representation is directly related to the canonical quantization in the electric field representation.

4.1 The Abelian group of loops

Although one could formulate Maxwell theory in terms of the full group of loops, it turns out that a subgroup of it is all that is needed due to the Abelian nature of the theory. We find it convenient to discuss in some detail the properties of this subgroup since they will help us to simplify the treatment of Maxwell theory.

Let us start by considering the elements of the group of loops of the following nature:

$$\kappa = \gamma \circ \eta \circ \gamma^{-1} \circ \eta^{-1}. \tag{4.1}$$

Generically, γ and η could be composed of an arbitrary number of loops $\gamma = \gamma_1 \circ \ldots \circ \gamma_n$, $\eta = \eta_1 \circ \ldots \circ \eta_n$. These kinds of loops are usually called commutators. It is easy to check that the set of all such loops and their products form a subgroup of the group of loops. We will denote it by $\mathcal{L}_{comm}$. One can immediately see that it forms a normal subgroup, i.e., given any element κ of $\mathcal{L}_{comm}$,

$$\gamma \circ \kappa \circ \gamma^{-1} \in \mathcal{L}_{comm} \qquad \forall \gamma \in \mathcal{L}. \tag{4.2}$$

Whenever one has a normal subgroup, one can define the quotient group. In order to do this we introduce an equivalence relation,

$$\gamma \sim \eta \iff \gamma \circ \eta^{-1} = \kappa \in \mathcal{L}_{comm}. \tag{4.3}$$

The reader can check that the relation is reflexive, symmetric and transitive. We denote the quotient group $\mathcal{L}_{Abel} = \mathcal{L}/\mathcal{L}_{comm}$. Its elements are the equivalence classes determined by the relation (4.3). Again it can be readily checked that the product of equivalence classes is independent of the representative element of the class chosen to perform the calculation.

The intuitive interpretation of the equivalence relation defined is that we have identified the commutators in the group of loops with the identity.

Therefore, if γ_1 and γ_2 belong to $\mathcal{L}_{Abel}$,

$$\gamma_1 \circ \gamma_2 = \gamma_2 \circ \gamma_1, \tag{4.4}$$

and $\mathcal{L}_{Abel}$ is an Abelian group.

As we saw in chapter 1, gauge theories are simply representations of the group of loops. Let us consider representations of the Abelian subgroup that we constructed. We therefore need matrices $H(\gamma)$, such that

$$H(\gamma_1)H(\gamma_2) = H(\gamma_1 \circ \gamma_2) = H(\gamma_2 \circ \gamma_1) = H(\gamma_2)H(\gamma_1) \tag{4.5}$$

for any pair of loops γ_1, γ_2. If one wishes to consider unitary representations of the group of loops, equation (4.5) can only hold if H is a unimodular complex number, i.e., an element of $U(1)$.

As we saw in section 1.4, any representation (sufficiently regular) of the group of loops can be written locally as

$$H_A(\gamma) = \exp\left(i \oint_\gamma dy^a A_a(y) \right), \tag{4.6}$$

where $A_a(y)$ is just a real number for the Abelian case we are considering and therefore

$$H_A(\gamma) = W_A(\gamma). \tag{4.7}$$

$W_A(\gamma)$ depends on the loop γ only through the circulation of A_a. This can be written using only the simplest of the loop coordinates introduced in chapter 2, the coordinate of order one,

$$\oint_\gamma dy^a A_a(y) = \int d^3z\, A_a(z) X^{az}(\gamma). \tag{4.8}$$

An interesting point is that the representation depends only on the information of the loop contained in the first order loop coordinate. This implies some strong differences with the general case. For instance, $W(\pi_x^y \circ \gamma_y \circ \pi_y^x \circ \eta_x) = W(\gamma_y \circ \eta_x)$, where γ is any loop basepointed at y and π_x^y is an arbitrary path and η_x is a loop basepointed at x. This implies that for an infinitesimal deformation

$$W(\pi_o^x \circ \delta u \delta v \delta \bar{u} \delta \bar{v} \circ \pi_x^0 \circ \gamma) = W(\delta u \delta v \delta u \delta v \circ \gamma). \tag{4.9}$$

Therefore loop derivatives are no longer path dependent but just point dependent,

$$\Delta_{ab}(\pi_o^x) \longrightarrow \Delta_{ab}(x) \qquad \forall \pi_o^x. \tag{4.10}$$

As a consequence, loop derivatives in the Abelian case commute,

$$[\Delta_{ab}(x), \Delta_{cd}(y)] = 0, \tag{4.11}$$

and the Bianchi identities can be expressed in terms of ordinary derivatives,

$$\partial_{[a}\Delta_{bc]}(x) = 0. \tag{4.12}$$

We will now study the classical Maxwell theory and the relation of the classical theory to quantities in terms of loops.

4.2 Classical theory

The classical canonical Maxwell theory can be expressed in terms of the canonical pair $\tilde{E}^a(x)$, $A_b(y)$,

$$\{A_b(y), \tilde{E}^a(x)\} = \delta_b^a \delta(x - y). \tag{4.13}$$

The only constraint of the theory is the Abelian Gauss law,

$$\partial_a \tilde{E}^a = 0. \tag{4.14}$$

The Hamiltonian of the theory is the sum of the squares of the electric and magnetic fields, integrated over space,

$$H = \int d^3x \tfrac{1}{2}\eta_{ab}(\tilde{E}^a\tilde{E}^b + \tilde{B}^a\tilde{B}^b), \tag{4.15}$$

where $\tilde{B}^a = \tilde{\eta}^{abc}F_{bc}$. Here η_{ab} is a flat Euclidean three-dimensional metric and from now on we will assume all indices are raised and lowered with it. The commutator of the electric field and the connection with the Hamiltonian gives the time evolution of the fields. These plus the Gauss law are equivalent to the usual four-dimensional Maxwell equations.

The Gauss law can be solved by considering only transverse electric fields, $\tilde{E}_T^a(x)$. The canonical theory can be reformulated entirely in terms of transverse fields (the transverse connection $A_a^T(x)$ is defined in terms of the fixed flat background metric), the canonical pair is then given in terms of Dirac brackets by

$$\{A_a^T(y), \tilde{E}_T^a(x)\} = \delta_{Tb}^a(x - y), \tag{4.16}$$

where the "transverse Dirac delta" is defined by

$$\delta_{Tb}^a(x - y) = \delta_b^a\delta(x - y) - \Delta^{-1}\partial^a\partial_b\delta^3(x - y), \tag{4.17}$$

where Δ^{-1} is the inverse of the Laplacian of the background metric on the three-manifold.

A usual simplification is to consider momentum space variables,

$$A_a^T(x) = \frac{1}{(2\pi)^{3/2}} \int d^3k \exp(i\vec{k}\cdot\vec{x})[q_1(\vec{k})e_a^1(\vec{k}) + q_2(\vec{k})e_a^2(\vec{k})], \tag{4.18}$$

$$E_T^a(x) = \frac{1}{(2\pi)^{3/2}} \int d^3k \exp(-i\vec{k}\cdot\vec{x})[p^1(\vec{k})e_1^a(\vec{k}) + p^2(\vec{k})e_2^a(\vec{k})], \tag{4.19}$$

where $e_A^a(\vec{k}), e_a^A(\vec{k})$ are transverse vectors and their dual one-forms in momentum space are normalized such that $k^a e_a^A = 0$, $e_a^a(\vec{k})e_a^B(\vec{k}) = \delta_A^B$, $e_a^A(\vec{k}) = (e_a^A(\vec{k}))^* = e_a^A(-\vec{k})$; also $q(-\vec{k}) = q^*(\vec{k})$ and $p(-\vec{k}) = p^*(\vec{k})$. These relations can be inverted to yield

$$q_A(\vec{k}) = \frac{1}{(2\pi)^{3/2}} \int d^3x \exp(-i\vec{k}\cdot\vec{x})e_A^a(\vec{k})A_a^T(x,) \qquad (4.20)$$

$$p^A(\vec{k}) = \frac{1}{(2\pi)^{3/2}} \int d^3x \exp(i\vec{k}\cdot\vec{x})e_a^A(\vec{k})E_T^a(x), \qquad (4.21)$$

with $A = 1, 2$.

The $q_A(\vec{k}), p^A(\vec{k})$ capture the two degrees of freedom of the electromagnetic field and describe the radiative modes corresponding to the two possible helicities of the photon. One can reformulate the theory in terms of these variables. The Poisson brackets are

$$\{q_A(\vec{k}), p^B(\vec{k'})\} = \delta_A^B \delta^3(\vec{k} + \vec{k'}). \qquad (4.22)$$

The Hamiltonian, written in terms of these basic variables, adopts the form of an infinite collection of harmonic oscillators, one for each $\vec{k}$,

$$H = \tfrac{1}{2} \int d^3k \left(p_A(\vec{k})p^A(-\vec{k}) + |k|^2 q_A(\vec{k})q^A(-\vec{k}) \right). \qquad (4.23)$$

Let us now introduce the two quantities,

$$a_A(\vec{k}) = \tfrac{1}{\sqrt{2}} \left(\sqrt{|k|}q_A(\vec{k}) + i\frac{1}{\sqrt{|k|}}p_A(\vec{k}) \right), \qquad (4.24)$$

$$a_A^*(\vec{k}) = \tfrac{1}{\sqrt{2}} \left(\sqrt{|k|}q_A(-\vec{k}) - i\frac{1}{\sqrt{|k|}}p_A(-\vec{k}) \right), \qquad (4.25)$$

with Poisson brackets

$$\{a_A(\vec{k}), a_B^*(\vec{k'})\} = -i\delta_{AB}\delta(k - k'), \qquad (4.26)$$

in terms of which the classical Hamiltonian reads

$$H = \int d^3k|k|a_C^*(\vec{k})a^C(\vec{k}). \qquad (4.27)$$

4.3   Fock quantization

The Fock quantization arises by considering the number representation for each harmonic oscillator of the Hamiltonian (4.23). Since there is a continuous infinite number of oscillators, one for each $\vec{k}$, it is convenient to consider quantization in a finite region of space ("a box") in order to have a countable infinity of modes $\vec{k}_i$. Then, the canonical commutation relations become

$$\{a_A(\vec{k}_i), a_B^*(\vec{k}_j)\} = -i\delta_{ij}.\delta_{AB} \tag{4.28}$$

The Hamiltonian becomes

$$H = \sum_{i=1}^{\infty} |\vec{k}_i| \, a_C^*(\vec{k}_i) a^C(\vec{k}_i). \tag{4.29}$$

One can introduce the Fock representation directly by considering the quantum representation of the algebra (4.28) in a space of functions of infinite pairs of integer variables $\Phi(\ldots, n_{i,c}, \ldots)$. Each variable represents the state of each harmonic oscillator for a given $\vec{k}_i$ and a given polarization. The representation of the algebra is as follows:

$$\hat{a}^*_C(\vec{k}_j)\Phi(\ldots, n_{i,D}, \ldots) = \sqrt{n_{j,C}} \, \Phi(\ldots, n_{j,D} - \delta_{CD}, \ldots), \tag{4.30}$$

$$\hat{a}_C(\vec{k}_j)\Phi(\ldots, n_{i,C}, \ldots) = \sqrt{n_{j,C} + 1} \, \Phi(\ldots, n_{j,D} + \delta_{CD}, \ldots), \tag{4.31}$$

where the wavefunctions vanish if any of their arguments are negative numbers.

The commutation relations can be immediately derived:

$$[\hat{a}_B(\vec{k}_i), \hat{a}^*_B(\vec{k}_j)] = \delta_{ij}\delta_{AB}. \tag{4.32}$$

The next step in the quantization program is to introduce an inner product. This can be readily done:

$$< \Phi | \Psi >=$$
$$\sum_{n_{1,1}=1}^{\infty} \sum_{n_{1,2}=1}^{\infty} \cdots \sum_{n_{j,1}=1}^{\infty} \sum_{n_{j,2}=1}^{\infty} \cdots \Phi(n_{1,1}, n_{1,2}, \ldots, n_{j,1}, n_{j,2}, \ldots)^* \times$$
$$\times \Psi(n_{1,1}, n_{1,2}, \ldots, n_{j,1}, n_{j,2}, \ldots). \tag{4.33}$$

In terms of this inner product the operators $\hat{a}_C(\vec{k}_j)$ and $\hat{a}^*_C(\vec{k}_j)$ satisfy the relations

$$\hat{a}_C^{\dagger}(\vec{k}_j) = \hat{a}^*_C(\vec{k}_j), \tag{4.34}$$

where † means adjoint in the operatorial sense. One can now define the Hermitian operator $N(\vec{k}_j, C)$ by

$$\hat{N}(\vec{k}_j, C) = \hat{a}_C^{\dagger}(\vec{k})\hat{a}_C(\vec{k}) \tag{4.35}$$

with no summation over C.

The explicit action of the operator $\hat{N}(\vec{k}_j, A)$ is given by

$$\hat{N}(\vec{k}_j, A)\Psi(n_{1,1}, n_{1,2}, \ldots, n_{l,1}, n_{l,2}, \ldots) =$$
$$n_{j,A}\Psi(n_{1,1}, n_{1,2}, \ldots, n_{l,1}, n_{l,2}, \ldots). \tag{4.36}$$

The reader can immediately notice the resemblance with the usual harmonic oscillator: $\hat{a}$ and $\hat{a}^\dagger$ are annihilation and creation operators and $\hat{N}$ is the number operator, and we have one of each per momentum $\vec{k}_j$ and polarization C. The usual commutation relations follow:

$$[\hat{N}(\vec{k}_i, C), \hat{a}_D^\dagger(\vec{k}_j)] = \delta_{ij}\delta_{CD}\hat{a}_D^\dagger(\vec{k}_j), \tag{4.37}$$

$$[\hat{N}(\vec{k}_i, C), \hat{a}_D(\vec{k}_j)] = -\delta_{ij}\delta_{CD}\hat{a}_D(\vec{k}_j). \tag{4.38}$$

Let us now introduce the quantum Hamiltonian. Rewriting (4.23) in terms of creation and annihilation operators, one gets

$$\hat{H} = \tfrac{1}{2}\sum_{j=1}^{\infty}|\vec{k}_j|(a_C^\dagger(\vec{k}_j)a^C(\vec{k}_j) + \tfrac{1}{2}) \tag{4.39}$$

and it should be realized that this corresponds to a different factor ordering than the natural one that we would have inferred from the classical expression (4.27). This expression is divergent even for the case we are considering (a finite box) since we are summing the zero point energy for each of the infinite excited modes. In order to make this expression finite, it is usual to subtract the zero modes through the procedure called "normal ordering" (denoted by enclosing expressions in colons) consisting in ordering the $a\dagger$ to the left,

$$: \hat{H} := \tfrac{1}{2}\sum_{j=1}^{\infty}|\vec{k}_j|(a_C^\dagger(\vec{k}_j)a^C(\vec{k}_j)). \tag{4.40}$$

Since $\hat{H}$ commutes with $\hat{N}(\vec{k}_i, C)$, $\forall i, C$, both operators could be diagonalized simultaneously. In the representation we are considering, this can be accomplished straightforwardly by determining the vacuum state. This is the state with minimal energy and it can be checked that such a state Φ_0 satisfies

$$\hat{a}_C(\vec{k}_j)\Phi_0 = 0 \quad \forall k, C. \tag{4.41}$$

Once this state is given, the whole space of "excited" states can be spanned by applying the "creation" operator $\hat{a}^\dagger$. One can interpret this construction in terms of particles: the application of the operator $\hat{a}_C^\dagger(\vec{k}_i)$ creates a photon with polarization C and three-momentum $\vec{k}_i$. This can be verified by computing the normal-ordered momentum operator $\hat{P}_b =: \int d^3x \tilde{\hat{E}}^a \hat{F}_{ab}$: in this state. It can be checked that $: \hat{H}^2 - \hat{P}_b\hat{P}^b := 0$ and therefore the photon is massless.

To diagonalize the Hamiltonian and number operators we introduce a basis of states labeled $|n_{1,1}, n_{1,2}, \ldots, n_{j,1}, n_{j,2}, \ldots >$, defined by

$$|n_{1,1}, n_{1,2}, \ldots, n_{j,1}, n_{j,2}, \ldots > = \prod_{j=1}^{\infty} \prod_{C=1}^{2} \frac{1}{\sqrt{n_{j,C}!}} (a_C^{\dagger}(\vec{k}_j))^{n_{j,C}} |0, \ldots, 0 >,$$
(4.42)

where $|0, \ldots, 0 >$ is the vacuum. Therefore

$$\hat{H} |n_{1,1}, n_{1,2}, \ldots, n_{j,1}, n_{j,2}, \ldots > =$$

$$\sum_{j=1}^{\infty} \sum_{C=1}^{2} n_{j,C} |\vec{k}_i| |n_{1,1}, n_{1,2}, \ldots, n_{j,1}, n_{j,2}, \ldots >, \quad (4.43)$$

$$\hat{N}(\vec{k}_j, C) |n_{1,1}, n_{1,2}, \ldots, n_{j,1}, n_{j,2}, \ldots > =$$

$$n_{j,C} |n_{1,1}, n_{1,2}, \ldots, n_{j,1}, n_{j,2}, \ldots > \quad (4.44)$$

and this is what is usually called the Fock basis.

It is useful to introduce a dual Fock basis through the relation,

$$< n_{1,1}, n_{1,2}, \ldots, n_{j,1}, n_{j,2}, \ldots |m_{1,1}, m_{1,2}, \ldots, m_{j,1}, m_{j,2}, \ldots > =$$

$$\prod_{j=1}^{\infty} \prod_{C=1}^{2} \delta_{m_{j,C}; n_{j,C}} \quad (4.45)$$

and this relation leads naturally to the inner product (4.33).

The Fock basis describes naturally states with a definite number of incoherent photons of definite energy and momentum. These states have vanishing expectation values for the field operators $\hat{E}^c$ and $\hat{A}_c$. They therefore present a description of electromagnetism that is not naturally associated with the classical one. To be able to make contact with the classical limit more easily it is convenient to introduce a basis of states in terms of which the expectation values of both $\hat{E}^c$ and $\hat{A}_c$ are non-vanishing. The elements of this basis are called the coherent states.

The coherent states form a basis labeled by arbitrary complex numbers $\alpha_{i,c}$, associated with each mode. Their definition is

$$\hat{a}_C(\vec{k}_j)|\alpha_{1,1}, \alpha_{1,2}, \ldots, \alpha_{j,1}, \alpha_{j,2}, \ldots > = \alpha_{j,C}|\alpha_{1,1}, \alpha_{1,2}, \ldots, \alpha_{j,1}, \alpha_{j,2}, \ldots >$$
(4.46)

and can be written in terms of the vacuum as

$$|\alpha_{1,1}, \alpha_{1,2}, \ldots, \alpha_{j,1}, \alpha_{j,2}, \ldots > =$$

$$\prod_{i=1}^{\infty} \prod_{C=1}^{2} \exp(-\tfrac{1}{2}|\alpha_{i,C}|^2) \exp(\tfrac{1}{2}\alpha_{i,C}\hat{a}_C^{\dagger}(\vec{k}_i))|0, \ldots, 0 > . \quad (4.47)$$

It should be noticed that the states introduced do not strictly belong to the Fock space but to its closure, due to the infinite summation. It can be checked that these states minimize the uncertainty in both the electric field and the connection and are therefore the closest to a "classical" configuration one can get.

Since we did not impose any restrictions on the eigenvalues of annihilation and creation operators while defining the coherent basis, it follows that the basis is *overcomplete*. A given state can be expanded in terms of this basis in infinitely many different ways. We will see later a connection between this overcompleteness and that of Wilson loops.

4.4   Loop representation

In order to introduce the loop representation let us first remind the reader of some aspects of the usual connection representation of the Maxwell theory. We can particularize the steps we presented in the previous chapter for the canonical quantization of Yang–Mills theories to the Maxwell case.

The connection representation is the most natural quantization since it is based on the straightforward quantization of the canonical algebra of connections and electric fields, taking a polarization based on wavefunctionals of the configuration variables.

Let us therefore start by picking a polarization in which wavefunctions are functionals of the connection $\Psi[A]$ and promote the connection and electric field to quantum operators,

$$\hat{\tilde{E}}^a \Psi[A] = -i \frac{\delta}{\delta A_a} \Psi[A], \qquad (4.48)$$

$$\hat{A}_a \Psi[A] = A_a \Psi[A]. \qquad (4.49)$$

Notice that we are considering functionals of the full (non-transverse) connection, so we will have to enforce the Gauss law as a quantum constraint,

$$\hat{\mathcal{G}} \Psi[A] = \partial_a \frac{\delta}{\delta A_a} \Psi[A] = 0, \qquad (4.50)$$

which tells us that $\Psi[A]$ has to be a gauge invariant function of A. We are imposing gauge invariance at a quantum level. This is different from what we did in the previous section where we solved the constraints at a classical level (reduced phase space quantization). Therefore there is potential for these two procedures to be inequivalent.

We can now formally write the quantum Hamiltonian,

$$\mathcal{H} \Psi[A] = \int d^3x \left(\eta^{ab} \frac{\delta}{\delta A_a} \frac{\delta}{\delta A_a} + \frac{1}{2} \eta^{ab} \eta^{cd} F_{ac} F_{bd} \right) \Psi[A], \qquad (4.51)$$

though it is clear that a detailed discussion of the first term requires a regularization.

One can solve the eigenvalue problem for this Hamiltonian (in terms of gauge invariant functions in order to satisfy the Gauss law) and determine the ground and excited states of the theory [63]. We will return to these issues in terms of other representations.

Let us now proceed to construct the loop representation. As described in the previous chapter one can introduce a loop representation either in terms of a non-canonical algebra of classical quantities or via a transform.

In the Abelian case one can immediately find a non-canonical algebra of gauge invariant operators in terms of which one can write all physical quantities by simply considering the Wilson loop and the electric field. In order to keep the construction as close as possible to that which we will later perform for the non-Abelian cases, let us introduce the operators

$$T(\eta) = W(\eta), \tag{4.52}$$

$$T^a(\eta_x^x) = \tilde{E}^a(x)W(\eta), \tag{4.53}$$

which satisfy the non-canonical algebra

$$\{T(\eta), T(\gamma)\} = 0, \tag{4.54}$$

$$\{T^a(\gamma_x^x), T(\eta)\} = -iX^{ax}(\eta)W(\eta \circ \gamma), \tag{4.55}$$

$$\{T^a(\gamma_x^x), T^b(\eta_y^y)\} = -iX^{ax}(\eta)T^b(\gamma \circ \eta) + iX^{by}(\gamma)T^a(\eta \circ \gamma). \tag{4.56}$$

A quantum realization of this algebra in a space of loop-dependent functions is

$$\hat{T}(\eta)\Psi(\gamma) = \Psi(\eta^{-1} \circ \gamma), \tag{4.57}$$

$$\hat{T}^a(\eta_x^x)\Psi(\gamma) = X^{ax}(\gamma)\Psi(\eta^{-1} \circ \gamma), \tag{4.58}$$

and the reader can check that this realizes correctly the Poisson algebra in terms of quantum commutators. A choice of factor ordering with the functional derivatives to the right has been made.

The loop transform is given by

$$\Psi(\gamma) = \int DA \exp\left(-i \int d^3z X^{az}(\gamma)A_a(z)\right)\Psi[A] \tag{4.59}$$

and due to the Abelian nature of the connection the integral can be rigorously defined [65].

If one considers operators $\hat{T}(\gamma), \hat{T}^a(\gamma_x^x)$ in the connection representation defined by

$$\hat{T}(\gamma)\Psi[A] \equiv W_A(\gamma)\Psi[A], \tag{4.60}$$

$$\hat{T}^a(\gamma_x^x)\Psi[A] \equiv \hat{\tilde{E}}^a(x)W_A(\gamma)\Psi[A], \tag{4.61}$$

one can check that applying the transform (4.59) one obtains the operators introduced in (4.57),(4.58).

In terms of the non-canonical algebra one can express the electric field and the field tensor in the following way:

$$\hat{F}_{ab}(x) = -i\Delta_{ab}(x)T(\gamma)|_{\gamma=\iota}, \tag{4.62}$$

$$\hat{\tilde{E}}^a(x) = T^a(\gamma_x^x)|_{\gamma=\iota}, \tag{4.63}$$

where ι is the identity loop. This allows a loop representation to be found naturally through equations (4.57),(4.58),

$$\hat{F}_{ab}(x)\Psi(\gamma) = -i\Delta_{ab}(x)\Psi(\gamma), \tag{4.64}$$

$$\hat{\tilde{E}}^a(x)\Psi(\gamma) = X^{ax}(\gamma)\Psi(\gamma). \tag{4.65}$$

Therefore there is a natural interpretation of loops as lines of electric flux in this representation.

One could arrive at these expressions by using the loop transform (4.59), integrating by parts and considering the action of the fields on Wilson loops in the connection representation,

$$\hat{F}_{ab}(x)W(\gamma) = F_{ab}(x)W(\gamma) = -i\Delta_{ab}(x)W(\gamma), \tag{4.66}$$

$$\hat{\tilde{E}}^a(x)W(\gamma) = X^{ax}(\gamma)W(\gamma) = \oint_\gamma dy^a \delta(x-y)W(\gamma). \tag{4.67}$$

The last expression ensures that the Gauss law is automatically satisfied in the loop representation (due to the transverse nature of the first order multitangent $X^{ax}(\gamma)$). This is a natural consequence of the fact that the loop representation is based on the quantization of an algebra of gauge invariant objects. Only gauge invariant quantities can be realized naturally in the loop representation. Gauge dependent objects could be introduced by means of the connection derivative defined in chapter 1. The gauge dependence is introduced through the path prescription used in the definition of the connection derivative.

The commutation relation of E and F,

$$[\hat{F}_{cd}(y), \hat{\tilde{E}}^a(x)] = -i\delta^a_{[c}\partial_{d]}\delta(x-y), \tag{4.68}$$

finds its natural counterpart in the expression of the action of the loop derivative on the loop coordinate that we introduced in chapter 2,

$$\Delta_{cd}(y)X^{ax}(\gamma) = \delta^a_{[c}\partial_{d]}\delta(x-y). \tag{4.69}$$

One can now realize the Hamiltonian in terms of loops. The magnetic field portion of it is given simply in terms of loop derivatives,

$$\eta_{ab}\hat{\tilde{B}}^a(x)\hat{\tilde{B}}^b(x)\Psi(\gamma) = -\tfrac{1}{2}\eta^{ac}\eta^{bd}\Delta_{ab}(x)\Delta_{cd}(x)\Psi(\gamma). \tag{4.70}$$

The electric field portion is given in terms of two loop integrals, which can be reexpressed as

$$\eta_{ab}\hat{\tilde{E}}^a(x)\hat{\tilde{E}}^b(x)\Psi(\gamma) = \eta_{ab}X^{ax}(\gamma)X^{bx}(\gamma)\Psi(\gamma). \qquad (4.71)$$

The Hamiltonian eigenvalue equation then reads

$$\hat{H}\Psi(\gamma) \equiv \int d^3x \left(-\tfrac{1}{4}\eta^{ac}\eta^{bd}\Delta_{ab}(x)\Delta_{cd}(x) + \tfrac{1}{2}\eta_{ab}X^{ax}(\gamma)X^{bx}(\gamma) \right) \Psi(\gamma)$$
$$= E\Psi(\gamma). \qquad (4.72)$$

The second term can be suggestively rewritten as

$$\int d^3x\, X^{ax}(\gamma)X^{bx}(\gamma)\Psi(\gamma) = \oint_\gamma dy^a \oint_\gamma dy'^b \delta(y-y')\eta_{ab}\Psi(\gamma), \qquad (4.73)$$

which is proportional (through a divergent factor that needs to be regularized) to the length of the loop. Therefore the eigenvalue equation can be qualitatively interpreted as a "Laplacian" in terms of the double loop derivative and a "quadratic potential" given by the length of the loop. Notice that the other term, involving the loop derivatives, is also potentially ill defined. If one considers wavefunctions such that their loop derivative is distributional a regularization may be needed. We will not discuss the details here since for the particular case of Maxwell theory the extended representation discussed in section 4.6 furnishes a natural setting to regularize the theory.

Let us now study the vacuum and excited states of this system. One possible avenue is to take this analogy with the Hamiltonian of a harmonic oscillator seriously and propose a "Gaussian" state of the form

$$\Psi_0(\gamma) = \exp\left(-\tfrac{1}{2} \oint_\gamma dy^a \oint_\gamma dy'^b K_{ab}(y-y') \right) \qquad (4.74)$$

$$\equiv \exp\left(-\tfrac{1}{2} X^{ax}(\gamma)X^{by}(\gamma)K_{ax\,by} \right) \qquad (4.75)$$

and insert this expression in the eigenvalue equation for the Hamiltonian to determine K_{ab}. This course has actually been pursued in reference [63]. Here, however, we will find the vacuum by introducing the creation and annihilation operators in the loop representation and finding the state annihilated by the annihilation operator. It will turn out that this construction yields the same vacuum as that of reference [63].

Both the creation and annihilation operators can be readily realized in loop space. To introduce them we need to realize the q and p operators, and therefore the $\hat{A}^T_a$ operator. To do so we use the relation in the classical theory

$$\partial^a F_{ab} = \Delta A^T_b, \qquad (4.76)$$

where Δ is the three-dimensional Laplacian, and realize this expression in terms of the loop derivative. Then,

$$\hat{A}_a^T(x)\Psi(\gamma) = -i\frac{1}{\Delta}\partial^b \Delta_{ab}(x)\Psi(\gamma). \tag{4.77}$$

In terms of this expression, the operator $\hat{q}^A(\vec{k})$ is,

$$\hat{q}^A(\vec{k})\Psi(\gamma) = \frac{1}{(2\pi)^{3/2}}\int d^3x \exp(-i\vec{k}\cdot\vec{x})e^{aA}(\vec{k})\frac{k^b}{|\vec{k}|^2}\Delta_{ba}(x)\Psi(\gamma). \tag{4.78}$$

The operator $\hat{p}^1(\vec{k})$ can be realized immediately,

$$\hat{p}^A(\vec{k})\Psi(\gamma) = \frac{1}{(2\pi)^{3/2}}\int d^3x e^{-i\vec{k}\cdot\vec{x}}e_a^A(\vec{k})X^{ax}(\gamma)\Psi(\gamma). \tag{4.79}$$

Therefore the creation and annihilation operators in the loop representation have the forms

$$\hat{a}_A^\dagger(\vec{k}) = \frac{1}{(2\pi)^{3/2}}\int d^3x \left(\exp(-i\vec{k}\cdot\vec{x})e_A^a(\vec{k})\frac{k^b}{|\vec{k}|^{3/2}}\Delta_{ba}(x) \right.$$
$$\left. -i\frac{1}{|\vec{k}|^{1/2}}\exp(-i\vec{k}\cdot\vec{x})e_{aA}(\vec{k})X^{ax}(\gamma) \right), \tag{4.80}$$

$$\hat{a}_A(\vec{k}) = \frac{1}{(2\pi)^{3/2}}\int d^3x \left(\exp(-i\vec{k}\cdot\vec{x})e_A^a(\vec{k})\frac{k^b}{|\vec{k}|^{3/2}}\Delta_{ba}(x) \right.$$
$$\left. +i\frac{1}{|\vec{k}|^{1/2}}\exp(-i\vec{k}\cdot\vec{x})e_{aA}(\vec{k})X^{ax}(\gamma) \right). \tag{4.81}$$

We now apply (4.81) to (4.74). The application of the first term in $\hat{a}$ yields,

$$-\frac{i}{(2\pi)^{3/2}}\int d^3x \exp(-i\vec{k}\cdot\vec{x})e_A^a(\vec{k})\sqrt{|k|}X^{by}(\gamma)K_{ax\,by}\Psi_0(\gamma). \tag{4.82}$$

We must now determine $K_{ax\,by}$ so that this terms cancels the second one. It can be straightforwardly checked that if one takes,

$$K_{ax\,by} = \frac{1}{(2\pi)^{3/2}}\int \frac{d^3q}{|q|}\exp\left(-i\vec{q}\cdot(\vec{x}-\vec{y})\right) \tag{4.83}$$

the two terms actually cancel. The expression for K is that of the homogeneous symmetric propagator of Maxwell theory.

It is now immediate to find the excited states, simply by operating with $\hat{a}^\dagger$ on the vacuum. The first excited state is given by

$$\Psi_1^{(A,\vec{k})}(\gamma) = \frac{1}{(2\pi)^{3/2}}\int d^3x \frac{1}{|\vec{k}|^{1/2}}\exp(-i\vec{k}\cdot\vec{x})e_{aA}(\vec{k})X^{ax}(\gamma)\Psi_0(\gamma). \tag{4.84}$$

This expression can be more compactly written in $\vec{k}$ space. Introducing the Fourier transform of the multitangent,

$$X^{ak}(\gamma) = \frac{1}{(2\pi)^{3/2}} \int d^3x \exp(-i\vec{k} \cdot \vec{x}) X^{ax}(\gamma), \qquad (4.85)$$

the first excited state is

$$\Psi_1^{(A,\vec{k})}(\gamma) = \frac{1}{|\vec{k}|^{1/2}} e_{aA}(\vec{k}) X^{ak} \Psi_0(\gamma). \qquad (4.86)$$

This state corresponds to a photon of momentum $\vec{k}$ and polarization A. The objects $X^{ak}(\gamma)$ are usually called "form factors" of the loop. The form factors are transverse,

$$k_a X^{ak}(\gamma) = 0. \qquad (4.87)$$

and therefore their only relevant components are the projections on the polarization vectors.

The n-photon state is given by,

$$\Psi_n^{(A_1,\vec{k}_1,...,A_n,\vec{k}_n)}(\gamma) = \left(\frac{1}{|\vec{k}_1|^{1/2}} e_{aA_1}(\vec{k}_1) X^{ak_1} \right.$$
$$\left. \cdots \frac{1}{|\vec{k}_n|^{1/2}} e_{aA_n}(\vec{k}_n) X^{ak_n} \right) \Psi_0(\gamma). \qquad (4.88)$$

An appealing fact is the form of the coherent states in this representation. They are given by

$$\Psi^{(\alpha)}(\gamma) = W(\gamma, A)\Psi_0(\gamma), \qquad (4.89)$$

where $W(\gamma, A)$ is the Wilson loop along the loop γ of a given connection A. It can be readily checked that these states are eigenvectors of the annihilation operator. When one operates with (4.81) on the state the first term (involving the loop derivative) acts both on the Wilson loop and on $\Psi_0(\gamma)$. The action on $\Psi_0(\gamma)$ cancels the contribution from the second term of (4.81) as we observed when deriving the vacuum. The action of the loop derivative on the Wilson loop gives the field tensor F_{ab} of the given connection, as we showed in chapter 1. The eigenvalue α is therefore given in terms of the connection as

$$\alpha = \frac{i}{|\vec{k}|^{3/2}} e_A^a(\vec{k}) k^b F_{ab}(k). \qquad (4.90)$$

The field tensor so introduced actually has a physical meaning. It corresponds to the expectation value of the spatial part of the Maxwell field tensor in the coherent state in question.

Up to now we have operated with the Hamiltonian in a formal fashion, ignoring the issues of regularization. As a result, the eigenfunctions we find are really ill defined. This can be readily seen from the expression of the vacuum (4.74) since the propagator diverges quadratically when $x \to y$.

A suitable regularization for the second term of the Hamiltonian is to replace the delta function by a function $f_\epsilon(y - y')$ such that

$$\lim_{\epsilon \to 0} f_\epsilon(y - y') = \delta(y - y'). \tag{4.91}$$

Explicitly,

$$f_\epsilon(y - y') = \frac{1}{(2\pi)^{3/2}} \int d^3q \, r(|q|\epsilon) \exp(i\vec{q} \cdot (\vec{y} - \vec{y'})), \tag{4.92}$$

where the function r is defined such that

$$\int_0^\infty r(x) dx = 0 \tag{4.93}$$

and explicit examples of such a function are

$$r(x) = (1 - x) \exp(-x), \tag{4.94}$$
$$r(x) = (1 - \tfrac{1}{2}x)\Theta(1 - x). \tag{4.95}$$

If one now repeats the procedure that led to the vacuum taking into account the regularization, one finds that the vacuum of the regularized Hamiltonian is also given by a Gaussian,

$$\Psi_0^\epsilon(\gamma) = \exp\left(-\tfrac{1}{2} \oint_\gamma dy^a \oint_\gamma dy'^b K_{ab}^\epsilon(x - y)\right), \tag{4.96}$$

where the regularized propagator is given by

$$K_{ab}^\epsilon(x - y) = \delta_{ab} \frac{1}{(2\pi)^{3/2}} \int d^3q \, \frac{r(\epsilon|q|)}{|q|} \exp(-i\vec{q} \cdot (\vec{y} - \vec{y'})), \tag{4.97}$$

where $r(x)$ is the function that we introduced while regularizing the Hamiltonian. Other regularizations for this same problem have been considered in reference [64].

Finally, we can introduce an inner product. We define a normalized form factor as

$$C^a(\vec{k}) = \frac{1}{|\vec{k}|^{1/2}} X^{ak}(\gamma), \tag{4.98}$$

in terms of which we introduce an inner product,

$$< \Phi_1(\gamma)|\Phi_2(\gamma) >= \int DCDC^* \, \Phi_1^*(C, C^*)\Phi_2(C, C^*). \tag{4.99}$$

The integrals on C and its complex conjugate are functional integrals. Note that the functional integrals defined above can only be computed in practice if one assumes that the normalized multitangents are arbitrary transverse fields, not necessarily associated with a loop. Therefore one is really going to an extension of the representation in order to perform it. We will return to these issues when we discuss the extended loop representation in section 4.6. The vacuum (4.74) is normalized with this inner product,

$$\int dC dC^* \Psi_0(C)^* \Psi_0(C) = \int DC DC^* \exp\left(-\int d^3k C^{*a}(\vec{k}) C^b(\vec{k}) \delta_{ab}\right) = 1.$$
(4.100)

Because in this representation excited states are proportional to the vacuum, the factor $\exp\left(-\int d^3k C^{*a}(\vec{k}) C^b(\vec{k}) \delta_{ab}\right)$ acts as a Gaussian measure in the inner product and the vacuum is simply represented as a constant and the excited states by the projections of form factors on the polarization vectors. We will see that a similar feature arises naturally in the Bargmann representation.

4.5 Bargmann representation

In 1962 Bargmann introduced a complex coordinatization for the harmonic oscillator. It is based on using as canonical coordinates $z \equiv q + ip$ and z^*, its complex conjugate. The resulting formulation is very elegant, wavefunctions are holomorphic, and the inner product is determined, fixing the reality of the relevant operators. This formulation has several analogous elements to Ashtekar's formulation of general relativity in which one of the canonical coordinates is complex and the other real. The hope is that similar analytic properties will help determine the inner product of quantum gravity. In this section we will present a Bargmann-like formulation of Maxwell theory in terms of both traditional variables and loops. This formulation naturally fixes the inner product to be the complex measure introduced a bit arbitrarily in the previous section. This treatment follows closely that of Ashtekar and Rovelli [65].

4.5.1 The harmonic oscillator

The canonical formulation of the harmonic oscillator is given in terms of coordinates q, p and the Hamiltonian is $H = p^2 + \omega^2 q^2$. Quantization is achieved through wavefunctions $\Psi(q)$ and the eigenvalue equation for the Hamiltonian is $(-\partial^2/\partial q^2 + \omega^2 q^2)\Psi(q) = E\Psi(q)$. The eigenstates of the system are given by a Gaussian in q times the Hermite polynomials.

Normally, as mentioned above, the Bargmann representation involves both real and complex coordinates. Discussion of the harmonic oscillator in those coordinates can be seen in reference [2] and in Bargmann's original paper [70]. Here, however, we will explore a fully complex representation which is better geared for comparison with what was done in reference [65] for the Maxwell case. One could also treat the Maxwell case in a mixed polarization and then it would more resemble Bargmann's original treatment.

Assume now that a complex coordinatization given by the variables and $z = \frac{1}{\sqrt{2}}(\omega q - ip)$ and $z^* = \frac{1}{\sqrt{2}}(\omega q + ip)$ is introduced. The Poisson bracket is $\{z, z^*\} = i\omega$. The variables satisfy reality conditions that say that they are complex conjugates of each other. One can then construct a representation of the canonical algebra on holomorphic functions $\Psi(z)$,

$$\hat{z}\Psi(z) = z\Psi(z), \tag{4.101}$$

$$\hat{z}^*\Psi(z) = \omega\frac{d\Psi(z)}{dz}. \tag{4.102}$$

An inner product is introduced that translates the reality conditions into operatorial relations:

$$\hat{z}^\dagger = \hat{z}^*, \tag{4.103}$$

$$\hat{z}^{*\dagger} = \hat{z}. \tag{4.104}$$

We will now use these relations to determine the inner product. Let us start with a generic inner product,

$$< \Phi|\Psi > = \int dz \int d\bar{z}\mu(z,\bar{z})\bar{\Phi}(z)\Psi(z), \tag{4.105}$$

and if one now requires that the operatorial relations be satisfied this fixes the measure uniquely to be

$$\mu(z,\bar{z}) = \exp(-z\bar{z}). \tag{4.106}$$

In terms of these variables the quantum Hamiltonian of the harmonic oscillator is

$$\hat{\mathcal{H}}\Psi(z) = \tfrac{1}{2}\omega\left(z\frac{\partial}{\partial z} + 1\right)\Psi(z) \tag{4.107}$$

where we have chosen a symmetric factor ordering in z and z^*. This ordering corresponds in the traditional variables to $\hat{\mathcal{H}} = \hat{p}^2 + \omega^2\hat{q}^2$. The vacuum is simply $\Psi_0(z) = 1$ and the excited states are polynomials in z. With the given measure, polynomial states are normalizable.

This is attractive because just by requiring the reality of the classical operators the inner product is uniquely fixed. Since Maxwell theory is just a collection of harmonic oscillators, it is immediate to construct a Bargmann representation. Since the reality conditions are a structure that is present in other theories (e.g. gravity) where other structures that one could use to build an inner product (e.g. Lorentz invariance) are absent, this gives some hope that a similar construction could yield the inner product for those theories. It is certainly reassuring that this construction at least yields the correct result for Maxwell theory as we will discuss in the next section.

4.5.2 *Maxwell–Bargmann quantization in terms of loops*

For the kind of calculation that we will perform in this section, it is convenient to introduce circular polarization. We express the fields as

$$A_a^T(x) = \frac{1}{(2\pi)^{3/2}} \int d^3k \exp(i\vec{k}\cdot\vec{x}) \left(q_1^{(c)}(\vec{k})m_a(\vec{k}) + q_2^{(c)}(\vec{k})m_a^*(\vec{k}) \right), \quad (4.108)$$

$$E_T^a(x) = \frac{-1}{(2\pi)^{3/2}} \int d^3k \exp(i\vec{k}\cdot\vec{x}) \left(p_1^{(c)}(\vec{k})m^a(\vec{k}) + p_2^{(c)}(\vec{k})m^{*a}(\vec{k}) \right), \quad (4.109)$$

where the complex polarization vectors satisfy*

$$k^a m_a(\vec{k}) = 0, \quad m^a(\vec{k})m_a(\vec{k}) = 0, \quad (4.110)$$

$$m_a(-\vec{k}) = -m^{*a}(\vec{k}), m^a(\vec{k})m_a^*(\vec{k}) = 1. \quad (4.111)$$

Given a conjugate pair A_a^T and E_T^a of Maxwell theory, one could decompose it into positive and negative frequency (for instance, by evolving it and decomposing the resulting spacetime solution). Examining the canonical commutation relations one finds that the positive frequency connection and the negative frequency electric field form a conjugate pair, given by,

$$A_a^+(x) = \frac{1}{\sqrt{2}}(A_a^T(x) + i\Delta^{-1/2}(E_T)^b\eta_{ab})$$

$$= \int \frac{d^3k}{(2\pi)^{3/2}|k|} \exp(i\vec{k}\cdot\vec{x})(\zeta_1(\vec{k})m_a(\vec{k}) + \zeta_2(\vec{k})m_a^*(\vec{k})), \quad (4.112)$$

$$\tilde{E}^{b-}(x) = \frac{1}{\sqrt{2}}(\tilde{E}_T^b(x) + i\Delta^{1/2}A_a^T\eta^{ab})$$

$$= \int \frac{d^3k}{(2\pi)^{3/2}} \exp(i\vec{k}\cdot\vec{x})(\zeta_1^*(-\vec{k})m_a(\vec{k}) + \zeta_2^*(-\vec{k})m_a^*(\vec{k})), \quad (4.113)$$

where $\zeta(\vec{k})_i = \frac{1}{\sqrt{2}}(|k|q_i^{(c)}(\vec{k}) - ip_i^{(c)}(\vec{k}))$. The definition of the ζs embodies exactly the same construction that we performed for the harmonic oscillator. The true degrees of freedom of the Maxwell field are now embodied in the two complex ζ fields. They provide a complex coordinatization on

---

* If one translates back to the language we used in section 4.2 by considering that the vector $m_a(\vec{k}) = \frac{1}{\sqrt{2}}(e_a^1(\vec{k}) + ie_a^2(\vec{k}))$ one finds that $e_a^1(\vec{k}) = -e_a^1(-\vec{k})$ as before but $e_a^2(\vec{k}) = e_a^2(-\vec{k})$. These conventions are also used by Bjorken and Drell [71].

the phase space of Maxwell theory. The canonical commutation relations for the ζs are

$$\{\zeta_B(\vec{k}), \zeta_C^*(\vec{k}')\} = i|k|\delta_{BC}\delta(\vec{k} - \vec{k}') \tag{4.114}$$

and we see the close relation between the ζ variables and the a, a^* variables that were introduced for the Fock representation.

Let us now quantize the theory by promoting the variables to quantum operators:

$$\hat{\zeta}_B(\vec{k})\Psi(\zeta) = \zeta_B(\vec{k})\Psi(\zeta), \tag{4.115}$$

$$\hat{\zeta}^*{}_B(\vec{k})\Psi(\zeta) = |k|\frac{\delta\Psi(\zeta)}{\delta\zeta_B(\vec{k})} \tag{4.116}$$

where the wavefunctions to be holomorphic functionals of the arguments.

One would like the fact that ζ and ζ^* are conjugate to each other translate itself into an operatorial relation of the kind

$$\hat{\zeta}_B^\dagger(\vec{k}) = \hat{\zeta}^*{}_B, \tag{4.117}$$

where by $\dagger$ we mean the operatorial adjoint under a suitable inner product. This relation implies that explicitly in terms of the inner product

$$< \Phi|\zeta_B(\vec{k})\Psi > = |k| \left\langle \frac{\delta}{\delta\zeta_B(\vec{k})}\Phi|\Psi \right\rangle. \tag{4.118}$$

To find an inner product that satisfies this condition, one can simply propose an explicit expression

$$< \Psi|\Phi > = \int d\zeta d\zeta_j^* \mu(\zeta, \zeta^*)\Psi^*(\zeta)\Phi(\zeta), \tag{4.119}$$

where $\mu(\zeta, \zeta^*)$ is the measure to be determined. It is easy to check that the condition (4.118) uniquely implies [70],

$$\mu(\zeta, \zeta^*) = \exp\left(-\int \frac{d^3k}{|k|}(|\zeta_1(\vec{k})|^2 + |\zeta_2(\vec{k})|^2)\right). \tag{4.120}$$

So we again get a Gaussian measure. Since the wavefunctions we are considering are holomorphic, we immediately conclude that this representation is essentially the same at the level of inner product and wavefunctions as the real connection representation that we introduced before. Again, we should notice that we found the Gaussian measure without any reference to Lorentz invariance. This therefore makes the method attractive for tackling cases in which such invariances are not present, such as in gravity.

In this representation, the normal ordered Hamiltonian is

$$\hat{\mathcal{H}} = \int d^3k |k| \sum_{B=1}^{2} \zeta_B(\vec{k}) \frac{\delta}{\delta \zeta_B(\vec{k})}, \tag{4.121}$$

The ground state, $\Psi(\zeta) = 1$ is equivalent to the Fock vacuum. A one-photon state with given polarization and momentum $\vec{k}_0$ is given by a linear function $\Psi = \zeta(\vec{k}_0)$. A generic one-photon state with given polarization is given by a superposition in momenta,

$$\Psi_1(\zeta) = \int \frac{d^3k}{|k|} f(\vec{k}) \zeta_1(\vec{k}) \tag{4.122}$$

with obvious generalizations for the n-photon states.

We now proceed to construct the loop representation. As usual we could proceed by quantizing an algebra of non-canonical loop-based gauge invariant quantities or via a loop transform. Since we have given examples of the first kind of construction before and in this particular case it leads to the same results, we will simply proceed with the transform. This will also allow us to show how the transform is explicitly defined for an Abelian theory. As we said before, for the Maxwell case the loop transform is well defined. In terms of the Bargmann coordinates, it reads

$$\Psi(\gamma) = \int \prod_B D\zeta_B D\zeta_B^* \exp\left(-\int \frac{d^3k}{|\vec{k}|} |\zeta_B(\vec{k})|^2\right) \exp\left(\oint dy^a A_a^+\right)^* \Psi(\zeta_B). \tag{4.123}$$

Notice that in the definition of the loop transform introduced in chapter 3 the complex conjugate of the Wilson loop appears. For the real case which we considered before this amounts to a change of sign due to the i that appears in the definition of the holonomy. Here it implies the complex conjugate of the connection,

$$\oint dy^a A_a^+ = \int \frac{d^3k}{|(2\pi)^{3/2}\vec{k}|} \left(\zeta_1(\vec{k}) m_a(\vec{k})(X^{ak})^* + \zeta_2(\vec{k}) m_a^*(\vec{k})(X^{ak})^*\right)$$

$$= \int \frac{d^3k}{|\vec{k}|} \sum_{B=1}^{2} \zeta_B(\vec{k}) X_B^{*k}, \tag{4.124}$$

where

$$X_1^{*k} = (2\pi)^{-3/2} m_a(\vec{k})(X^{ak})^*, \quad X_2^{*k} = (2\pi)^{-3/2} m_a^*(\vec{k})(X^{ak})^*. \tag{4.125}$$

Also, as we said in chapter 3, the introduction of a loop transform requires the introduction of an inner product in terms of connections. Since we have the Gaussian inner product given by the reality conditions, we use it in the definition of the transform.

Therefore the expression for the loop transform for this particular case
is given by

$$\Psi(\gamma) = \int \prod_B D\zeta_B \, D\zeta_B^* \, \exp\left(-\int \frac{d^3k}{|\vec{k}|} |\zeta_B(\vec{k})|^2\right)$$

$$\times \exp\left(\int \frac{d^3k}{|\vec{k}|} \sum_{B=1}^{2} \zeta_B^*(\vec{k}) X_B^k(\gamma)\right) \Psi(\zeta). \qquad (4.126)$$

Let us now evaluate this explicitly for some states. Generically the
n-photon states are going to be polynomials in ζ. It is easy to transform
such states. Simply expand the exponential $\exp(\zeta_B^* X_B)$ and note that
the $\zeta^n/\sqrt{n!}$ are an orthonormal basis with the Gaussian measure. Then
the loop transform of any state $\Psi(\zeta) = \sum_n C_n(\zeta)^n$ is simply given by
$\Psi(\gamma) = \sum_n C_n(X)^n$ with immediate generalizations for states depending
on several ζ_Bs. The vacuum, in particular, is $\Psi(\gamma) = 1$ and the one-photon
state with helicity B and momentum $\vec{k}$ given by

$$\Psi_1(\gamma) = X_B^k(\gamma). \qquad (4.127)$$

With this we end the discussion of this representation. Let us now
compare the results obtained with the loop representation constructed
from real variables. The first thing to notice is that the use of the reality
conditions in the Bargmann case fixes a non-trivial inner product in terms
of connections and therefore a non-trivial measure in the loop transform.
Historically, this was not done with the real loop representation since
the intention was to recover the Fock space structure (which, in turn, is
determined by Poincaré invariance). However, it is very easy to check that
if one constructs a connection representation for the real case in terms of
q and p and requires the quantum operators $\hat{q}$ and $\hat{p}$ to be real, the inner
product given by the trivial measure in q, p appears as a result. This, in
turn, implies the trivial measure in the As which is the one we used in
section 4.4 to compute the loop transform.

The appearance in the Bargmann case of a non-trivial measure in the
inner product and the loop transform implies certain important differences in the two representations. To start with, the vacuum is just a
constant. The Gaussian factor that appeared in the real representation is
"absorbed in the non-trivial measure". Although one may consider this
point irrelevant from a practical point of view, it has implications in the
rigorous definition of the space of states. In fact, while in the real case
we needed the introduction of a regularization to have a well defined vacuum and space of states, in the Bargmann case the states are well defined
without the introduction of a regularization.

Have we gained something from nothing? That is, can we forget the
regularization issues altogether by considering a non-trivial measure in

the loop transform? The answer is negative. If one wishes to complete the quantization in the loop representation, one would like to introduce an inner product in terms of loops, as was done in section (4.4). If one does so in the Bargmann case, one notices that now a non-trivial Gaussian measure in the Fs appears in the loop representation. This measure coincides exactly with the expression of the vacuum in the real case. If one wants to define an inner product only in terms of loops, the expression of the measure is illdefined. If one wants to proceed as in section 4.4 and "extend" the inner product to all Xs then the difficulties disappear at the price of extending the notion of loops.

Let us now study the extended representation.

4.6 Extended loop representation

We will now explore the consequences of introducing an "extended loop representation", a representation based on the loop coordinates introduced in chapter 2. We will immediately see that such a representation presents computational economy, technical cleanliness and also allows us to view in a conceptually different way the problem of loop quantization. We will see that regularization difficulties are better dealt with in terms of loop coordinates. We will also see that we are also able to determine the classical canonical theory that underlies the loop representation. In the particular case of Maxwell theory we will see that the extended loop representation coincides with the electric field representation. This, however, does not generalize to non-Abelian fields and in those cases the extended loop representation is a new representation that contains the loop representation as a limiting case. As a bonus we will find a way of writing the action for electromagnetism purely in terms of loops. This version of the action is amenable to lattice Monte Carlo techniques and has the potential to offer new insights into non-perturbative QED problems. The fact that so much is gained in the Maxwell case by going to an extended loop representation clearly suggests that a similar avenue should be pursued in the non-Abelian cases and especially gravity.

Let us start by replacing in our formalism the usual loop holonomy by its extended counterpart in terms of the loop coordinates,

$$H_A(\gamma) \to H_A(\mathbf{X}) = \exp\left(i \int d^3x\, A_{ax}\, X^{ax}\right). \qquad (4.128)$$

Because of the Abelian nature of the theory we only need the first order multitensor, which can be simply viewed as a divergence-free vector density on the three-manifold,

$$\partial_{ax} X^{ax} = 0. \qquad (4.129)$$

One can now introduce a loop coordinate representation by means of the transform

$$\Psi(\mathbf{X}) = \int DA\,\Psi[A] \exp\left(-ig\int d^3x\, A_a(x) X^{ax}\right). \qquad (4.130)$$

In this representation, wavefunctions are functionals of the smooth vector density X. In terms of this representation we can realize the operators $\hat{F}_{ab}$, $\hat{\tilde{E}}^{ax}$ and the Wilson loop $\hat{W}_A(\mathbf{X})$ through

$$\hat{W}_A(\mathbf{X}_0)\,\Psi(\mathbf{X}) = \Psi(\mathbf{X} - \mathbf{X}_0), \qquad (4.131)$$

$$\hat{E}^{ax}\,\Psi(\mathbf{X}) = X^{ax}\Psi(\mathbf{X}), \qquad (4.132)$$

$$\hat{F}_{ab}(x)\,\Psi(\mathbf{X}) = i\,\partial_{[a}\frac{\delta}{\delta X^{b]x}}\,\Psi(\mathbf{X}). \qquad (4.133)$$

As a consequence, the quantum Hamiltonian reads

$$\hat{H}\,\Psi(\mathbf{X}) = \int d^3x\left[\frac{1}{2}\hat{X}^{ax}\hat{X}^{ax} + \frac{1}{4}(\partial_{[a}\hat{P}_{b]x})^2\right]\Psi(\mathbf{X}), \qquad (4.134)$$

where

$$\hat{P}_{bx} = i\,\frac{\delta}{\delta X^{bx}}. \qquad (4.135)$$

From these equations, one realizes, making the identifications

$$\hat{P}_{bx} \longrightarrow A_{bx}, \qquad (4.136)$$

$$X^{ax} \longrightarrow \tilde{E}^{ax}, \qquad (4.137)$$

that the representation we have just introduced is nothing but the electric field representation of electromagnetism, and the vector density X^a is just the electric field. This is in agreement with the picture that we introduced before in which the loops played the role of lines of electric flux.

A remarkable fact is that one can go back to the loop representation through the substitution $X^{ax} \to X^{ax}(\gamma)$. For instance, if one finds a physical state in the extended representation one can find a physical state in the loop representation by evaluating it on multitangents (since it is a function of multitensors, it has a definite value for multitangents). Care should be exercised in general since multitangents are distributional and limits could be ill defined. For the particular case of Maxwell theory it can easily be checked that the converse property also holds: if one replaces multitangents by multitensors in the physical states of the loop representation, one obtains the physical states for the extended representation. This does not, in general, hold for non-Abelian fields.

Using this correspondence we can immediately write the expression for the vacuum in the extended loop representation,

$$\Psi(\mathbf{X}) = \exp\left(-\frac{1}{2}\int d^3x\int d^3y\, X^{ax}X^{ay}D_1(x-y)\right). \qquad (4.138)$$

Here we observe a crucial feature of the extended representation. While the vacuum in terms of loops is, as we pointed out, a singular divergent quantity that only makes sense after a regularization procedure has been introduced, the vacuum in the extended representation is automatically well defined. It is an analogous situation to the one that appears in classical electrostatics: if one tries to formulate the theory in terms of point charges one needs to regularize it, whereas the theory is automatically well defined if one considers smooth charge distributions. It is natural to expect that a similar behavior will appear in non-Abelian theories and quantum gravity. This is one of the main features that make the extended representation attractive. The loop representation only appears as a singular limit, in the same spirit as the electrostatics of point charges appears through a limiting procedure from the electrostatics of smooth charge distributions.

The existence of the extended loop representation is an illustration of a property pointed out by Ashtekar and Isham [73]: that there exist possibly non-equivalent representations of quantum theories. One can introduce an inner product in the loop representation in terms of extended loop coordinates that allows a Fock interpretation as we did in section 4.4. This is the natural inner product in the extended representation and corresponds exactly to the inner product one introduces in the connection representation to implement the reality conditions of the theory. One can also introduce a representation in terms of usual loops and a discrete inner product which seems to describe naturally the states of a Type II superconductor [73].

Since we have a theory written in terms of usual smooth tensorial quantities with a well defined Hamiltonian it is immediate (in this simple Abelian case) to introduce a *classical action* in terms of which one can formulate the theory. This is by no means trivial. Whereas usually the loop representation has been viewed as a "mysterious" construction that either arises indirectly via a transform or through an unusual non-canonical quantization, the extended representation teaches us that one can actually find a *canonical* classical theory in terms of which a straightforward quantization leads to the loop representation. This construction can actually be generalized to the non-Abelian cases, although it presents more subtleties than the Abelian case we are examining here.

Let us therefore write the classical action which yields the quantum theory corresponding to the extended loop representation,

$$ S = \int dt \left\{ P_{ax}\dot{X}^{ax} - \left[\frac{1}{2} X^{ax}X^{ax} + \frac{1}{4}(\partial_{[a}P_{b]x})^2 \right] + \lambda_x X^{ax}_{,a} \right\}. \quad (4.139) $$

We immediately recognize the action for classical electromagnetism if we identify the loop coordinate with the electric field and the momentum with the connection as we did before.

This action could be rewritten in terms of loops,

$$S = \int dt \left\{ \int_{\gamma_t} dy^a \dot{A}_a(y) + \tfrac{1}{2} \int d^3x F_{ab}(x) F_{ab}(x) \right.$$
$$\left. + \int_{\gamma_t} dy^a \int_{\gamma_t} dy'^a f_\epsilon(y - y') \right\}, \tag{4.140}$$

where f_ϵ is a regularization of the delta function and the loops γ_t belong to the surface $t = $ constant. This action could also be presented in second order form (modulo regularization difficulties). It could also be regularized by considering the theory on a lattice. This has been pursued in detail in reference [72] and it has been found to lead to the usual Kogut–Susskind formulation [74].

4.7   Conclusions

The example discussed in this chapter, due to its simplicity, allows us to illustrate in an explicit fashion several properties that are important for the program of quantization of the gravitational field and cannot be proved for that case.

We have shown that the language of loops is adequate to describe the free quantum Maxwell field. We have shown that the use of loops is inherently associated with regularization difficulties which can be cured by considering the extended loop representation. The loop representation is totally equivalent in this case to the traditional Fock quantization. The Wilson loop functional appears as naturally related to coherent states. The loop transform in this case is rigorously defined through the inner product in the connection representation. This inner product can be determined through the reality conditions of the theory, as we proved for the Bargmann case. We also showed that the loop representation can also be constructed for a complex coordinatization of phase space similar to the one that the Ashtekar variables introduce for gravity.

In the next chapter we will discuss the quantization in terms of loops of non-Abelian fields.

5

Yang–Mills theories

5.1 Introduction

Since the unification of the electromagnetic and weak interactions through the Glashow–Salam–Weinberg model [75], Yang–Mills theories [76] have been widely accepted as correctly describing elementary particle physics. This belief was reinforced when they proved to be renormalizable [77, 78]. Moreover, the discovery of color symmetry as the underlying gauge invariance associated with strong interactions raised the possibility that all interactions of nature could possibly be cast as Yang–Mills theories. This spawned interest in grand unified models and some partial successes were achieved in this direction.

A crucial ingredient in the description of elementary particle physics through gauge theories is the maintenance of the gauge invariance of physical results and the underlying theory and this is also crucial in order to be able to prove renormalizability.

The success of the electroweak model is yet to be achieved by the quark model of strong interactions. The reason is that perturbative techniques, which were adequate for the electroweak model, are only appropriate in the high energy regime of strong interactions. This motivated the interest in non-perturbative techniques, especially to prove the existence of a confining phase. A great effort took place in the late 1970s and suggestive arguments were put forward but a rigorous proof of quark confinement is still lacking.

In several of these attempts the use of loops played an important role. Loops were used in a variety of contexts and approaches including the one we are focusing on in this book, the loop representation. In this chapter we will also briefly highlight some of the aspects of other approaches which seem of most interest for gravitational physicists. We are forced to omit, for reasons of space, many other valuable constructions.

The first gauge invariant, path dependent formulation of a gauge theory was Mandelstam's reformulation of QED [8]. Mandelstam later extended his formulation to the Yang–Mills case and applied it to the development of Feynman diagrammatic rules [9]. This was the first time the Feynman rules for non-Abelian gauge theories had been found through canonical quantization. They had been established in the S-matrix approach by Feynman [79] and DeWitt [80] and in the functional approach by Fadeev and Popov [81]. The main feature of the Mandelstam approach was to avoid using gauge dependent quantities, introducing instead path dependent field variables $F_{ab}(P)$ for the field where P is a path going from a basepoint to the point of interest: translating to the language introduced in chapter 1

$$F_{ab}(\pi) = \lim_{\gamma \to \iota} \Delta_{ab}(\pi_o^x) H_\gamma[A]. \tag{5.1}$$

These quantities satisfy the identities induced by those of the loop derivative that we introduced in chapter 1 and the Yang–Mills equation of motion, $D_a F^{ab}(P) = 0$, where D^a is the Mandelstam covariant derivative. Notice that the aim of this approach was to develop a perturbative formulation and in that respect it was successful.

Another approach was that of Polyakov [83, 82]. This was based on the hope that holonomies for Yang–Mills theories could satisfy equations similar to those of non-linear σ models, which, in turn, are integrable. This is based on what happens in $2 + 1$ dimensions. The basic variable is a derivative of the holonomy

$$F_\mu(s, \gamma) = \frac{\delta H_\gamma[A]}{\delta \gamma^\mu(s)} H_\gamma^{-1}[A] \tag{5.2}$$

and the formalism assumes a parametrization has been picked for the loop and extra equations are added to impose invariance under reparametrizations. The equations of motion are

$$\frac{\delta F_\mu(s, \gamma)}{\delta \gamma^\nu(t)} - \frac{\delta F_\mu(t, \gamma)}{\delta \gamma^\nu(s)} + [F_\mu(s, \gamma), F_\nu(t, \gamma)] = 0, \tag{5.3}$$

$$\frac{d\gamma^\mu(s)}{ds} F_\mu(s, \gamma) = 0, \tag{5.4}$$

$$\frac{\delta F_\mu(s, \gamma)}{\delta \gamma^\mu(s)} = 0. \tag{5.5}$$

The first equation is the usual vanishing of a curvature that appears in non-linear σ models. The second equation is related with the invariance under reparametrizations of the holonomy and the last equation is a consequence of the Yang–Mills dynamical equation.

This approach had several difficulties. Even in the three-dimensional case, the equations presented are not exactly the same as those of a traditional non-linear σ model. In a traditional non-linear σ model, the first equation would involve a partial derivative with respect to a coordinate. In the present case, this means that one is really dealing with an infinite number of components, one per each point in parameter space, as is expected in loop space. The third equation, which in the usual case is a divergence, should be summed over all components (integrated over s), but then, it is not true that the Yang–Mills equations follow. This difficulty was recognized by Polyakov [84]. Moreover the situation becomes more complicated if one considers the four-dimensional case, since in that case it is not even clear how to reformulate the fields as σ models. Other technical difficulties appear, mainly related to the parametrization dependence [85]. In particular it was shown that when the equations are rewritten in a parametrization independent way (using the techniques discussed in chapter 1) extra terms appear, which break the resemblance with the σ model.

The plan of this chapter is as follows. We will discuss in some detail in the next section an alternative approach, due to Polyakov and Migdal. We then devote a section to the loop representation of Yang–Mills theories, discussing the $SU(2)$ and $SU(N)$ cases. We end with a section on some ideas relating loops to confinement.

5.2   Equations for the loop average in QCD

The approach we are about to discuss originated in an idea of Polyakov [82] and was later developed by Makeenko and Migdal [12, 86]. We only present a sketch of the main ideas here, in part because we will use similar techniques in the context of Chern–Simons theory in chapter 10. We refer the reader to the review article by Migdal [11].

The basic idea is as follows. The expectation value of the Wilson loop functional in (Euclidean) four dimensions (i.e, the loop exists in a four-dimensional space) operates as a generating functional of the Green functions of the theory, as can be simply seen by considering its successive loop derivatives at different points,

$$\Delta_{\mu_1\nu_1}(\pi_o^{x_1})\dots\Delta_{\mu_n\nu_n}(\pi_o^{x_n}) <W(\gamma)>_{\gamma=\iota}=<\text{Tr}[F_{\mu_1\nu_1}(x_1)\dots F_{\mu_n\nu_n}(x_n)]> .$$
(5.6)

Notice that to write the right-hand side as point dependent a prescription for the paths π has been chosen (as we discussed at the end of chapter 1).

The right-hand side of expression (5.6) is the n-point function of the theory. This was the insight of Polyakov. Now consider the action of the field equations on the expectation value of the Wilson loop functional,

$$D^\mu \Delta_{\mu\nu}(\gamma_o^x) <W(\gamma)> = \int DA \, \exp(-S_{YM}) D^\mu \Delta_{\mu\nu} W_A(\gamma)$$

$$= \int DA \, \exp(-S_{YM}) \mathrm{Tr}[D^\mu F_{\mu\nu}^i(x) \mathbf{X}^i U_A(\gamma_x^x)]$$

$$= -\int DA \, \frac{\delta}{\delta A_\nu^i} \exp(-S_{YM}) \mathrm{Tr}[\mathbf{X}^i U_A(\gamma_x^x)]$$

$$= \int DA \, \exp(-S_{YM}) \frac{\delta}{\delta A_\nu^i} \mathrm{Tr}[\mathbf{X}^i U_A(\gamma_x^x)]$$

$$= \int DA \, \exp(-S_{YM}) \oint_\gamma dy^\mu \delta^4(x-y)$$

$$\times \mathrm{Tr}[\mathbf{X}^i U_A(\gamma_x^y) \mathbf{X}^i U_A(\gamma_y^x)] \qquad (5.7)$$

where S_{YM} is the Yang–Mills action $\frac{1}{2}\int d^4x \, \mathrm{Tr}[\mathbf{F}_{\mu\nu}\mathbf{F}^{\mu\nu}]$ and $\mathbf{X}^i$ are the generators of the group.

Let us now particularize the gauge group to $SU(N)$. This allows us to use the identity

$$\mathbf{X}_{AB}^i \mathbf{X}_{CD}^i = (\delta_{AD}\delta_{BC} - \tfrac{1}{N}\delta_{AB}\delta_{CD}) \qquad (5.8)$$

and to reexpress the above result as

$$D^\mu \Delta_{\mu\nu}(\gamma_o^x) <W(\gamma)> =$$

$$\oint_\gamma dy^\mu \delta^4(x-y)(<W(\gamma_x^y)W(\gamma_y^x)> - \frac{1}{N} <W(\gamma)>). \, (5.9)$$

Notice that this equation couples the expectation value of the Wilson loop functional with the expectation value of products of Wilson loops. In general one would therefore need to consider similar equations for $<W(\gamma_1)\dots W(\gamma_n)>$. However, in the particular case of $SU(2)$ or $\lim_{N\to\infty} SU(N)$ it is enough to consider only the expectation value of one Wilson loop functional. In the $SU(2)$ case this is justified since one can reexpress any product of Wilson loop functionals in terms of a single Wilson loop. In the $N \to \infty$ case it can be shown [12] that,

$$<W(\gamma_1)\dots W(\gamma_n)> = <W(\gamma_1)> \dots <W(\gamma_n)> +O(N^{-2}), \quad (5.10)$$

due to the fact that the leading Born terms correspond to the sum of all planar diagrams [87]. The Makeenko–Migdal equation can be rewritten for this particular case as

$$D^\mu \Delta_{\mu\nu}(\gamma_o^x)\phi(\gamma) = \oint_\gamma dy^\mu \delta^4(x-y)\phi(\gamma_x^y)\phi(\gamma_y^x), \qquad (5.11)$$

where

$$\phi(\gamma) = <W(\gamma)> . \qquad (5.12)$$

This equation is reminiscent of that of a $\lambda\phi^3$ scalar field theory. Notice that the equation is only non-trivial if one considers intersecting loops. For smooth loops the right-hand side of the equation becomes $\phi(\gamma)$ and the solution to the equation can be found and coincides with the vacuum state of Maxwell theory in terms of loops that we introduced in the previous chapter. That is, the non-Abelian character of the theory is lost if one does not consider intersecting loops, a fact we will also see reflected in the Hamiltonian case.

It can be shown [88] that the Makeenko–Migdal equations generate all planar diagrams in perturbation theory in a regularized fashion (if one regularizes the equation), although they are not renormalized and no concrete proposal has been found for an equation that could take care of the renormalization.

This approach offered the promise of reformulating QCD entirely in terms of free color fields, which raised the hope that confinement could be understood. Moreover, it makes it possible to express the expectation values of the observables through integrals in loop space. The diagrams are automatically free of infrared catastrophes since one works only with gauge invariant quantities. Finally, Migdal [12] gave a heuristic argument that showed that the behavior of the Wilson loop is consistent with the asymptotic area law typical of confinement.

Several obstacles hampered further development of this approach. To begin with, the expectation values considered are divergent and need to be renormalized, as can be seen from their perturbative study [88]. The equation was initially written [12] in terms of a functional derivative, which led to some technical problems, though a later more geometric reformulation was accomplished [89]. Historically, for a long time the structure and completeness of Mandelstam identities were not understood for the different gauge groups. Although the conceptual simplicity introduced by expressing observables as integrals in loop space was appealing, it is also the case that one does not know how to compute such integrals.

But the main obstacle in this approach is the fact that not a single solution of the Makeenko–Migdal equation has ever been found in four dimensions. Progress has been made in the two-dimensional case [91, 90, 93] and also with variational techniques [92]. It would be interesting to test whether the ideas of the extended representation we present in this book can be used to tackle this problem. Another interesting aspect is that relations have been found between the Makeenko–Migdal equation in the large N limit and equations for string theory [94].

Most of the advantages and disadvantages of Migdal's approach are shared by other loop formulations. The appeal of the loop representation, which is the main theme of this book, lies elsewhere. On the one hand, loop representations based on Hamiltonian approaches deal with three-dimensional loop equations in a realistic case instead of four-dimensional ones as in the Migdal construction. Moreover, they are better suited for a canonical description of quantum gravity. In particular one can solve the diffeomorphism constraint rather easily, as we will see in chapter 8. We will also see in chapter 10 that by applying exactly the same construction that we presented here for the Yang–Mills case to the Chern–Simons action, one can find the connection between the expectation value of a Wilson loop and the Jones polynomial of knot theory.

5.3   The loop representation

Constructing a loop representation for a Yang–Mills theory is a straightforward matter with the concepts introduced in chapter 3.

We will first discuss the $SU(2)$ case as an illustration of the quantization of a non-canonical algebra. We then discuss the $SU(N)$ case via a transform.

5.3.1 $SU(2)$ Yang–Mills theories

Let us construct the quantum theory for $SU(2)$ through the quantization of a non-canonical algebra of loop dependent operators. A quantization could be achieved (formally) using the loop transform, leading to the same quantum theory.

Let us consider a gauge theory with $SU(2)$ gauge group. The connections are group-valued $A_a = A_a^i \mathbf{X}^i$ where the Xs are elements of the $su(2)$ algebra. We define the following quantities

$$T^0(\gamma) = \mathrm{Tr}[U(\gamma)] = \mathrm{Tr}\left[P \exp\left(ig \oint_\gamma dy^a A_a\right)\right], \qquad (5.13)$$

$$T^a(\gamma_x^x) = \mathrm{Tr}[U(\gamma_o^x)\tilde{\mathbf{E}}^a(x)U(\gamma_x^o)], \qquad (5.14)$$

in the same spirit as those defined in chapter 3, except that we are making explicit the dependence on the coupling constant g of the theory.

Classically, these quantities satisfy an algebra under Poisson brackets, which we discussed in chapter 3,

$$\{T^a(\gamma_x^x), T(\eta)\} = \frac{i}{2}\sum_{\epsilon=-1}^{1} \epsilon X^{ax}(\eta)T(\gamma \circ \eta^\epsilon), \qquad (5.15)$$

$$\{T^a(\gamma_x^x), T^b(\eta_y^y)\} = -\frac{i}{2}\sum_{\epsilon=-1}^{1} \epsilon X^{ax}(\eta)T^b(\eta_y^x \circ (\gamma_x^x)^\epsilon \circ \eta_x^y),$$

$$+\frac{i}{2}\sum_{\epsilon=-1}^{1} \epsilon X^{by}(\gamma)T^a(\gamma_x^y \circ (\eta_y^y)^\epsilon \circ \gamma_y^x), \qquad (5.16)$$

where η^ϵ represents either η or η^{-1}.

The best picture of these relations can be obtained by considering the $T^a(\gamma_x^x)$ to be represented by a loop with a "hand" at the point x. Then the commutators are only non-zero when the hand on one of the loops grabs the other loop (which implies the loops must intersect). The effect of "grabbing" one loop with the hand of the other is to insert the accompanying loop at the point of intersection.

This algebra is closed, as seen above, but it is insufficient in the sense that one cannot express all observables of interest in Yang–Mills theories in terms of it, in particular, the Hamiltonian. Therefore one cannot base a quantum theory simply on finding a representation of this algebra. One should consider a larger algebra including the objects with n insertions defined by formula (3.102) that were discussed in chapter 3. The algebra of T^0 and T^a has for this reason been called the "small algebra" [38]. A generic Poisson bracket between higher order Ts is given by

$$\{T^{b_1 \ldots b_n}(\gamma_{x_1}^{x_2}, \ldots, \gamma_{x_n}^{x_1}), T^{a_1 \ldots a_m}(\eta_{y_1}^{y_2}, \ldots, \eta_{y_m}^{y_1})\} = \frac{i}{2} \sum_{\epsilon=-1}^{1}$$

$$\left\{ -\sum_{k=1}^{m} \epsilon X^{b_k \, x_k}(\eta) T^{b_1 \ldots \slashed{b}_k \ldots b_m, a_1 \ldots a_n}(\eta_{y_1}^{y_2}, \ldots, \eta_{y_{k-1}}^{x_k} \circ (\gamma_{x_k}^{x_k})^\epsilon \circ \eta_{x_k}^{y_k}, \ldots, \eta_{y_m}^{y_1}) \right.$$

$$\left. + \sum_{k=1}^{n} \epsilon X^{a_k \, y_k}(\eta) T^{a_1 \ldots \slashed{a}_k \ldots a_n, b_1 \ldots b_m}(\gamma_{x_1}^{x_2}, \ldots, \gamma_{x_{k-1}}^{y_k} \circ (\eta_{y_k}^{y_k})^\epsilon \circ \gamma_{y_k}^{x_k}, \ldots, \gamma_{x_m}^{x_1}) \right\},$$

$$(5.17)$$

where $\slashed{b}_k$ means that the index is not present.

The Hamiltonian for a Yang–Mills theory was introduced in chapter 3, and is given by

$$\tilde{\tilde{\mathcal{H}}} = \int d^3x \, \text{Tr}(\tfrac{1}{2}(\tilde{\mathbf{E}}^a \tilde{\mathbf{E}}^b \eta_{ab} + \tilde{\mathbf{B}}^a \tilde{\mathbf{B}}^b \eta_{ab})). \tag{5.18}$$

The two terms in this expression have different properties. The $\mathbf{B}^2$ term can actually be written in terms of a $T^0(\gamma)$ as

$$\text{Tr}(\tilde{\mathbf{B}}^a(x)\tilde{\mathbf{B}}^b(x))\eta_{ab} = -\tfrac{1}{2} \lim_{\eta \to \iota} \eta^{ac}\eta^{bd} \Delta_{ab}(\pi_o^x)\Delta_{cd}(\pi_o^x)T^0(\gamma), \tag{5.19}$$

i.e., by taking a double loop derivative and then shrinking the loop dependence to a point. The right-hand side of this formula does not depend on the path choice, as can be seen explicitly by considering

$$\Delta_{ab}(\pi_o^y)\Delta_{cd}(\pi_o^x)T^0(\gamma) = \text{Tr}[U(\pi_o^y)F_{ab}(y)U(\pi_y^o)U(\pi_o^x)F_{ab}(x)U(\pi_x^o)U(\gamma)] \tag{5.20}$$

and taking the limit in which $x \to y$ and the limit of γ shrinking to a point giving,

$$\Delta_{ab}(\pi_o^x)\Delta_{cd}(\pi_o^x)T^0(\gamma) \mid_{\gamma \to \iota} = Tr[F_{ab}(x)F_{cd}(x)], \tag{5.21}$$

which is independent of the path prescription.

The term involving two electric fields cannot be written in terms of T^0 or T^a, one needs a T^{ab}. Classically this can be seen from

$$\mathrm{Tr}(\tilde{\mathbf{E}}^a(x)\tilde{\mathbf{E}}^b(x))\eta_{ab} = \lim_{\gamma \to \iota} \eta_{ab}T^{ab}(\gamma_x^y, \gamma_y^x) \qquad (5.22)$$

and in the limit $\gamma \to \iota$ the points x and y coincide.

We now proceed to propose a quantization of the classical non-canonical algebra. We consider a space of wavefunctions of loops $\Psi(\gamma)$ as discussed in section 3.5.3 and define the action of the operators as

$$\hat{T}^0(\eta)\Psi(\gamma) \equiv \Psi(\gamma \circ \eta) + \Psi(\gamma \circ \eta^{-1}), \qquad (5.23)$$

$$\hat{T}^a(\eta_x^x)\Psi(\gamma) \equiv -\tfrac{1}{2} \sum_{\epsilon=-1}^{1} \epsilon \oint_\gamma dy^a \delta(x-y)\Psi(\gamma \circ \eta^\epsilon). \qquad (5.24)$$

The action of the T^2 is defined as [39]

$$\hat{T}^{ab}(\eta_x^y, \eta_y^x)\Psi(\gamma) = \tfrac{1}{4}X^{ax}(\gamma)X^{by}(\gamma)[\Psi(\gamma_x^y \circ \bar{\eta}_y^x, \gamma_y^x \circ \bar{\eta}_x^y)$$
$$+\Psi(\gamma_x^y \circ \eta_y^x, \gamma_y^x \circ \eta_x^y) + \Psi(\gamma_x^y \circ \bar{\eta}_y^x \circ \bar{\gamma}_x^y \circ \eta_y^x) + \Psi(\gamma_y^x \circ \bar{\eta}_x^y \circ \bar{\gamma}_y^x \circ \eta_x^y)]. \qquad (5.25)$$

This last expression could be rearranged in terms of wavefunctions of a single loop using the Mandelstam identity (3.119).

This representation for the T operators yields a quantum commutator algebra that reproduces, to first order in $\hbar$, the classical Poisson algebra of the T operators that we introduced in chapter 3*.

We are now in a position to give a quantum representation of the Hamiltonian of the theory. We have written the Hamiltonian of the theory (5.18) and it is the sum of two terms, one electric and one magnetic. The electric portion was given as a limit of a T^2, so we can now find the corresponding quantum representation, by taking the limit in equation (5.25) in which we shrink the loop η to a point. We will study the action of this operator at a point in the loop γ where there is an intersection (of arbitrary order). The action on a regular point of γ can be obtained as a limit (or by direct calculation). The result is

$$\lim_{\gamma \to \iota} T^{ab}(\eta_x^y \circ \eta_y^x)\Psi(\gamma) = 2X^{ax}(\gamma)X^{by}(\gamma)(\tfrac{1}{4}\Psi(\gamma_x^y \circ \gamma_y^x) + \tfrac{1}{2}\Psi(\gamma_x^y \circ \bar{\gamma}_x^y)). \qquad (5.26)$$

In the case of an intersection, the points x and y lie on any "petal" of the loop and therefore the portions γ_x^y and $\bar{\gamma}_y^x$ refer to the various combinations of petals contained between x and y. For the case of a

---

* We are taking $\hbar = 1$. The orders of $\hbar$ can be restored by noting that momenta are first order in $\hbar$, for instance, $T^1 = O(\hbar)$, $T^2 = O(\hbar^2)$, etc.

regular point of the loop, one simply takes the loop and shrinks the points x to y and therefore $\gamma_x^y \to \gamma$ and $\gamma_y^x \to \iota$. The result then is

$$\lim_{\eta \to \iota} T^{ab}(\eta_x^y \circ \eta_y^x)\eta_{ab}\Psi(\gamma) = \tfrac{3}{4}X^{ax}(\gamma)X^{by}(\gamma)\Psi(\gamma). \tag{5.27}$$

We also see that smooth loops are eigenstates of the electric part of the Yang–Mills Hamiltonian. This allows us to think of Wilson loops as lines of electric flux.

The complete Hamiltonian for an $SU(2)$ Yang–Mills theory in the loop representation is given by

$$\begin{aligned}
\hat{\mathcal{H}}\Psi(\gamma) = \Big[& -\tfrac{1}{4}\int d^3x\, \eta^{ac}\eta^{bd}\Delta_{ab}(x)\Delta_{cd}(x) \\
& + \tfrac{1}{4}\oint_\gamma dy^a \oint_\gamma dy'^b \eta_{ab}\delta^3(y-y')\Big]\Psi(\gamma) \\
& + \tfrac{1}{2}\oint_\gamma dy^a \oint_\gamma dy'^b \eta_{ab}\delta^3(y-y'))\Psi(\gamma_y^{y'} \circ \bar{\gamma}_{y'}^y). \tag{5.28}
\end{aligned}$$

The path dependence of the loop derivatives has been dropped since we showed above that they are prescription independent. Notice that if the loop γ does not have intersections, the last term is equal to the second one and the equation is identical (up to constants) to the one obtained for Maxwell theory (4.72). Therefore it is clear that wavefunctions must have support on intersecting loops if the theory is to capture the full non-Abelian nature of the fields.

As in the case of Maxwell theory, the Hamiltonian is singular and needs to be regularized and renormalized. As we pointed out in the case of Maxwell theory, in principle *all terms* in the Hamiltonian require a regularization. In the case of Maxwell theory we knew how to compute the vacuum and this suggested a suitable regularization of wavefunctions and operators. In the non-Abelian case, unfortunately, we do not know a single solution of the Hamiltonian eigenvalue equation and the issue of regularization and renormalization is largely unexplored. The eigenvalue equation has been extensively studied in the lattice in different approximations, leading to results for the energy density, gluon mass spectrum and other observables which coincide with those obtained with more standard methods. We will return to the lattice treatment in the next chapter.

5.3.2 *SU(N) Yang–Mills theories*

The loop representation for $SU(N)$ Yang–Mills theories can be built along similar lines to the ones we followed in the previous section for the $SU(2)$ case. We will see that the main difference consists in the fact that one needs to consider wavefunctions of multiloops. The classical "small" algebra of T^0 and T^1 can be readily generalized to the $SU(N)$ case,

$$\{T^0(\gamma), T^0(\eta)\} = 0, \tag{5.29}$$

$$\{T^0(\gamma), T^a(\eta_x^x)\} = ig \oint_\gamma dy^a \delta(y-x)$$
$$\times \left(T^0(\gamma_o^y \circ \eta_x^x \circ \gamma_y^o) - \tfrac{1}{N} T^0(\gamma) T^0(\eta)\right), \tag{5.30}$$

$$\{T^a(\gamma_x^x), T^b(\eta_y^y)\} = -ig \oint_\eta dz^a \delta(z-x)$$
$$\times \left(T^b(\eta_y^z \circ \gamma_x^x \circ \eta_z^y) - \tfrac{1}{N} T^0(\gamma) T^b(\eta_y^y)\right)$$
$$+ig \oint_\gamma dz^b \delta(z-y)$$
$$\times \left(T^a(\gamma_x^z \circ \eta_z^y \circ \gamma_z^x) - \tfrac{1}{N} T^0(\eta) T^a(\gamma_x^x)\right). \tag{5.31}$$

Up to now, representations of the large algebra have not been studied for $SU(N)$ with $N > 2$. The approach we will take will be to consider the representation of the elements of the small algebra with the addition of a T^{ab}, which is sufficient to write the Hamiltonian. This is clearly not enough: the Poisson bracket of a T^{ab} with a T^{cd} gives rise to T^{abc}s. A quantization of the full algebra can always be performed such that this Poisson bracket is represented by a correct commutation relation. A direct constructive procedure to obtain such an algebra would be to perform a usual canonical quantization in terms of Es and As and to consider the Ts as derived quantities. The resulting quantum algebra will coincide with Poisson bracket algebra up to factor ordering differences. We should remind the reader that in the quantization process Poisson brackets are replaced by i times $\hbar$ and the factor ordering ambiguities are of order $\hbar^2$ or higher. Since the T algebra for $SU(N)$ has not been explicitly computed up to now, we will proceed with the constructive technique we just outlined. It would be interesting to check explicitly that this technique yields the same result as consideration of the full T algebra.

We will choose the following ordering prescription for the T^1 operator in terms of the canonical operators,

$$\hat{T}^a(\gamma_x^x) = \text{Tr}\left(H(\gamma_x^x)\hat{\tilde{\mathbf{E}}}^a\right). \tag{5.32}$$

The operator algebra of $\hat{T}^1$ and $\hat{T}^0$ reproduces the same classical Poisson algebra described above, where the brackets are replaced by $\hbar$ times the commutators.

We now represent this algebra in terms of loop-based operators and wavefunctions, giving rise to the loop representation. As we mentioned in section 3, we need to consider functions of multiloops, via the transform

$$\Psi(\gamma_1, \ldots, \gamma_n) = \int d\mathbf{A}\, W_{\mathbf{A}}^*(\gamma_1) \cdots W_{\mathbf{A}}^*(\gamma_n) \Psi[\mathbf{A}]. \qquad (5.33)$$

It turns out to be convenient to consider a certain combination of products of Wilson loops in the transform which is slightly different than the one presented in equation (5.33). We will consider the following transform:

$$\Psi(\gamma_1, \ldots, \gamma_n) = \int d\mathbf{A}\, M_N^*(\gamma_1, \ldots, \gamma_n) \Psi[\mathbf{A}], \qquad (5.34)$$

where the functionals $M_N(\gamma_1, \ldots, \gamma_n)$ were introduced in section 3.4.1. This does not imply any loss of generality since the product of N Wilson loops can be reconstructed from the Ms.

We now define the action of the T^0 operator, which is given by,

$$\hat{T}^0(\gamma)\Psi(\eta_1, \ldots, \eta_N) = \Psi(\bar{\gamma} \circ \eta_1, \ldots, \eta_n) + \Psi(\eta_1, \bar{\gamma} \circ \eta_2, \ldots, \eta_N)$$
$$+ \Psi(\eta_1, \ldots, \bar{\gamma} \circ \eta_N). \qquad (5.35)$$

Notice that this expression only involves wavefunctions of N entries, since due to the Mandelstam identities (3.39) $M_{N+1} = 0$ and consequently $\Psi(\eta_1, \ldots, \eta_{N+1})$ vanishes identically. The fact that we are dealing with a special group (determinant equal 1) implies that

$$\Psi(\eta \circ \eta_1, \eta \circ \eta_2, \ldots, \eta \circ \eta_N) = \Psi(\eta_1, \ldots, \eta_N) \qquad (5.36)$$

and therefore by considering $\eta = \bar{\eta}_i$ one immediately concludes that the wavefunctions are really only functions of $N - 1$ loops. We will continue using wavefunctions with N entries for convenience.

We complete the small T algebra by representing the $\hat{T}^1$ operator,

$$\hat{T}^a(\gamma_x^x)\Psi(\eta_1, \ldots, \eta_N) = \sum_{k=1}^{N} \oint_{\eta_k} dy^a \delta^3(x - y)$$
$$\times [\Psi(\eta_1, \ldots, (\eta_k)_o^y \circ \gamma_x^x \circ (\eta_k)_y^o, \ldots, \eta_N) - \tfrac{1}{N}\hat{T}^0(\gamma)\Psi(\eta_1, \ldots, \eta_N)]. \qquad (5.37)$$

To represent the Hamiltonian, we first recall that the magnetic part is given by equation (5.20). Using the formula for $\hat{T}^0(\gamma)$, (5.35), one can immediately realize the action of the magnetic part of the Hamiltonian,

$$\int d^3x \eta_{ab} \mathrm{Tr}(\hat{\tilde{\mathbf{B}}}^a \hat{\tilde{\mathbf{B}}}^b)\Psi(\eta_1, \ldots, \eta_n) =$$
$$-\tfrac{1}{4}\sum_{i=1}^{N} \int d^3x \eta^{ac}\eta^{bd}\Delta_{ab}^{(i)}(x)\Delta_{cd}^{(i)}(x)\Psi(\eta_1, \ldots, \eta_n), \qquad (5.38)$$

where the loop derivatives $\Delta_{ab}^{(i)}(x)$ are path independent and act only on the ith entry of the loop.

For the electric part, we need to study the action of the operator

$$\hat{\mathcal{E}} = \int d^3x \text{Tr}(\hat{\tilde{\mathbf{E}}}{}^a\,\hat{\tilde{\mathbf{E}}}{}^b)\eta_{ab} \tag{5.39}$$

on the functionals M_N in the connection representation. To perform this calculation it is useful to consider the following identity,

$$\hat{\mathcal{E}}\hat{T}^0(\eta_1) \times \ldots \times \hat{T}^0(\eta_N) = \sum_{i=1}^{N-1} \left\{ \hat{T}^0(\eta_1) \times \ldots [\hat{\mathcal{E}}, \hat{T}^0(\eta_i)] \ldots \times \hat{T}^0(\eta_N) \right.$$
$$\left. + \hat{T}^0(\eta_1) \times \ldots \times \hat{T}^0(\eta_{N-1})\hat{\mathcal{E}}\hat{T}^0(\eta_N) \right\} \tag{5.40}$$

and recalling that the commutator of the electric part of the Hamiltonian with $\hat{T}^0$ is

$$[\int d^3x \text{Tr}(\hat{\tilde{\mathbf{E}}}{}^a\,\hat{\tilde{\mathbf{E}}}{}^b)\eta_{ab}, \hat{T}^0(\gamma)] = -\oint_\gamma dy^a \hat{T}^b(\gamma_y^y)\eta_{ab}$$
$$+\frac{1}{2}\oint_\gamma dy^a \oint_\gamma dy'^b \eta_{ab}\delta^3(y-y')[\hat{T}^0(\gamma_y^{y'})\hat{T}^0(\gamma_{y'}^y) - \frac{1}{N}\hat{T}^0(\gamma)] \tag{5.41}$$

and

$$\hat{\mathcal{E}}\hat{T}^0(\gamma) = \frac{1}{2}\oint_\gamma dy^a \oint_\gamma dy'^b \eta_{ab}\delta^3(y-y')[\hat{T}^0(\gamma_y^{y'})\hat{T}^0(\gamma_{y'}^y) - \frac{1}{N}\hat{T}^0(\gamma)]. \tag{5.42}$$

one can construct explicitly the action of the Hamiltonian constraint in the $SU(N)$ case. Very little is known about this operator in the continuum though some progress has been made in the $SU(3)$ case in the lattice [95] and in the $SU(N)$ case in $1+1$ dimensions [96].

Representations in terms of multiloops also appear in the context of general relativity coupled to gauge fields [97].

5.4 Wilson loops and some ideas about confinement

In his pioneering work, which stimulated most of the interest in the use of loop variables in the treatment of non-Abelian gauge theories, Wilson [48] introduced the idea that the trace of the holonomy could act as an order variable for the theory and could therefore be used to study phase transitions.

The intuitive picture behind this is the following. Consider a Yang–Mills theory coupled to fermions (quarks) and consider the creation and subsequent annihilation of a quark–antiquark pair. Assuming the usual interaction term in the Hamiltonian of the type $\vec{\mathbf{J}} \cdot \vec{\mathbf{A}}$, and neglecting

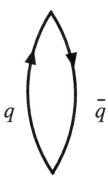

Fig. 5.1. Creation of quark–antiquark pairs viewed as Wilson loops. Creation of a free pair is suppressed if the expectation value of the Wilson loop is a decreasing function of the area of the loop

vacuum polarization effects, one expects such a process to have a weight proportional to the holonomy of the connection $\mathbf{A}_a$ along the closed path formed by the quark–antiquark creation and annihilation process. There are other weight factors independent of the connection and also a weight factor given by the free action of the field.

In order that quarks exist as separate final-state particles it must be possible to consider quark–antiquark processes in which the quark and antiquark lines are well separated, at least when the points of creation and annihilation are far apart. The behavior of the expectation value of the Wilson loop under the separation of the quark–antiquark lines will therefore determine if it is possible for quarks to exist as final states.

For instance, if the expectation value of the Wilson loop turns out to go as $\exp(-l)$, where l is the length of the loop (perturbative and lattice calculations suggest this result for non-confining theories) one sees that one could separate the quark and antiquark lines at will without increasing the length of the loop. This implies that creating well separated quark–antiquark pairs is as likely as creating pairs close together. Therefore the theory is not confining. For instance, for QED an explicit calculation can be performed and $< W >= \exp[-\oint_\gamma dy^\mu \oint dz^\nu D_{\mu\nu}(y-z)]$ where the loop γ is four-dimensional and $D_{\mu\nu}$ is the free propagator of the theory. If one regularizes the calculation one can see that this is proportional to $\exp(-l)$ with l the length of the loop (for details of the regularization see reference [62]).

On the other hand, if $< W >\sim \exp(-a)$, where a is the area of the loop, a process with the quark–antiquark lines shown in figure 5.1 far away from each other is suppressed with respect to one in which the lines are close together and therefore the theory exhibits confinement.

These qualitative considerations have been extensively verified in the lattice. It is immediate to confirm them in strong coupling expansions [48, 74, 98], in Monte Carlo simulations [98] and in perturbative calculations.

We therefore see that quark confinement can be thought of as the appearance of confinement of Wilson loops. Since we have extensively argued that Wilson loops can be thought of as lines of electric flux, this gives an image of quark confinement in which lines of electric flux are confined. This is reminiscent of what happens in superconductivity, except that in that case, the confinement refers to lines of magnetic flux. It seems therefore that a system can have two possible confining regimes, one electric and the other magnetic. Each confining regime will be characterized by an order parameter. We argued above that the Wilson loop acted as an order parameter for electric confinement. What could such a parameter be for magnetic confinement? We will now discuss a proposal by t'Hooft for such a parameter and its implications for the loop representation.

t'Hooft [99] introduced a quantity that can be viewed as an order (actually he refers to it as a disorder) parameter for a Yang–Mills theory. The idea can be illustrated by means of the following example in $(2+1)$ dimensions.

Consider an $SU(N)$ Yang–Mills theory in $2+1$ dimensions coupled to an $SU(N)$ Higgs field such that the gauge symmetry is spontaneously and completely broken. Both the Yang–Mills connection $\mathbf{A}_a$ and the Higgs field $\mathbf{H}(x)$ are invariant under gauge transformations generated by the center of the group $SU(N)$, $Z(N)$. A generic element of $Z(N)$ is given by $\exp(2\pi i n/N)$ with n an integer. A system like this admits a classical solitonic solution of the following kind. Consider a region R in two-dimensional space surrounded by a region B. In region B symmetry is spontaneously broken, the Higgs field having acquired a "constant" non-zero value. Being an element of the group, "constant" means that there exists a gauge transformation at each point that relates the value of the fields to a certain fixed value, $\mathbf{H}_\Omega(x) = \mathbf{\Omega}(x)\mathbf{H}_0\mathbf{\Omega}(x)^{-1}$. Consider a closed curve in B that surrounds R. Since B is not simply connected, it could happen that by going around the curve, $\mathbf{\Omega}$ becomes multivalued, i.e., $\mathbf{\Omega}_{2\pi} = \exp(i2\pi n/N)\mathbf{\Omega}_0$. We say that the field has a winding number n in such a configuration. The presence of this multivaluedness in the field implies that the configuration is stable. If it were not, it could be radiated away, the final configuration would have $n = 0$ and this could not be achieved from a state with given n in a continuous fashion.

Let us now consider an operator that, starting from a regular field configuration a configuration (such that there exists everywhere a single-valued gauge transformation that maps the field to a constant), will create a configuration like the one we discussed above. To simplify, we will shrink the region R to a single point, at which the gauge transformation mapping the field to a constant is singular. Let us call such a point x_0.

We now define an operator $\phi(x_0)$ that materializes a singular gauge transformation that changes the winding number of the fields,

$$\hat{\phi}(x_0)\Psi[A, H] = \Psi[A_\Omega(x_0), H_\Omega(x_0)], \tag{5.43}$$

where $\Omega(x_0)$ the gauge transformation singular at x_0 such that for every oriented curve $c(\theta)$ that surrounds x_0 once

$$\Omega(x_0)_{\theta=2\pi} = \Omega(x_0)_{\theta=0} \exp(2\pi i/N). \tag{5.44}$$

Let us now consider a state in the physical space of states $\Psi_{ph}(A, H)$. Such states are gauge invariant under regular gauge transformations. Under the singular transformations we are considering here

$$\hat{\phi}(x)\hat{\phi}(y)\Psi_{ph}[A, H] = \hat{\phi}(y)\hat{\phi}(x)\Psi_{ph}[A, H]. \tag{5.45}$$

This statement is self-evident: the resulting gauge transformation only depends on the singularity structure. If one takes a curve surrounding x it will detect the multivaluedness induced by $\hat{\phi}(x)$ and similarly for a curve surrounding y. A curve that surrounds both singularities will detect the combined winding number. All this is independent of the order in which the singularities were added.

The point of this construction was to introduce the operator $\hat{\phi}(x)$. We will now show that this operator plays the role we wanted: that of a "disorder" parameter for the theory. In order to see this, let us study the commutation relation of this operator with the Wilson loop. Acting on a physical state

$$\hat{\phi}(x)W_\gamma[A]\Psi[A, H] = W_\gamma[A_\Omega(x)]\Psi[A_\Omega(x), H_\Omega(x)] \tag{5.46}$$

and noting that $W_\gamma[A_\Omega(x)] = \exp(2\pi i n(\gamma)/N)W_\gamma[A]$, where $n(\gamma)$ is the number of times γ winds around x_0, and $\Psi[A_\Omega(x), H_\Omega(x)] = \phi(x)\Psi[A, H]$ we get

$$\hat{\phi}(x)W_\gamma[A] = \exp(2\pi i n(\gamma)/N)W_\gamma[A]\hat{\phi}(x). \tag{5.47}$$

Let us now consider a basis $|\phi>$ in which the operator $\phi(x)$ is diagonal. In such a basis, the operator $W_\gamma[A]$ introduces a jump of magnitude $\exp(2\pi i n(\gamma)/N)$ in the operator $\phi(x)$ if the point x is within γ. This implies that the Wilson loop acts as a creation operator for a domain inside of which the operator $\phi(x)$ has a different value. Using the natural association of Wilson loops with lines of electric field one can view the domain in which $\phi(x)$ jumps in value as delimited by a closed line of electric field.

This argument can be extended to the case of a theory coupled to fermions (quarks) and in this case one should consider an operator built with a holonomy along an open path with quarks at the ends. By reasoning analogous to that above one can view this open path as a confined line of electric field joining the quark–antiquark pair. Trying to separate the pair requires that the line of confined electric flux be stretched and since the flux is constant, the energy needed to separate the pair is proportional to the distance between the particles. This is a signal of confinement in the theory.

Let us now outline how to generalize the above reasoning to $3 + 1$ dimensions. In this case the point x_0 at which the gauge transformation was singular becomes a closed line η. Any gauge transformation along a curve γ that is linked with η will be multivalued. The order of multivaluedness is related to the linking number of both curves.

The commutation relation in three dimensions between the Wilson loop and the generalization of $\phi(x_0)$ to three dimensions (which is usually referred to as the t'Hooft operator $B(\eta)$ is

$$W_\gamma B_\eta = B_\eta W_\gamma \exp\left(\frac{2\pi i}{N} GL(\gamma, \eta)\right) \qquad (5.48)$$

where $GL(\gamma, \eta)$ is the Gauss linking number of the two curves. The Bs commute among themselves.

The physical results that arise from this picture are that either W or B can exhibit behavior dependent on the area or the length of the loop. According to the possible combinations, four different phases can be identified for the theory. A physical discussion of the four phases in the context of QCD can be found in reference [99], where it is argued that the only relevant phases in the case of pure gauge theories (no fermions) are either electric or magnetic confinement.

The phase in which electric field lines are confined is called the confining phase. From an energetic point of view, this phase is characterized by a degeneracy of the vacuum. This is due to the fact that the Hamiltonian commutes with the operator B and therefore it does not cost extra energy to add magnetic field lines. Electric field lines carry an energy proportional to their length.

The explicit form of the B operator in the connection representation is complicated. t'Hooft [99] was able to find an explicit form for this operator in the lattice and Mandelstam [100] discussed its form in the continuum case. It is remarkable that in the loop representation these operators can be realized in a rather straightforward manner [13].

Recalling the action of the Wilson loop on a state of an $SU(N)$ Yang–Mills theory in the loop representation,

$$\hat{T}^0(\gamma)\Psi(\eta_1, \ldots, \eta_N) = \Psi(\bar\gamma \circ \eta_1, \ldots, \eta_n) + \Psi(\eta_1, \bar\gamma \circ \eta_2, \ldots, \eta_N)$$
$$+ \Psi(\eta_1, \ldots, \bar\gamma \circ \eta_N), \qquad (5.49)$$

we can define the operator $\hat{B}(\gamma)$,

$$\hat{B}(\gamma)\Psi(\eta_1,\ldots,\eta_N) = \exp\left(\frac{2\pi i}{N}\sum_{k=1}^{N} GL(\gamma,\eta_k)\right)\Psi(\eta_1,\ldots,\eta_N), \quad (5.50)$$

where

$$GL(\gamma,\eta_k) = \frac{1}{4\pi}\oint_\gamma dx^a \oint_{\eta_k} dy^b \epsilon_{abc}\frac{(x-y)^c}{|x-y|^3} \qquad (5.51)$$

is the Gauss linking number of γ and η_k. This topological invariant measures how many times the loop η_k "threads through" the loop γ (for more details see chapter 10).

It is straightforward to study the commutation relations of this operator with the Yang–Mills Hamiltonian. Because the Gauss linking number is a topological invariant, it commutes with the portion of the Hamiltonian with two loop derivatives, since adding an infinitesimal loop does not change the value of the topological invariant. This point really requires a regularization since the Hamiltonian adds an infinitesimal loop at all points in the manifold and could introduce divergences. The electric part of the Hamiltonian also commutes with the $\hat{B}$ operator. This can be seen by recalling that the effect of the electric part of the Hamiltonian on a wavefunction of N loops is to produce a wavefunction with $N+1$ loops produced by fissions of loops at their self-intersections, as shown in equations (5.28),(5.42). Computing the linking number before or after the fission gives the same result and the operators commute. The reader may be interested in what happens if one characterizes the action of the Hamiltonian purely in terms of N loops using the Mandelstam identity, as we did in the $SU(2)$ case, equation (5.28). In this case some portions of the loop are rerouted and some of the linking numbers — which depend on the orientation — may change sign. However, the result remains unchanged, because the operator $\hat{B}$ takes values on $Z(N)$ and this makes the operator compatible with the Mandelstam identities.

So we see that the operator $\hat{B}$ commutes with the Hamiltonian. We will now study the commutation relation of the Wilson loop with the Hamiltonian. In the loop representation the Wilson loop becomes the $\hat{T}^0(\gamma)$ operator and let us assume that we are considering loops γ that are smooth. Consider the commutator of the $\hat{T}^0(\gamma)$ operator with the electric part of the Hamiltonian, equation (5.42). Its action on a state in the loop representation can be computed using (5.35),(5.37). From (5.42) one can see that there are two terms. The first one is proportional to the length of the loop (double integral along γ) and the second one, taking into account (5.37), involves an integral along γ and another integral along the loop that appears in the argument of the wavefunction. If one considers long loops, the first term (proportional to the length) dominates the other term, which involves intersections of γ with the argument of the wavefunction. The commutator of $\hat{T}^0(\gamma)$ with the magnetic part of the Hamiltonian vanishes, which can immediately be seen from (5.38) since the loop derivatives act specifically on the arguments of the wavefunction and the loop dependence of the $\hat{T}^0$ is transparent.

We therefore see that it is simple to prove that the loop representation naturally describes the confining phase of Yang–Mills theory. There is a natural representation of the disorder operator $\hat{B}$ and adding an electric field line has an energetic cost proportional to the length of the line being added, which is one of the signs of confinement. It should be remarked that these arguments show that there exists a confining phase in which it is energetically expensive to create Wilson loops. However, one could conceive different phases, where the distribution of loops is dense in space and then the dominating term in the expressions considered above, instead of being that of the length of the loop could be the one involving intersections. However, the fact that $\hat{B}$ *always* commutes with the Hamiltonian suggests that Yang–Mills is always in a confining phase.

5.5 Conclusions

We have considered various loop-based approaches to Yang–Mills theories. We have emphasized the use of Hamiltonian techniques and the loop representation, which we constructed explicitly for $SU(2)$ and $SU(N)$ Yang–Mills theories. As the reader may have perceived, the treatment of Yang–Mills theories in the language of loops in the continuum has only a formal character and little progress has actually been made towards understanding the non-perturbative physics of QCD. Only qualitative arguments, like the ones we introduced in the previous section, shed some light on the various physical processes of non-Abelian gauge theories. On the other hand, the gauge invariant description of Yang–Mills theories based on holonomies has found application in attempts to set the theory in a more mathematically rigorous basis. For instance, it may be possible to define an infinite-dimensional measure rigorously in the space of connections modulo gauge transformations in terms of the loop algebra [40]. Progress in this respect has also been made in lower dimensions [96, 102, 103]. In this chapter we have concentrated on pure Yang–Mills theories. Coupling to fermions and Higgs fields can be introduced in the loop representation but again most results are only formal. In the next chapter we will return to the issue of matter couplings in Yang–Mills theories.

6
Lattice techniques

6.1 Introduction

As we mentioned in the previous chapter, the definition of Yang–Mills theories in the continuum in terms of loops requires a regularization and the resulting eigenvalue equations are, in the non-Abelian case, quite involved. Lattice techniques appear to be a natural way to deal with both these difficulties. First of all since on a lattice there is a minimum length (the lattice spacing), the theory is naturally regularized. An important point is that this is a gauge invariant regularization technique. Secondly, formulating a theory on a lattice reduces an infinite-dimensional problem to a finite-dimensional one. It is set naturally to be analyzed using a computer.

Apart from these technical advantages, the reader may find interest in this chapter from another viewpoint. In terms of lattices one can show explicitly in simple models many of the physical behaviors of Wilson loops that we could only introduce heuristically in previous chapters.

Lattice gauge theories were first explored in 1971 by Wegner [104]. He considered a usual Ising model with up and down spins but with a local symmetry. He associated a spin to each link in the lattice and considered an action that was invariant under a spin-flip of all the spins associated with links emanating from a vertex. He noted that this model could undergo phase transitions, but contrary to what happens with usual Ising models, his model did not magnetize. The absence of the magnetization posed him with the problem of distinguishing the phases of the theory. That lead him to introduce correlation functions associated with loops ("loop correlation functions") and to find laws of area and perimeter very much in the same spirit as the ones introduced in the previous chapter.

In a similar fashion, usual gauge theories can be introduced in the lattice, associating to each link an element of the corresponding gauge group.

Many lattice formulations of a certain theory may be written down. This is totally analogous to trying to discretize a differential equation in the sense that many discretized versions of a single equation in the continuum may exist.

The first application of lattice techniques to Yang–Mills theories is due to Wilson [48], who showed how to quantize a gauge theory on a lattice using path integral techniques. Making a Wick rotation to a Euclidean spacetime, he showed that the computation of the Green functions of the field theory essentially coincides with the computation of the correlation functions of a Euclidean four-dimensional statistical mechanics system. Wilson also noticed that the lattice theory admits a strong coupling regime in which there are no free quarks, i.e., confinement appears explicitly. The strong coupling expansion is not completely satisfactory since it does not preserve Lorentz invariance. Kogut and Susskind [49] were the first to introduce a Hamiltonian formulation for lattice gauge theories. In this case space is discretized but time is retained as a continuous variable. They studied the $SU(2)$ theory, which then becomes a quantum mechanical problem, and they studied the strong coupling expansion, which becomes the usual time independent perturbation theory.

Exploiting the connection pointed out above between gauge theories and four-dimensional statistical mechanical problems has allowed the introduction of Monte Carlo techniques for the covariant description of lattice field theories. These computational techniques were developed in the 1950s and a widely used practical implementation is due to Metropolis *et al.* [105]. In the context of lattice gauge theories these techniques were first applied by Wilson [106] and further developed by Creutz [107]. The application of these methods has allowed a concrete prediction of the mass spectra of the physical excitations of the theory and has been implemented on supercomputers yielding values of elementary excitations within 10% error of experimental measurements.

The main limitation of lattice approaches is that the number of degrees of freedom increases very rapidly with lattice size. The situation is worse in higher dimensions and when the theory is coupled to fermions. Progress in lattice approaches to gauge theories is therefore more dependent on the development of new analytical techniques and the identification of the relevant degrees of freedom than on the development of faster computers. We have argued in previous sections that loops are natural objects for describing gauge theories in a gauge invariant fashion. This raises hopes that the loop representation could be a useful tool for addressing some of the difficulties that arise in lattice formulations. The lattice context is very useful for putting in a concrete and rigorous setting many of the formal results discussed in the previous chapters and for gaining an intuitive feel for the loop representation. Loop representations on the lattice have been developed for the $Z(2)$ model [108] and more recently some progress has been made for non-Abelian gauge theories. Concrete calculations in $2 + 1$ dimensions have been performed for $SU(2)$ [109, 110] and $SU(3)$ [95].

One is usually interested in the continuum limit of a lattice gauge theory. This involves shrinking the separation between lattice points to zero and increasing the number of lattice points to infinity in such a way that distances are conserved. The theory on the lattice involves interactions between the variables in different lattice sites. These interactions give rise to correlations. If the correlations are short range with respect to sites, when taking the continuum limit the correlations vanish for nonvanishing lengths. Therefore, in order to have a non-trivial continuum limit, a lattice model needs to allow a regime (at least for some value of the coupling constants) such that the system becomes scale-free and long range correlations appear. These regimes correspond to second class phase transitions in the statistical mechanics sense, and it is in this regime that the continuum limit is usually taken.

The organization of this chapter is as follows. In the following section we discuss as a toy model the $Z(2)$ model on the lattice. We analyze it in the covariant and Hamiltonian versions in terms of the usual variables and then study the loop representation. In the following section we repeat the analysis for $SU(2)$ and in the last section we discuss the inclusion of fermions in an open-path representation.

6.2 Lattice gauge theories: the $Z(2)$ example

Wegner [104] introduced the Ising lattice gauge theories in 1971. He was interested in building a model similar to the Ising model but which did not exhibit spontaneous magnetization and which had a non-trivial phase structure. He wanted to study how to characterize the phases of a model without local order parameters.

The treatment of this section will follow closely the presentation due to Kogut [74], to which we refer the reader for further details.

6.2.1 Covariant lattice theory

Consider a cubic lattice in a three-dimensional Euclidean spacetime. We label the lattice sites by a triplet of integers n and the unit vectors on the lattice which we characterize by unit vectors along the lattice directions. The lattice is oriented and the unit vectors will have a $+$ or $-$ sign in front according to their orientations. Notice that each link corresponds to two possible arrangements of (n, μ), since $(n, \mu) = (n + \mu, -\mu)$. At each link

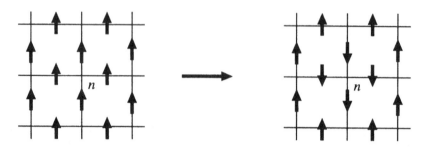

Fig. 6.1. The local gauge transformations of the Wegner $Z(2)$ model.

we associate an Ising spin $\sigma = \pm 1$ and therefore each configuration of the system is associated with an assignment of a spin orientation to each link of the lattice.

Consider now a transformation (which we will call "a local gauge transformation") such that it flips all the spins connected with one site in the lattice. An example is shown in figure 6.1.

We now consider the following action for the model

$$S(\sigma) = -\beta \sum_{n,\mu,\nu} \sigma(n,\mu)\sigma(n+\mu,\nu)\sigma(n+\mu+\nu,-\mu)\sigma(n+\mu,-\nu), \quad (6.1)$$

given by the sum of the products of all spins around each elementary plaquette of the lattice. β is the coupling constant of the model. This action is invariant under the gauge transformations introduced above. This can be readily checked noticing that a gauge transformation at n simultaneously changes the sign of $\sigma(n,\mu)$ and $\sigma(n+\mu,-\nu)$. Notice that any product of spins around any loop on the lattice will be invariant. We readily see how for this simple model the ideas of loop and holonomy play an important role.

If one attempts to define an order parameter for this model in the same spirit as the one defined for the Ising model — the magnetization — one finds that the statistical mean value of such an order parameter identically vanishes. This is a particular case of the result due to Elitzur [111] that states that taking the statistical mean value of a local gauge dependent quantity averages it over the gauge orbits. For a compact Lie group (or a discrete group like $Z(2)$) this means that the mean value vanishes.

Wegner proposed the idea of considering as an order parameter for the model the gauge invariant quantity

$$W_\gamma = \prod_{l \in \gamma} \sigma(l), \quad (6.2)$$

which represents the product of all the spins situated at the links l that compose the closed loop γ on the lattice. This idea appears natural in view of what we discussed in the previous chapter, but it should be remembered that historically it appeared before Wilson's proposal. It was the first time that a "Wilson loop" was proposed as an order parameter for a gauge theory.

The statistical mean value of the operator W_γ is given by

$$< W_\gamma >= \frac{1}{Z} \sum_{\sigma_l} W_\gamma \exp(-S(\sigma)), \tag{6.3}$$

where $Z = \sum_{\sigma_l} \exp(-S(\sigma))$ is the partition function. The summations on σ_l above mean summing over all possible spin configurations on the lattice (i.e., assignments of spin values to the links). Notice that in this context one can reinterpret β as the inverse temperature in a statistical model. Wegner proved that at small values of β (high temperatures), $< W_\gamma >\sim \exp(-\text{area}(\gamma))$. At low temperatures — large values of the coupling constant — it decreases as $\sim \exp(-\text{length}(\gamma))$ and therefore the expectation value of this operator allows us to distinguish the high and low temperature phases.

Let us discuss the proof of the area behavior via a high temperature expansion. We start by considering the identity

$$\exp(-S(\sigma)) = \exp(\beta \sigma \, \sigma \, \sigma \, \sigma)$$
$$= \cosh(\beta) + \sigma \, \sigma \, \sigma \, \sigma \sinh(\beta) \tag{6.4}$$
$$= (1 + \sigma \, \sigma \, \sigma \, \sigma \tanh(\beta)) \cosh(\beta), \tag{6.5}$$

valid for $\sigma \in Z(2)$. Therefore

$$< W_\gamma >= \frac{\sum_{\sigma_l} \prod_\Box (1 + \sigma \, \sigma \, \sigma \, \sigma \tanh(\beta)) \prod_{l \in \gamma} \sigma(l)}{\sum_{\sigma_l} \prod_\Box (1 + \sigma \, \sigma \, \sigma \, \sigma \tanh(\beta))}, \tag{6.6}$$

where symbolically the product of four sigmas represents a product along a plaquette like the one considered in equation (6.3). The product over $\Box$ means over all plaquettes in the lattice.

In order to evaluate the above expression one should recall that,

$$\sum_{\sigma_l} \sigma = 0, \qquad \sum_{\sigma_l} \sigma^2 = 2. \tag{6.7}$$

Therefore the only contributions that survive in the numerator are those in which each link is traversed at least twice, in particular the links of the loop γ. This means that the interior of the loop has to be filled by the plaquettes in the product. This ensures that each link in the loop and all links in the internal plaquettes are traversed twice. Notice that in three dimensions this could be accomplished by many configurations of plaquettes, not just planar ones. Similarly in a compact lattice the same effect could be achieved in the exterior of the loop. We will see immediately that all those possibilities are suppressed and the minimal area surrounded by the loop gives the dominant term. In order to see this, recall that $\tanh(\beta) \ll 1$ and therefore the leading contribution to the numerator will be of order $(\tanh \beta)^N$ where N is the minimal number of plaquettes that fill the loop γ and, therefore, to leading order

$$< W_\gamma >= (\tanh(\beta))^N + \cdots = \exp(N \ln(\tanh(\beta))) + \cdots. \qquad (6.8)$$

Since the minimal area inside the loop is given by N times the area of the elementary plaquette we get the area law

$$< W_\gamma >= \exp(-f(\beta)\text{area}), \qquad (6.9)$$

where the leading term in the expansion of $f(\beta)$ is $-\ln(\tanh(\beta))$.

A similar perturbative expansion for the behavior at low temperature gives a dependence proportional to the length of the loop. We refer the reader to reference [74] for details.

6.2.2   The transfer matrix method

A lattice Hamiltonian version of a quantum gauge theory can be introduced in two different ways. One would be simply to consider the theory in a Hamiltonian fashion in the continuum and to propose a discretization on a lattice. There is another method, which is commonly used, in which one discretizes the covariant theory on a Euclidean spacetime lattice and then takes the continuum limit in the time direction to end with a Hamiltonian formulation. This procedure is called the transfer matrix formalism and was introduced by Schulz *et al.* [113]. Why would one proceed in this way? It turns out that for several theories it is more immediate to write a discretized version of the covariant theory and it exhibits in a clearer fashion the symmetries of the theory. For statistical models that do not come from a discretization of a continuum theory (like the example we are considering), the transfer matrix method is the only way to construct a Hamiltonian theory from the covariant one. The Hamiltonian version of a statistical theory can only agree with the covariant version at critical points since it is a partial continuum limit (in the time direction) of the latter.

To illustrate the transfer matrix method, let us consider it for a simple mechanical system, a particle in a potential. We follow the treatment due to Creutz [112]. The transfer matrix method is based on the close analogy between the Euclidean path integral formulation of quantum mechanics and statistical mechanics. The idea is the following. One starts with a theory, the covariant lattice version of it giving a statistical mechanical system with a discretized time. One then writes the partition function in terms of a product of elements of a certain matrix. This matrix is then reinterpreted, in the limit in which the discrete time intervals go to zero, as the matrix elements of the evolution operator of a Hamiltonian quantum theory. The vacuum energy of this theory may be identified with the free energy of the statistical mechanical system. The propagator can be identified with the correlation functions and the mass gap with the inverse of the correlation length.

We will now prove for the simplified case of a particle in a potential the relation between the partition function and the energy of a quantum Hamiltonian. Let us consider the Lagrangian of a particle in a potential $V(x)$

$$\mathcal{L} = \tfrac{1}{2}m\dot{x}^2 + V(x), \tag{6.10}$$

where the $+$ sign is due to the Euclidianization. Then the path integral is given by

$$Z = \int Dx(t)\, \exp(-S), \tag{6.11}$$

where the integral is over all possible trajectories $x(t)$ from an initial configuration at t_0 to a configuration at t_N. We will perform this integral in a lattice in which space is continuous but time is discrete, divided into intervals spaced by $a = (t_N - t_0)/N$. The discretized version of the action is

$$S = a \sum_i \left[\frac{1}{2}m \left(\frac{x_{i+1} - x_i}{a} \right)^2 + V(x_i) \right]. \tag{6.12}$$

The functional integral is now precisely defined as a multiple integral,

$$Z = \int \prod_1^{N-1} dx_i \, \exp(-S). \tag{6.13}$$

Notice that if one considers periodic boundary conditions in time and sums for all $x(t_0) = x(t_N)$, Z becomes the partition function of a statistical mechanic system.

We will now see that evaluating this partition function is equivalent to solving a quantum mechanical Hamiltonian system. To see this, let us write the partition function

$$Z = \int \prod_{i=1}^{N} dx_i \, T(x_{i+1}, x_i), \tag{6.14}$$

where $T(x, x')$ are the elements of the transfer matrix, given by,

$$T(x, x') = \exp\left(-\frac{m}{2a}(x - x')^2 - \frac{a}{2}(V(x') + V(x)) \right). \tag{6.15}$$

Consider now a Hilbert space $\Psi(x)$ with the usual inner product, in which we define the position and translation operators,

$$\hat{x}\Psi(x) = x\Psi(x), \tag{6.16}$$

$$\exp(ib\hat{p})\Psi(x) = \Psi(x+b). \tag{6.17}$$

We wish to identify the elements of the transfer matrix as the matrix element of an operator in the position representation,

$$T(x,x') = <x|\hat{T}|x'>. \tag{6.18}$$

It can be seen that the operator,

$$\hat{T} = \int db \, \exp\left(-\frac{aV(\hat{x})}{2}\right) \exp\left(-\frac{b^2 m}{2a} - ib\hat{p}\right) \exp\left(-\frac{aV(\hat{x})}{2}\right) \tag{6.19}$$

$$= \sqrt{2\pi a/m} \, \exp\left(-\frac{aV(\hat{x})}{2}\right) \exp\left(-\frac{a\hat{p}^2}{2m}\right) \exp\left(-\frac{aV(\hat{x})}{2}\right) \tag{6.20}$$

gives the desired result. In the limit in which the lattice spacing is small, one can rewrite the operator as

$$\hat{T} = \sqrt{\frac{2\pi a}{m}} \, \exp(-a\hat{H} + \mathcal{O}(a^2)), \tag{6.21}$$

and one recognizes the usual Hamiltonian operator,

$$\hat{H} = \frac{\hat{p}^2}{2m} + V(\hat{x}). \tag{6.22}$$

In terms of the $\hat{T}$ operator, the partition function can be written as

$$Z = \text{Tr}(\hat{T}^N) = \text{Tr}(\exp(-\hat{H}(t_N - t_0))). \tag{6.23}$$

It is immediate to establish the relation between the partition function and the vacuum energy of the associated Hamiltonian formulation. Suppose we are in a basis that diagonalizes $\hat{H}$ with eigenvalues E_j. Then,

$$Z = \sum_j \exp(-E_j(t_N - t_0)), \tag{6.24}$$

and in the limit in which the time interval is large, the dominant term is given by $\exp(-E_0 t)$.

6.2.3 Hamiltonian lattice theory

Let us now apply the transfer matrix method to the $Z(2)$ theory. For this purpose we consider an asymmetric lattice with time spacing τ and spatial spacing a. Let β_τ be the coupling constant of the plaquettes that contain time-like links $\Box_t$ and β the constant associated with the purely spatial plaquettes $\Box_s$. The action can be split into two types of terms

$$S = -\beta_\tau \sum_{\Box_t} \sigma\,\sigma\,\sigma\,\sigma - \beta \sum_{\Box_s} \sigma\,\sigma\,\sigma\,\sigma. \qquad (6.25)$$

We now fix the gauge in such a way that the spins associated with all the temporal links take the value $+1$ (this is called the "temporal gauge" and is analogous to $A_0 = 0$ in a gauge theory). The contribution of the plaquettes that include time-like links can then be rewritten as

$$\beta_\tau \sum_{n,\mu} \sigma(n,\mu)\sigma(n+\tau,\mu), \qquad (6.26)$$

where the sum is over all the spatial links, i.e., μ is a space-like unit vector of the lattice. This can be immediately rewritten — apart from an irrelevant additive constant — as

$$S = \tfrac{1}{2}\beta_\tau \sum_{n,\mu}(\sigma(n+\tau,\mu) - \sigma(n,\mu))^2 - \beta \sum_{\Box_s} \sigma\,\sigma\,\sigma\,\sigma. \qquad (6.27)$$

If we now denote by σ^i the spatial spins associated with the spatial surface $t = t_i$, the partition function can be rewritten

$$Z = \sum_{\substack{\text{spatial spin} \\ \text{configurations}}} \prod_i \exp\left(\frac{\beta_\tau}{2}\sum_{l_s}(\sigma^{i+1}(l) - \sigma^i(l))^2 + \beta \sum_{\Box_s} \sigma\,\sigma\,\sigma\,\sigma\right).$$

$$(6.28)$$

We can write the partition function as

$$Z = \sum_k \prod_i T(k_{i+1}, k_i), \qquad (6.29)$$

where k_i is a spatial configuration of spins and T is the transfer matrix. If the lattice has Q spatial links at $t = t_i$ there will be 2^Q different configurations and therefore the transfer matrix will be of dimension $2^Q \times 2^Q$. Following the procedure outlined in the previous subsection, we can introduce a Hilbert space of functions that depend on the spin configurations on a given spatial surface $\Psi(k_i) = \Psi(\sigma^i_1, \ldots, \sigma^i_Q)$. The inner product is given by $< \psi|\phi > = \sum_k \psi^*(k)\phi(k)$. The product of two given configurations is $< k|k' > = \delta_{\sigma_1,\sigma'_1} \cdots \delta_{\sigma_Q,\sigma'_Q}$.

Let us now write the diagonal matrix elements of the transfer matrix among the configurations at time t_i and the configurations at time t_{i+1}. We denote the diagonal elements as "zero flip", meaning that all the spins are unchanged,

$$T(0\,\text{flips}) \equiv T(k_i, k_i), = \exp\left(\beta \sum_{\Box_s} \sigma\,\sigma\,\sigma\,\sigma\right), \qquad (6.30)$$

where we see that all the terms of the first summation of the action (6.29) vanish since the configurations are the same.

Let us consider the contribution when the configuration at t_i differs by one spin from that at t_{i+1}, which we denote as "one flip",

$$T(1\,\text{flip}) = \exp\left(-2\beta\tau + \beta\sum_{\square_s}\sigma\,\sigma\,\sigma\,\sigma\right). \tag{6.31}$$

We see that only one term contributes to the first summation in the action, the one corresponding to the lattice site where the spin has been flipped. In general,

$$T(n\,\text{flips}) = \exp\left(-2n\beta\tau + \beta\sum_{\square_s}\sigma\,\sigma\,\sigma\,\sigma\right). \tag{6.32}$$

We would now like to adjust the parameters β and τ such that in the limit $t_{i+1}-t_i \to 0$, the transfer matrix becomes $\exp(\tau H) \sim 1-\tau H$ with H the Hamiltonian. This immediately leads us to the following conclusions (in the limit),

$$\beta \sim \tau, \tag{6.33}$$

$$\exp(-2\beta\tau) \sim \tau, \tag{6.34}$$

which leads us to identify $\beta = \lambda\exp(-2\beta\tau)$, where λ is a constant. Therefore from the expression of the elements of the transition matrix we can infer the elements of the quantum Hamiltonian,

$$H(0\,\text{flips}) \equiv H(k_i, k_i) = \left(\lambda\sum_{\square_s}\sigma\,\sigma\,\sigma\,\sigma\right), \tag{6.35}$$

whereas

$$H(1\,\text{flips}) = 1 + \mathcal{O}(t^2). \tag{6.36}$$

Let us give a representation for the action of the Hamiltonian operator. We represent an upward pointing spin by the two-dimensional vector $(1,0)$ and the downward pointing spin by $(0,1)$. The operator that produces a spin-flip is the Pauli matrix $\sigma_1 = \begin{pmatrix} 0 & 1 \\ 1 & 0 \end{pmatrix}$, whereas the diagonal operator in this basis is $\sigma_3 = \text{diag}(1,-1)$. Therefore the Hamiltonian can be written as

$$H = -\sum_{l_s}\sigma_1 - \lambda\sum_{\square_s}\sigma_3\,\sigma_3\,\sigma_3\,\sigma_3. \tag{6.37}$$

The operator that materializes the gauge transformations in this model can be written as

$$\Omega(n) = \prod_{l_n}\sigma_1, \tag{6.38}$$

where l_n are the spatial links that emanate from the lattice site n. It can be readily checked that this operator commutes with the Hamiltonian.

We see that via the transfer matrix method we have obtained from a $(d+1)$-dimensional covariant theory a quantum Hamiltonian theory in d dimensions. We will now show that one can build a loop representation for such a theory.

6.2.4   Loop representation

In the same spirit as we discussed in the section of Yang–Mills theory, we introduce a set of operators that are gauge invariant and that are based on loops,

$$T^0(\gamma) = \prod_{l \in \gamma} \sigma_3^{(l)}, \qquad (6.39)$$

where γ is a loop on the lattice. We only need to consider loops without repeated links, since $(\sigma_3)^2 = \text{diag}(1,1)$, and therefore the repeated links do not contribute to the product. Also note that $T^0(\gamma) = T^0(\gamma^{-1})$. These identities reflect the fact that the group in question is $Z(2)$ and could be viewed as "Mandelstam identities" for this simple case.

The second invariant operator is

$$T^1(l) = \sigma_1, \qquad (6.40)$$

where we see that the operator T^1 depends on a particular link. Notice the similarity with the construction for Maxwell theory, where we could have taken as operators the Wilson loop and the electric field. There we decided to multiply the electric field by the holonomy to keep the similarity with the non-Abelian case. Here we decide to make the variable T^1 loop independent. The commutator of the two operators is

$$[T^1(l), T^0(\gamma)] = 2T^1(l)T^0(\gamma) \sum_{l' \in \gamma} \delta_{ll'}. \qquad (6.41)$$

Defining $X_l(\gamma)$ as $\sum_{l' \in \gamma} \delta_{ll'}$ (a lattice analogue of the first rank multitensor $X^{ax}(\gamma)$ in the continuum) we get the T algebra

$$[T^0(\gamma), T^0(\eta)] = 0, \qquad (6.42)$$

$$[T^1(l), T^1(l)] = 0, \qquad (6.43)$$

$$[T^1(l), T^0(\eta)] = 2X_l(\gamma)T^1(l)T^0(\gamma). \qquad (6.44)$$

Notice that one could define in analogy with Maxwell theory an "electric field operator" $E(l)$,

$$E(l) = 1 - \tfrac{1}{2}T^1(l). \qquad (6.45)$$

As in the case of Maxwell theory, where we defined a special group of loops to reflect the symmetries of the theory, one can define an Abelian group of loops that reflects the symmetries of $Z(2)$. In order to do this, one starts from the Abelian group of loops defined in section 4.1 and identifies the squares of loops with the identity. The elements of this group therefore satisfy

$$\gamma_1 \circ \gamma_2 = \gamma_2 \circ \gamma_1, \qquad (6.46)$$

$$\gamma \circ \gamma = 1, \qquad (6.47)$$

$$\gamma_1 = \gamma_1 \circ (\bar{\gamma}_1)^2 = \bar{\gamma}_1. \qquad (6.48)$$

We now give a representation of the T algebra in terms of a space of wavefunctions dependent on loops $\Psi(\gamma)$

$$T^0(\eta)\Psi(\gamma) = \Psi(\gamma \circ \eta), \qquad (6.49)$$

$$T^1(l)\Psi(\gamma) = (1 - 2X_l(\gamma))\Psi(\gamma). \qquad (6.50)$$

The electric field has the usual action for an Abelian theory,

$$E(l)\Psi(\gamma) = X_l(\gamma)\Psi(\gamma). \qquad (6.51)$$

The Hamiltonian in this representation can be written in terms of the elementary operators,

$$\hat{H}\Psi(\gamma) = -\lambda \sum_{\square_s} T^0(\square_s) - \sum_{l_s} T^1(l), \qquad (6.52)$$

and is called the Wegner Hamiltonian. To make contact with usual gauge theories it is common to define a Hamiltonian which differs from Wegner's by a constant, $\hat{H} = \lambda^{-1}(H_{\text{Wegner}} + \Lambda)$ where Λ is the product of the number of sites in the lattice and the dimension of the space. The modified Hamiltonian then reads

$$\hat{H}\Psi(\gamma) = \mu \sum_l E(l) - \sum_{\square_s} \sigma\sigma\sigma\sigma, \qquad (6.53)$$

where $\mu = -2/\lambda$ and we see the appearance of an "electric" and a "magnetic" piece. This is clarified by studying the action on a state,

$$\hat{H}\Psi(\gamma) = -\sum_{\square_s} \Psi(\square_s \circ \gamma) + \mu l(\gamma)\Psi(\gamma), \qquad (6.54)$$

where $l(\gamma)$ is the length of the loop. Usually one would expect a length squared in the term arising from the electric part. However, for the $Z(2)$ case loops are only traversed once.

This Hamiltonian can be put on a computer and used to study the vacuum energy, observables and mass spectra of the theory [114] and the results can be compared with those obtained with other methods. There is a great wealth of knowledge about this model because — in three dimensions — it is the dual of the Ising model. Duality in this context means that one can associate to the lattice a dual lattice in which to each cube one associates a lattice site and to each plaquette a link and rewrite the action of the Wegner model as the action of an Ising model on the dual lattice [74]. This allows us to import all the knowledge about the Ising model to the Wegner model.

Let us now sketch one of the approximation techniques that appear naturally in the loop representation on the lattice. It is called the collective variables technique.

Given the known behavior at long lengths of the Wilson loop, the length and area of a loop appear as natural variables with which to study the theory. The length and area of loops are two examples of possible collective variables that characterize loops in the asymptotic region of large loop lengths. One could refine this picture by introducing other variables that give additional information about the loop, such as information about corners [114].

Let us therefore propose a description of the model in which we consider wavefunctions that are a function of the length of the loop, the area of the loop and a variable c that codes the information about the number and kinds of corners of the loop $\Psi(l, A, c)$.

The Hamiltonian in terms of these variables can be constructed from the action of the Hamiltonian in the loop representation (6.52). The action of the electric part is trivial, it just multiplies the wavefunction and the value of the length. The magnetic part adds a plaquette. The effect of this will depend on where the Hamiltonian is acting. If the plaquette is completely exterior to the loop, the area is increased by one unit, the length is increased by four units and four corners are added. If the plaquette is completely inside the loop, the area decreases by one unit, and the length and number of corners increase by four units. In the case where the plaquette shares a link with the loop and there are no corners adjacent the length increases by two units, the area by (plus or minus) one and the number of corners by four units. All actions are weighted by a factor that states how many plaquettes with the action of interest are possible. The action of the Hamiltonian is therefore coded in a finite difference equation involving the three variables of the problem and the total number of plaquettes in the lattice.

One can search for solutions of the finite difference equation minimizing the energy per plaquette. Since one is aware of possible exponential behaviors with length and area — as we argued in the previous chapter — one can propose a solution $\Psi(l, A, c) = Y^A X^l Z^c$ with X, Y, Z constants. The typical behaviors of the energy and the Y constant as a function of the coupling constant μ is shown in figure 6.2. Notice that between $\mu = 0$ and a certain critical value (weak coupling regime) the constant $Y = 1$ and therefore the wavefunction does not depend on the area. For $\mu > \mu_0$ the variable $Y < 1$ and therefore the wavefunction decreases exponen-

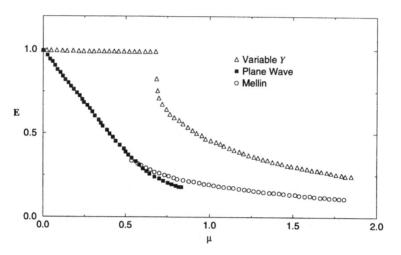

Fig. 6.2. Dependence of the energy per plaquette and the variable Y as a function of the coupling constant μ in the Wegner model.

tially with the area. For the energy we show two curves, the one in the weak coupling regime was obtained as discussed in the text, the one in the strong coupling regime was obtained by a Mellin transform method [108, 114]. Notice that there is a discontinuity suggesting a second order phase transition for the value $\mu = 0.68$ which agrees to within 2% of the value observed with other methods.

So we see that the use of the loop representation allows a very natural and intuitive action for the Hamiltonian constraint. Collective variables representing qualitative properties of the loop allow an understanding of the phase structure and the observables of the theory with small computing power. Unlike statistical methods, such as Monte Carlo simulations, they produce analytic results. The results can be plotted with a computer as we chose to do, but they are available in analytic form.

These initial results should act as an encouragement for further work on the use of collective variables on the lattice and the development of other approximation methods in the loop representation.

6.3   The $SU(2)$ theory

We will now discuss a more realistic lattice gauge theory, related to a continuum theory. Many of the techniques developed here are also applicable to other cases of direct physical interest, such as QED and QCD.

Let us start by setting up a lattice version of the connection representation. We assign to each link in the lattice an element of the $SU(2)$ gauge group $\mathbf{U}(l)$. This element is the parallel transport operator along the link. Notice the parallel with Wegner's model. Each link variable $\mathbf{U}(l)$ is not gauge invariant, the gauge transformations act at each site. We will be able to construct gauge invariant quantities by taking the product of link variables along closed contours. Notice that the variables $\mathbf{U}(l)$ are just the natural counterparts of the holonomies along open paths that we discussed in chapter 1. A point to notice is that in the continuum descriptions of gauge theories one usually takes variables defined on the algebra of the gauge group whereas the fundamental variables on the lattice take values in the group.

6.3.1 *Hamiltonian lattice formulation*

A field configuration is determined by an assignment of $SU(2)$ matrices to each link on the lattice. The assignment depends on an orientation of the lattice: if the link l has associated with it the matrix $\mathbf{U}(l)$, the reversed link has $\mathbf{U}(l)^{-1}$. We denote by $\mathbf{A}(l)$ the element of the algebra associated with $\mathbf{U}(l) = \exp(iag\mathbf{A}(l))$ where a is the lattice spacing and g is the coupling constant of the theory. We want to introduce a variable canonically conjugate to U that in the limit of zero lattice spacing plays the role of the electric field and that in general has the same transformation properties under gauge transformations that the electric field has in the continuum case. We introduce the variable $\mathbf{E}(l)$ which takes values in the $SU(2)$ algebra and which has Poisson brackets,

$$\{\mathbf{U}_A^B(l), \mathbf{U}_C^D(l')\} = 0, \tag{6.55}$$

$$\{\mathbf{E}_j(l), \mathbf{U}_A^B(l')\} = -i\delta_{l,l'}(X_j)_A^C \mathbf{U}_C^B(l), \tag{6.56}$$

$$\{\mathbf{E}_j(l), \mathbf{E}_k(l')\} = \sqrt{2}\epsilon_{jkm}\mathbf{E}_m(l)\delta_{l,l'}, \tag{6.57}$$

where as usual the indices j refer to components in a basis of generators of the algebra, $(\mathbf{X}_j)_A^B$, l and l' are in the positive orientation and $\delta_{l,l'} = 1$ if $l = l'$ and zero otherwise.

The continuum limit of these variables (when the lattice spacing a goes to zero) is defined as

$$\lim_{a\to 0} \mathbf{A}(l) = \lim_{a\to 0} \mathbf{A}(n, \mu_b) = \mathbf{A}_b(n), \tag{6.58}$$

$$\lim_{a\to 0} \mathbf{U}(l) = \lim_{a\to 0}(\mathbf{1} + ia\mathbf{A}(l)) = \mathbf{1}, \tag{6.59}$$

$$\lim_{a\to 0} \mathbf{E}_j(l) = \lim_{a\to 0} \mathbf{E}_j(n, \mu_b) = \lim_{a\to 0} a^2\mathbf{E}_j^b(n), \tag{6.60}$$

where $\mathbf{A}(n, \mu_b)$ is the value of the field at the lattice position n where the link l starts and in the direction of the vector μ_b along the link l (the μs were a triad of vectors along the lattice directions and b is a triad index).

With this definition we can check that the Poisson brackets correspond to the usual canonical brackets of Yang–Mills theory,

$$\lim_{a \to 0} \frac{1}{a^3} \{\mathbf{E}_j(n, \mu_b), \mathbf{U}(n', \mu_c)\} = -i \lim_{a \to 0} \frac{1}{a^3} \delta_{n,n'} \delta_{bc} \mathbf{X}^j, \qquad (6.61)$$

and in the limit,

$$\{\mathbf{E}_b^j(x), \mathbf{A}_c(x')\} = -\delta(x - x') \delta_{b,c} \mathbf{X}^i, \qquad (6.62)$$

where x and x' are the coordinate positions of the lattice sites n and n' respectively. Following a similar calculation one can show that the Poisson bracket of the $\mathbf{E}$ variables leads to the usual vanishing Poisson bracket of electric fields.

We now write the Gauss law and the Hamiltonian for the classical theory. The Gauss law is

$$\mathcal{G}_j(n) = \sum_{l_n} \mathbf{E}_j(l_n) = 0, \qquad (6.63)$$

where l_n are all the links emanating from the site n. In the limit of vanishing lattice spacing it can be checked that this equation gives rise to the usual divergence of E. To perform this limit the reader should be aware of the commutation relation of $\mathbf{E}_j(\bar{l})$ where $\bar{l}$ is the link l with a reversed orientation. This Poisson bracket is derived from the ones introduced above and recalling that $\mathbf{U}(\bar{l}) = \mathbf{U}^{-1} = \mathbf{U}^\dagger(l)$ where $\dagger$ is the conjugate transpose matrix. The result is

$$\{\mathbf{E}_j(l), \mathbf{U}_A^B(l')\} = -i\delta_{l,l'} \mathbf{U}_A^C(l)(X_j)_C^B. \qquad (6.64)$$

It is possible to show that the Gauss law generates infinitesimal canonical transformations associated with gauge transformations on the lattice,

$$\mathbf{U}(n, \mu) \to \mathbf{V}(n)\mathbf{U}(n, \mu)\mathbf{V}^\dagger(n + \mu), \qquad (6.65)$$

$$\mathbf{E}_j(n, \mu) \to \mathbf{V}(n)\mathbf{E}_j(n, \mu)\mathbf{V}^\dagger(n). \qquad (6.66)$$

The Hamiltonian is

$$\mathcal{H} = \frac{g^2}{2} \sum_{l>0} \mathbf{E}_j(l)\mathbf{E}_j(l) - \frac{1}{g^2} \sum_{\square} \mathrm{Tr}(\mathbf{U}(\square)), \qquad (6.67)$$

where $l > 0$ means all the positive oriented links of the lattice and $\mathbf{U}(\square)$ is the product of the four U variables associated with the links of the elementary plaquettes over which the sum runs. We have set $a = 1$ as is customary in the lattice. If not, the electric part would be altered by a factor $1/a$ and the magnetic part by a factor a. In the continuum limit, the electric piece immediately reproduces the continuum electric field squared. The magnetic part is more complicated. In order to recover the usual magnetic part one should subtract a negative constant proportional to the number of plaquettes. The Hamiltonian that we consider is therefore bounded from below but is not positive-definite.

To construct a quantum theory in the $\mathbf{U}$ representation we introduce wavefunctions $\Psi(\mathbf{U})$ where $\mathbf{U}$ denotes a configuration of the system that assigns to each element on the lattice an $SU(2)$ matrix. The variables $\mathbf{U}$ and $\mathbf{E}$ become multiplicative and purely derivative operators in this space. This space is endowed with a natural inner product

$$< \Psi|\Phi > = \int \prod_{l>0} d\mathbf{U}_l \Psi^*(\mathbf{U})\Psi(\mathbf{U}). \qquad (6.68)$$

The measure is, for every lattice link, the Haar measure associated with $SU(2)$ [98]. In this space the operators associated with $\mathbf{U}$ and $\mathbf{E}$ have the following action:

$$\hat{U}(l)_A^B \Psi(U) = \mathbf{U}(l)_A^B \Psi(U), \qquad (6.69)$$

$$\hat{E}_j(l)\Psi(U) = -(X_j)_A^B \mathbf{U}(l)_B^C \frac{\partial}{\mathbf{U}(l)_A^C}\Psi(\mathbf{U}), \qquad (6.70)$$

and the quantum expressions for the Hamiltonian and Gauss law can be constructed straightforwardly.

6.3.2 Loop representation in the lattice

Following the same steps that we used when discussing the Wegner model, it is immediate to introduce gauge invariant variables for the $SU(2)$ theory on the lattice. Loop representations on the lattice have been considered by several authors [115, 110, 109]. In this section we will follow closely the treatment of reference [110].

Let us consider an algebra of classical gauge invariant quantities on the lattice defined by[*]:

$$T^0(\gamma) = \tfrac{1}{2}\mathrm{Tr}\left[\prod_{l\in\gamma} \mathbf{U}(l)\right] \equiv \tfrac{1}{2}\mathrm{Tr}\left[\mathbf{U}(\gamma)\right], \qquad (6.71)$$

$$T_l^1(\gamma) = \tfrac{1}{2}\mathrm{Tr}\left[\mathbf{U}(\gamma_n^n)\mathbf{E}(l)\right], \qquad (6.72)$$

where as usual $l = (n,\mu)$ and $\mathbf{U}(\gamma_n^n)$ denotes the product of $\mathbf{U}(l)$ with l links in γ starting at n and ending at n. The Poisson algebra is

$$\{T^0(\gamma), T^0(\eta)\} = 0, \qquad (6.73)$$

---

[*] Our conventions in this chapter for the T variables differ from those in the rest of the book by a factor $\tfrac{1}{2}$. We do this in order to facilitate the comparison with the particle physics literature which usually includes that factor in the definition of the Wilson loops.

$$\{T^0(\gamma), T_l^1(\eta)\} = \frac{i}{2}g \sum_{l' \in \gamma} \delta_{l,l'} \left[T^0(\gamma_n^n \circ \eta_{n'}^{n'}) - T^0(\gamma)T^0(\eta) \right], \qquad (6.74)$$

$$\{T_l^1(\gamma), T_{l'}^1(\eta)\} = \frac{i}{2}g \sum_{l'' \in \gamma} \delta_{l,l''} \left[T_{l'}^1(\eta_{n'}^{n''} \circ \gamma_n^n \circ \eta_{n''}^{n'}) - T^0(\gamma)T_{l'}^1(\eta) \right]$$
$$- \frac{i}{2}g \sum_{l'' \in \eta} \delta_{l',l''} \left[T_l^1(\gamma_n^{n''} \circ \eta_{n'}^{n'} \circ \gamma_{n''}^n) - T^0(\eta)T_l^1(\gamma) \right],$$

$$(6.75)$$

where $\delta_{l,l'} = \pm 1$ if l and l' are the same link, the sign depending on the orientation of links, and being zero otherwise.

We now proceed to quantize the theory. We need to realize the algebra of classical quantities that we just introduced on a space of functions of loops on the lattice. As we argued in chapter 3, for the case of $SU(2)$ it is sufficient to consider wavefunctions of a single loop. This was due to the fact that the Mandelstam identities (for $SU(2)$) allow us to express any product of Wilson loops as a linear combination of Wilson loops. In this chapter however, we will not take advantage of this fact and we will consider a representation in terms of wavefunctions of multiloops ("clusters" in lattice notation).

The reason for this is that the use of multiloops will lead us naturally to calculational techniques that are more economical and efficient from the point of view of the lattice. In short we will trade aesthetics (having a single loop) for calculational efficiency. For instance, we may consider the action of the magnetic term of the Yang–Mills Hamiltonian in the loop representation. In terms of a representation based on multiloops its action is to add a plaquette to the loops in the argument of the wavefunction. In terms of functions of a single loop, one obtains reroutings at intersections (as is typical of the use of the Mandelstam identities).

As we discussed in chapter 5, in order to represent the Hamiltonian of Yang–Mills theory, it is not enough to consider the "small" T algebra formed by T^0 and T^1. As in the discussion of $SU(N)$ in the continuum, we will not study the representation of the "large" T algebra but just of the T^0 and T^1 supplemented with the Hamiltonian constraint. As we argued in the continuum, one can always find a representation of the large T algebra such that the representation we introduce for the T^0, T^1 and Hamiltonian is reproduced simply by considering the expression of the Ts in terms of E and U and choosing a factor ordering (E to the right in this particular case). The action of the operators on the space of wavefunctions of multiloops is

$$\hat{T}^0(\eta)\Psi(\gamma_1, \ldots, \gamma_n) = \Psi(\eta, \gamma_1, \ldots, \gamma_n), \qquad (6.76)$$

$$\hat{T}_l^1(\eta)\Psi(\gamma_1,\ldots,\gamma_n) = \frac{g}{2}\sum_{k=1}^{n}\sum_{l'\in\gamma_k}\delta_{l,l'}$$

$$\times\left(\Psi(\gamma_1,\ldots,\eta_n^n\circ(\gamma_k)_{n'}^{n'},\ldots,\gamma_n) - \Psi(\eta,\gamma_1,\ldots,\gamma_n)\right)(6.77)$$

To realize the electric part of the Hamiltonian we proceed in the same way as in chapter 5, recalling the commutation relation of the electric part of the Hamiltonian,

$$\hat{\mathcal{E}} = \frac{g^2}{2}\sum_{l>0}\hat{E}_j(l)\hat{E}_j(l), \tag{6.78}$$

with the Wilson loop (which can be derived in the U representation),

$$[\hat{\mathcal{E}},\hat{T}^0(\gamma)] = g^2\sum_{l,l'\in\gamma}\delta_{l,l'}[\hat{T}^0(\gamma_n^{n'})\hat{T}^0(\gamma_{n'}^n) - \tfrac{1}{4}T^0(\gamma)] - g\sum_{l\in\gamma}\hat{T}_l^1(\gamma). \tag{6.79}$$

From here we can read off the realization of the electric part of the Hamiltonian on a wavefunction, which consists of four distinct contributions $\mathcal{E} = \mathcal{E}_1 + \mathcal{E}_2 + \mathcal{E}_3 + \mathcal{E}_4$, given by

$$\hat{\mathcal{E}}_1\Psi(\gamma_1,\ldots,\gamma_n) = g^2\sum_{j=1}^{n}L(\gamma_j)\Psi(\gamma_1,\ldots,\gamma_n), \tag{6.80}$$

$$\hat{\mathcal{E}}_2\Psi(\gamma_1,\ldots,\gamma_n) = -\frac{g^2}{4}\sum_{i,j=1}^{n}\Lambda(\gamma_i,\gamma_j)\Psi(\gamma_1,\ldots,\gamma_n), \tag{6.81}$$

where $L(\gamma)$ is the number of links in the loop γ, $\Lambda(\gamma,\eta) = \sum_{l\in\gamma}\sum_{l\in\eta}\delta_{l,l'}$ (sometimes called the "quadratic length" in the case $\gamma = \eta$),

$$\hat{\mathcal{E}}_3\Psi(\gamma_1,\ldots,\gamma_n) = \frac{g^2}{2}\sum_{k<j=1}^{n}\sum_{l\in\gamma_k}\sum_{l'\in\gamma_j}\delta_{l,l'}$$

$$\times\Psi(\gamma_1,\ldots,\gamma_{i-1},(\gamma_k)_n^n\circ(\gamma_j)_{n'}^{n'},\ldots,\gamma_{j-1},\gamma_{j+1},\ldots\gamma_n),(6.82)$$

$$\hat{\mathcal{E}}_4\Psi(\gamma_1,\ldots,\gamma_n) = g^2\sum_{j=1}^{n}\sum_{l,l'\in\gamma_j}\delta_{l,l'}$$

$$\times\Psi(\gamma_1,\ldots,\gamma_{j-1},(\gamma_j)_n^{n'},(\gamma_j)_{n'}^n,\gamma_{i+1},\ldots,\gamma_n). \tag{6.83}$$

The magnetic part of the Hamiltonian,

$$\hat{\mathcal{B}} = -\frac{1}{g^2}\sum_{\square}\mathrm{Tr}(\hat{U}(\square)), \tag{6.84}$$

has also a very simple action,

$$\hat{\mathcal{B}}\Psi(\gamma_1,\ldots,\gamma_n) = -\frac{1}{g^2}\sum_{\square}\Psi(\square,\gamma_1,\ldots,\gamma_n), \tag{6.85}$$

and the loop could have been inserted in any entry. The order of loops in the multiloop is irrelevant.

We therefore see that the Hamiltonian in the lattice has a beautifully simple geometric action. The term $\mathcal{E}_1$ measures the length of the loop, the term $\mathcal{E}_2$ measures "overlapping" and quadratic effects in the length. The term $\mathcal{E}_3$ fuses two loops if they have an intersecting point and $\mathcal{E}_4$ takes any loop with a self-intersection and produces a fission. The magnetic terms simply add a loop. For the case $SU(N)$ the action of the Hamiltonian is exactly the same (apart from factors dependent on the dimensionality of the group) [110].

We will now discuss methods for treating the Hamiltonian on the lattice.

6.3.3 Approximate loop techniques

The cluster approximation techniques we are to describe are based on the combination of strong coupling expansions and collective variables. Let us therefore start with a brief discussion of the strong coupling expansion in terms of loops. If one takes the limit $g \to \infty$, the magnetic term in the Hamiltonian of Yang–Mills theory drops out and the Hamiltonian eigenvalue problem can be solved exactly. Remembering that all the terms in the electric part of the Hamiltonian are proportional to loop lengths, it is immediate to realize that the vacuum is a ket with zero loops $|0>$. The energy of the vacuum vanishes. The first excited state is given by a plaquette excitation $|\square>$. The second involves at most two plaquettes and so on. In this approximation the magnetic term is considered a perturbation of the electric term. The effect of the magnetic term is to add a plaquette. Therefore, in the perturbative expression of the vacuum in the strong coupling regime, terms involving many plaquettes are suppressed by a power of $1/g^4$.

A cluster is a set of loops in a finite region of space. The quantum states we will consider will be based on sets of clusters, which we assume to be far apart from each other.

Examples of clusters are:

- a single plaquette,

- two plaquettes nearby,

- a rectangle,

- a plaquette traversed twice.

The idea of this approximation is based on the action of the Hamiltonian we described above. Since the clusters are assumed to be far away

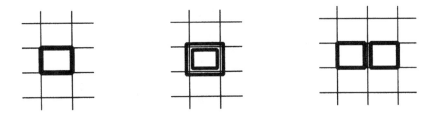

Fig. 6.3. The three clusters considered in the example

from each other, the Hamiltonian never connects them. The approximation is based on truncating the basis of all possible states (all possible clusters) and considering a finite number of clusters and the action of the Hamiltonian in the truncated basis.

Let us consider a concrete example of the action of the Hamiltonian of $SU(2)$ theory in $2+1$ dimensions on a particular set of clusters. We truncate the basis of clusters to only three types of cluster. This is clearly a toy model only and later we will give a procedure for constructing approximations to any desired order.

Cluster type 1 is a plaquette. Cluster type 2 is a plaquette traversed twice and type 3 consists of two plaquettes with a common link in a plane, as shown in figure 6.3. The three elements of the cluster basis considered are

$$\text{type 1}: \qquad T^0(\square)|0>, \qquad (6.86)$$

$$\text{type 2}: \qquad T^0(\square)T^0(\square)|0>, \qquad (6.87)$$

$$\text{type 3}: \qquad T^0(\square)T^0(\square')|0>, \qquad (6.88)$$

where $\square'$ has a common link with $\square$. The states spanned by this basis are denoted by

$$|n_1, n_2, n_3 >, \qquad (6.89)$$

where n_i indicates the number of clusters of type i present. The lattice position of the clusters is immaterial (as long as the clusters are far apart) since the action of the Hamiltonian is local and sums over all clusters.

Let us study the action of the Hamiltonian. For convenience we rescale it by a factor $g^2/2$. The magnetic term adds a plaquette. Its action can be written as

$$g^2\hat{\mathcal{B}}|n_1, n_2, n_3 >= n_1|n_1 - 1, n_2 + 1, n_3 >$$
$$+4n_1|n_1 - 1, n_2, n_3 + 1 > +5n_2|n_1, n_2 - 1, n_3 >$$
$$+8|n_1, n_2, n_3 - 1 > +(P - 5n_1 - 5n_2 - 8n_3)|n_1 + 1, n_2, n_3 >, \quad (6.90)$$

where P is the number of plaquettes in the lattice. The first term corresponds to the addition of a plaquette on top of one of the single plaquettes and therefore producing a cluster of type 2. The second term introduces a plaquette adjacent to one of the clusters of type 1, forming a cluster of type 3. The third and fourth terms destroy clusters of type 2 and type 3 respectively and produce a state rigorously out of the basis of clusters considered. The last term corresponds to the addition of a plaquette without contact with any of the existing clusters. We see that the action of the magnetic part of the Hamiltonian does not leave the basis invariant. There are two possibilities: either one ignores this fact and the calculation becomes valid only in the strong coupling limit, or one tries to encode the missing information in an extra set of variables (collective variables).

The electric part of the Hamiltonian gives a diagonal contribution that can be written as

$$\hat{\mathcal{E}}_1 + \hat{\mathcal{E}}_2 |n_1, n_2, n_3> = \frac{g^4}{2}(3n_1 + 4n_2 + \tfrac{13}{2})|n_1, n_2, n_3 > . \tag{6.91}$$

The remaining terms are the fission and fusion terms. The fission terms do not contribute because we are not considering loops with self-intersections at this order of approximation. The fusion terms give a non-diagonal contribution,

$$\hat{\mathcal{E}}_3 |n_1, n_2, n_3> = \frac{g^4}{2}(4n_2|n_1, n_2, n_3 > -2n_2|n_1, n_2 - 1, n_3 >$$
$$-\tfrac{1}{2}n_3|n_1, n_2, n_3 - 1 >). \tag{6.92}$$

The first and second term originate in the action of the fusion terms on a cluster of type 2; this leads to a $\hat{T}^0(\square \circ \square)$ which can be rearranged using the Mandelstam identities into a linear combination of a cluster of type 2 and the vacuum. The last term comes from the fusion of the two plaquettes present in the clusters of type 3 and leads to a rectangle, which does not appear at the present order.

We are now in a position to cast the problem of finding the vacuum and the excited states of $SU(2)$ Yang–Mills theory in terms of a finite difference equation. We start from

$$< \Psi|H|n_1, n_2, n_3> = E < \Psi|n_1, n_2, n_3 >, \tag{6.93}$$

and get

$$n_1[5\Psi(n_1 + 1, n_2, n_3) - \Psi(n_1 - 1, n_2 + 1, n_3)$$
$$-4\Psi(n_1 - 1, n_2, n_3 + 1) + \tfrac{3}{2}g^4\Psi(n_1, n_2, n_3)]$$
$$n_2[5\Psi(n_1 + 1, n_2, n_3) - 5\Psi(n_1, n_2 - 1, n_3)$$
$$+4g^4\Psi(n_1, n_2, n_3) - g^4\Psi(n_1, n_2 - 1, n_3)]$$
$$n_3[8\Psi(n_1 + 1, n_2, n_3) - 8\Psi(n_1, n_2, n_3 - 1)$$
$$+\tfrac{13}{4}g^4\Psi(n_1, n_2, n_3) - \tfrac{1}{4}g^4\Psi(n_1, n_2, n_3 - 1)] =$$
$$P[\Psi(n_1 + 1, n_2, n_3) + \epsilon\Psi(n_1, n_2, n_3)], \tag{6.94}$$

where P is the number of plaquettes and $\epsilon = E/P$ is the energy per plaquette.

We now propose a power-law solution for the vacuum (as discussed above),

$$\Psi_0(n_1, n_2, n_3) = x_1^{n_1} x_2^{n_2} x_3^{n_3}, \tag{6.95}$$

with x_i constants, that leads to a system of non-linear equations,

$$(5x_1 + \tfrac{3}{2}g^4)x_1 - x_2 - 4x_3 = 0, \tag{6.96}$$

$$(5x_1 + 4g^4)x_2 - g^4 - 5 = 0, \tag{6.97}$$

$$(8x_1 + \tfrac{13}{4}g^4)x_3 - \frac{g^4}{4} - 8 = 0, \tag{6.98}$$

and the relation $\epsilon_0 = -x_1$. Note that in spite of the fact that one started from a strong-coupling approximation, if one takes $g = 0$, then $x_1 = x_2 = x_3 = 1$, which corresponds to the exact solution of the system when $g = 0$. This last fact can be better seen in the U representation. In such a representation the magnetic part is just a multiplication by $-T^0(\Box)$, which has a minimum at -1, and this implies that T^0 is 1 and therefore it corresponds to a configuration in which each link has associated the element $U = 1$ (up to gauge). This implies that the vacuum in the loop representation is $\Psi(\gamma) = 1$ for any γ in the weak coupling limit, which is the result we found above.

The excited states are found by trial of ansätze of the form

$$(1 + a_1 n_1 + a_2 n_2 + a_3 n_3)x_1^{n_1} x_2^{n_2} x_3^{n_3}, \tag{6.99}$$

which resemble the kind of polynomial construction we performed for the excited states of Maxwell theory.

We now return to the general discussion of the cluster approximation technique. A generic state will be characterized by a list of clusters, $|n_1, \ldots, n_k, \ldots >$. We now propose a recursive ordering among clusters. This idea of order is associated with the different orders in the strong coupling approximation. The zeroth order will be the vanishing loop, the first order is a single plaquette. The nth order is obtained from the $(n-1)$th one through the action of the Hamiltonian on the basis of $(n-1)$th clusters. We restrict the action of the Hamiltonian to the addition of a plaquette immediately adjacent to the existing one. The combined action of the electric and magnetic terms will give rise to loops of large area and disconnected (but close). It is simple to see that through this procedure one can obtain any loop on the lattice. Therefore, there exists a natural approximation scheme that consists in considering clusters up to a certain order. In particular, the previous example is an approximation of order 2. This proposal for the construction of a basis of clusters has the drawback that one may consider clusters that are equivalent under the Mandelstam identities, that in the conventions used in this section read (in the loop representation)

$$2\Psi(\ldots, \gamma_i, \gamma_j, \ldots) = \Psi(\ldots, \gamma_i \circ \gamma_j, \ldots) + \Psi(\ldots, \gamma_i \circ \bar{\gamma}_j, \ldots). \quad (6.100)$$

An interesting point is that if one is only interested in the energy spectrum of the Hamiltonian, one can ignore completely the Mandelstam identities. The only consequence (apart from the obvious one of working on a larger set of states) is that the level of degeneracy of each energy eigenstate is increased (since one is considering as independent states that are not). But the point is that the energy levels are unchanged. This statement is an exact one as long as one does not truncate the basis. With a truncation one should be careful because clusters apparently of higher order could be Mandelstam rearranged to a lower order. An explicit calculation for the $SU(2)$ case in $2+1$ dimensions taking into account the complete set of Mandelstam identities [109] shows no appreciable difference in the energy levels from those in which the Mandelstam identities were taken into account only partially [110], as we shall discuss later.

Although it is difficult to exhaust entirely the content of the Mandelstam identities to reduce the set of states, it is possible to use them partially in a simple way to curb the number of states considerably. The Mandelstam identity for two adjacent loops sharing a common link with the same orientation implies that the state can be completely rewritten in terms of loops in which the link appears to be traversed only once. This means that one can automatically eliminate from the set of states those which include links traversed twice or more in the same direction.

A technique that considerably improves the performance of the cluster approximation is the use of collective variables. The idea is to supplement the information one has about the clusters with the use of certain variables, similar to the ones we discussed for the Wegner model. An example of a collective variable is the length of a loop L. The value of a collective variable Q for a certain configuration of clusters $|n_1, \ldots, n_k>$ is given by $Q = \sum_{i=1}^{k} n_i Q_i$ where the coefficients Q_i are the values of the collective variable for the cluster i (for instance, it could be the length of the cluster i). One usually normalizes the variables such that $Q_1 = 1$. The wavefunctions $\Psi(n_1, \ldots, n_k)$ can be reexpressed as $\Psi'(Q, n_2, \ldots, n_k)$, and therefore to search for the vacuum one chooses the ansatz $x_1^Q y_2^{n_2} \ldots y_k^{n_k}$ with $x_1, y_2, \ldots, y_k$ are constants. In spite of the fact that this description may seem quite similar to the one presented before, the use of the collective variable Q allows us to take into account when the action of the Hamiltonian goes out of the space of clusters considered. Note that in searching for the vacuum one solves a system of non-linear equations. It can be seen that the equations associated with the collective variables are non-linear and that all the others are linear.

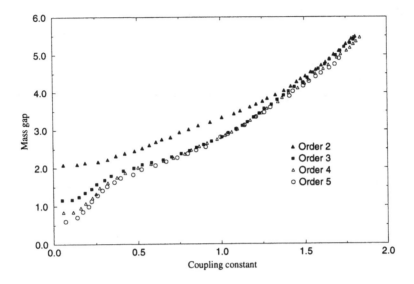

Fig. 6.4. Energy of the first excited state minus the vacuum energy (mass gap) as a function of the coupling constant. The calculation was performed using clusters up to order five and one collective variable.

The use of collective variables allows the approximation to remain meaningful to a certain extent in the weak coupling regime, as can be seen in figure 6.4. This figure corresponds to the use of a single collective variable, appropriately chosen to fit the analytic behavior of the energy in the weak coupling regime. The collective variable considered is $Q = ((N+1)L - \Lambda)/4N$, where N is the order of the approximation, L the length and Λ the quadratic length.

In the figure one can see the transition between strong and weak coupling regimes around $g^2 = 1$. If one does not use collective variables, the approximation breaks down around $g^2 = 2$ ([109] figure 8). In the weak coupling regime the mass gap should go to zero as $g \to 0$ linearly in g^2 (as can be seen in perturbation theory [116]). We see that there are signs of convergence to the expected behavior for the higher cluster orders. The slope for the best approximation (fifth order) is 4.03 [110], whereas strong coupling calculations predict a value of 4.4 ± 0.5[117, 118] and the Monte Carlo result is 4.06 ± 0.6 [119].

6.4   Inclusion of fermions

Yang–Mills theories arise in nature as the theories of the vector bosons
that mediate the interactions of fermionic matter fields. An example
would be the photons that allow particles with electric charge to interact.
Another example could be the interaction of colored particles, such as
quarks, via the exchange of $SU(3)$ gluons. In all these examples the
particles that interact via the Yang–Mills fields are charged fermions.
Therefore to study interactions of gauge theories with matter we need to
incorporate fermions in the discussion. We have already mentioned that
fermions give rise to open paths in the language of loops both in chapter
1 and in the discussion of the ideas about confinement. We are also
going to present a discussion of the interaction of fermions with gravity
in chapter 9. In this section we do not intend to develop in great detail
the discussion of gauge theories interacting with fermions in the loop
representation. We simply want to introduce very briefly some techniques
that have been developed to deal with fermions interacting with gauge
fields. These techniques are formulated on the lattice and that is the
rationale for including the discussion of fermions in this chapter.

The action of a gauge field interacting with a charged fermion is given
by

$$S = S_{\text{Free}} + \bar{\psi}^I (\delta_I^J \gamma^\mu (\partial_\mu - ig(A_\mu)_I^J))\psi_J, \qquad (6.101)$$

where ψ_I is a group-valued four-component Dirac spinor. We have omit-
ted the Dirac indices, as is usually done. γ_μ are the four Dirac matrices
and $\bar{\psi}^I = (\psi^I)^\dagger \gamma^0$ where $(\psi^I)^\dagger$ is the complex conjugate of the transpose
of ψ_J viewed as a four-component vector. The indices I, J are those of a
representation of the gauge group.

If one constructs the Hamiltonian theory of this action one finds that
the canonical variables are ψ_I and its canonically conjugate momentum is
$(\psi^I)^\dagger$. From here one can quantize and arrive at a "connection representation" in which wavefunctions are labeled by the Yang–Mills connection
and the spinor field ψ_I, $\Psi[A, \psi]$.

One would like to find an analogue of the loop representation. In order
to do this, one wants to introduce a transform in which one expands the
wavefunctions of the connection representation in terms of a basis of gauge
invariant quantities. The natural quantities that arise in this context are
holonomies along open paths with fermions at their ends,

$$W(\pi_x^y) = (\psi^I)^\dagger H(\pi_x^y)_I^J \psi_J, \qquad (6.102)$$

where we have used the letter W to stress the analogy with the Wilson
loop. Notice that W has two Dirac indices which we omit, since we do not

assume a contraction between $(\psi^I)^\dagger$ and ψ_J on the Dirac indices, which would be too restrictive.

Notice that we immediately face a difficulty in the sense that the above quantity depends on both the configuration variable and its conjugate momentum. We therefore cannot use it to expand the wavefunctions in the connection representation in terms of it. There do not exist, in general, natural invariants associated with a single open path that are functions only of the configuration variables. For example, in the Abelian case (QED) the only gauge invariant quantity has the form (6.102).

For particular gauge groups there are different alternatives for tackling this problem. For instance, one can make contractions (if the gauge group is special, like in $SU(N)$) with the Levi-Civita symbol in the gauge group and obtain gauge invariant quantities that only depend on the configuration variables. For example, for $SU(2)$ one can construct an object depending on a single path

$$W(\pi_x^y) = \psi_I(x)\epsilon^{IJ}H(\pi_x^y)_J^K\psi_K(y) \tag{6.103}$$

and we will see in chapter 10 a detailed discussion of the resulting representation in the case of fermions interacting with gravity.

Unfortunately for $SU(N)$ with $N > 2$ the Levi–Civita symbol has more than two indices and one is forced to consider more than a single path in order to construct an invariant. For instance, for $SU(3)$,

$$W(\pi_x^y,\eta_x^z,\gamma_x^w) = \epsilon_{LMN}H(\pi_x^y)_L^I H(\eta_x^z)_M^J H(\gamma_x^y)_N^K\psi_I(y)\psi_J(z)\psi_K(w). \tag{6.104}$$

The above object corresponds naturally to physical excitations of the theory in the confining phase. It represents a baryon constructed as three quarks at the ends of three gluon lines that join at the point x.

Evidently, constructing a representation in terms of the above objects is more complicated than we expected. It also leads to completely different representations, even at the most basic kinematical level, for the different gauge groups. Notice also that the above construction does not work for the simplest case, that of a $U(1)$ gauge theory.

A possibility for solving this problem, which has not been explored, would be to decompose the Dirac spinors in their up and down components and construct a representation with wavefunctions that are functionals of the connection, $(\psi_{\rm up}^I)^\dagger$ and $\psi_{I\,\rm down}$. One can then construct gauge invariants that only depend on the configuration variables that are based on a single open path.

All this has led to the use of a different approach for the inclusion of fermions in the loop representation of gauge theories, inspired by the last observation about decomposing the Dirac spinor into its different components [121, 122]. The resulting procedure makes use of the staggered fermion technique introduced by Susskind in the context of lattice gauge theories [120]. This technique arose as a solution of the "fermion doubling" problem that is present in lattice gauge theories.

We will only present a discussion of this technique in a simplified context, that of a free theory in one spatial dimension. In that case, Dirac spinors have only two components (ψ_1, ψ_2). The spinors can be group-valued, but since we are not considering interactions this does not play any role, so we drop the group index in the subsequent discussion. The Dirac matrices in one dimension are given by $\gamma^0 = \sigma_3$ and $\alpha = \gamma^0\gamma^1 = \sigma_1$, where σ_i are the Pauli matrices and the (massless) Dirac equation is

$$\frac{\partial\psi}{\partial t} = -\alpha\frac{\partial\psi}{\partial x}. \tag{6.105}$$

If one now considers a one-dimensional lattice of spacing a the discretized equation is

$$\dot{\psi}(n) = -\frac{i\alpha}{2a}(\psi(n+1) - \psi(n-1)). \tag{6.106}$$

The solution of the continuum equation is given by plane waves of the form $\exp(ik_1 x - k_0 t)$, which lead to the eigenvalue problem $k_0\psi = k\alpha\psi$. The solution of the eigenvalue problem leads to a dispersion relation $k_0 = \pm k_1$. The discrete equation, on the other hand, has solutions of the form $\exp(ik_1 na - k_0 t)$. In this case k_1 takes a discrete set of values $|k_1| = \pi m/Na$ where N is the number of lattice sites and $m \leq N$. This corresponds to a Brillouin zone of $|k| \leq \pi/a$. The resulting eigenvalue problem is

$$k_0\psi = \alpha\frac{\sin(ka)}{a}\psi, \tag{6.107}$$

which leads to a dispersion relation $k_0 = \pm\sin(k_1 a)/a$.

There are two values of $k_1 a$ that lead to a continuum limit, $k_1 a = 0$ and $k_1 a = \pm\pi$. For a given value of k_0 close to zero, there are two values of k_1 allowed by the dispersion relation, each close to the two values of $k_1 a$ that lead to continuum limits. In one case the corresponding k_1 is positive and in the other negative. This is the root of the fermion doubling problem in the lattice: in the continuum limit one gets two fermions moving in opposite directions.

The staggered fermion technique consists in putting the different components of the Dirac spinor in different lattice positions. For the one-dimensional case we are considering this amounts to putting the two components in alternating positions in the lattice. In $3 + 1$ dimensions it is considerably more complicated, since one has to double each dimension of the lattice and therefore there is an eight-fold increase in the components.

The end result is that the up components lie at the even sites and the down components at the odd sites. See reference [120] for more details.

After staggering the lattice position of the two components of the Dirac fermion, the discretized Dirac equation reads in components,

$$\dot{\psi}_1(n) = \frac{1}{2a}(\psi_2(n+1) - \psi_2(n-1)), \tag{6.108}$$

$$\dot{\psi}_2(n) = \frac{1}{2a}(\psi_1(n+1) - \psi_1(n-1)). \tag{6.109}$$

We now introduce a field $\phi(n)$ defined by

$$\phi(n) = \left\{ \begin{array}{l} \psi_1(n) \text{ for } n \text{ even}, \\ \psi_2(n) \text{ for } n \text{ odd}, \end{array} \right. \tag{6.110}$$

in terms of which we can rewrite the Dirac equation as

$$\dot{\phi}(n) = \frac{1}{2a}(\phi(n+1) - \phi(n-1)). \tag{6.111}$$

We therefore see that the resulting equation (6.111) is equivalent to the original Dirac equation but with a double lattice spacing. This translates in terms of the momentum space into a reduction of the Brillouin zone to half its original size, i.e, $|k| \leq \pi/(2a)$. This excises from the dispersion relation the second continuum limit point.

With this idea in hand, we are now in a position to return to the main argument which was to define gauge invariant quantities depending only on configuration variables to introduce a geometric formulation for Yang–Mills theories interacting with fermions. To do that one considers as configuration variables the Dirac fields $\psi(y_{\text{odd}})$ at the odd sites and their conjugate momenta $\psi(x_{\text{even}})^\dagger$ at the even sites. One introduces the following quantities:

$$W(\pi_x^y) = \psi(x_{\text{even}})^\dagger H(\pi_x^y)\psi(y_{\text{odd}}), \tag{6.112}$$

in terms of which one can define a transform to a representation purely in terms of paths. In this representation one can now realize the action of physical excitations, such as the baryonic excitation (6.104). New Mandelstam identities arise relating baryonic excitations and open-path mesonic excitations. We will not present the details here, the reader is referred to reference [121].

6.5 Conclusions

We have seen several examples of the formulation of gauge theories in the lattice in terms of loop representations. It was shown that practical calculations of excitation energies and observables are feasible in the language of loops. The main advantage is that the formulation is gauge invariant and the action of the operators admits very simple geometric formulations in the lattice. In the case of the inclusion of fermions through the use of open paths one does not need to introduce Grassmann variables, which leads to computational economies. The main drawback is that the basis of loops grows very rapidly with the lattice size and since the description is Hamiltonian one does not have at hand statistical methods, like the Monte Carlo techniques, to deal efficiently with a large number of degrees of freedom. The use of cluster techniques, as we have seen, allows us with relative simplicity to obtain a complete approximate description of the phase diagram of theories, although it is not a systematic approximation procedure.

7

Quantum gravity

7.1   Introduction

There have been many different attempts to provide a quantum description of gravitational phenomena. Although there is at present no immediate experimental evidence of quantum effects of the gravitational field, it is expected on general grounds that at sufficiently high energies quantum effects may be relevant. The fact that quantum field theories in general involve virtual processes of arbitrarily high energies may suggest that an understanding of quantum gravity may be needed to provide a complete picture of quantum fields. Ultraviolet divergences arise as a consequence of an idealization in which one expects the field theory in question to be applicable up to arbitrarily high energies. It is generally accepted that for high energies gravitational corrections could play a role. On the other hand, classical general relativity predicts in very general settings the appearance of singularities in which energies, fields and densities become intense enough to suggest the need for quantum gravitational corrections.

In spite of the many efforts invested over the years in trying to apply the rules of quantum mechanics to the gravitational field, most attempts have remained largely incomplete due to conceptual and technical difficulties. There are good reasons why the merger of quantum mechanics and gravity as we understand them at present is a difficult enterprise. We now present a brief and incomplete list of the issues involved. The reader should realize that every one of these problems is to some extent currently being actively investigated by several groups and some of these difficulties could eventually be overcome.

- It is not clear which theory of gravity to start from at a classical level. The fact that general relativity is the simplest viable theory does not necessarily mean it is appropriate for quantization. Some people argue that a successful theory of gravity should also incorporate all other interactions in nature in a unified fashion.

• The rules of quantum mechanics, as we know them at present, may not be applicable to systems without a defined notion of time, as is the case for generally covariant theories of gravity.

• It is not clear that a continuous description of spacetime and fields will be enough to provide a framework to quantize gravity. It may be that the description provided by general relativity is an "effective" macroscopic theory with an underlying, more fundamental theory. As in the case of the Fermi model, quantizing the effective theory can be misleading.

• There is a tendency to incorporate into quantum descriptions of the gravitational field issues related to the quantization of the universe as a whole ("quantum cosmology"). As a consequence it is not clear what the measurement process exactly is and how to define observers and measurable quantities for the theory.

As well as these more fundamental problems, several attempts to quantize the gravitational field have encountered more specific difficulties. Again, we present just a brief list and many of these difficulties are currently being studied by several researchers.

• Attempts based on perturbation theory, in which one starts with a fixed background metric and quantizes deviations from it have led to nonrenormalizable theories. This has sometimes been perceived as a pathology of the classical theory of gravity chosen, and has motivated the study of quantizations of theories other than Einstein's, most notably higher order theories, supersymmetric theories and theories based on strings. Another point of view is to notice that these attempts ignore the rich non-linear, geometric and topological nature of general relativity. This suggests that from the beginning they offered little hope of dealing appropriately with the fundamental difficulties listed above. It is therefore not entirely surprising that they encounter difficulties at some point.

• In recent years there has been great interest in considering string theories as the fundamental theory of particles and gravity. Apart from possibly being able to unify all interactions, string theory was expected to be perturbatively much better behaved than regular field theories based on point particles. In spite of this better behavior, which makes each term in the perturbation expansion finite, the series diverges rather badly. Again one could view this as a failure of perturbative techniques and it is still possible that a non-perturbative theory of strings could yield the correct quantum theory of gravity.

• The use of path integral quantization techniques has been advocated for gravity since it is naturally covariant and allows us to consider in a dynamical fashion the geometric and topological nature of gravity. With the exception of some mini-superspace examples, several technical difficulties have prevented the application of these techniques to gravity.

Among them is the lack of understanding of the appropriate gauge invariant measure of integration in the path integral, the unboundedness of the Einstein action and the inapplicability of Wick rotation techniques without a notion of time.

• Canonical techniques have been applied to quantum gravity for quite some time. All the discussion in this book will be focused on this kind of approach and we will discuss in detail some of the difficulties that appear. Among the difficulties is the choice of a natural time variable in the theory, the construction of gauge invariant observables, the imposition of an appropriate Hilbert space structure compatible with regularized constraints enforcing gauge invariance and the fact that the spacetime topology is fixed.

• Other more radical approaches try to start from theories that are fundamentally different from general relativity or other field theories, usually with some degree of discreteness, and try to recover the usual theories in some limit. The main difficulty is that they are faced with the burden of checking that all desirable features of the usual field theories are reproduced and that no unexpected behaviors are introduced.

In this book we will concentrate on a very specific approach to quantum gravity: we will apply canonical quantization techniques to general relativity.

The use of canonical quantization techniques is suggested by the results on Yang–Mills theories that we introduced in chapter 5. As we saw, one can obtain considerable progress in the canonical formulation using loop variables. Although there has been recent progress on the use of loop techniques at a covariant level, most of the emphasis up to now has been on canonical approaches. The choice of general relativity (in four spacetime dimensions) as the theory of gravity to be quantized is based on the fact that it is the simplest purely geometric theory available and it should serve well as a testbed for quantization techniques, even if it ultimately is superseded by another theory.

The canonical approach to quantum general relativity had been considered extensively in the past and had several complications. As we will see, general relativity is a constrained system and the constraint equations turned out to be unmanageable at a quantum level. The situation changed a few years ago with the introduction of a new set of variables that has allowed a significant amount of progress. In particular, the new variables cast general relativity in a form that is similar to that of a Yang–Mills theory and is therefore quite suited to the techniques we have developed in this book.

The outline of this chapter is as follows. In the first section we recall the traditional Hamiltonian formulation of general relativity. In the next section we work out the new canonical formulation. In the last section we use the new Hamiltonian formulation to quantize canonically the theory as if it were a Yang–Mills theory, obtaining a connection representation. This will be the starting point for the development of the loop representation in the next chapter.

7.2 The traditional Hamiltonian formulation

7.2.1 Lagrangian formalism

General relativity is a theory of gravity in which the gravitational interaction is accounted for by a deformation of spacetime. The fundamental variable for the theory is the spacetime metric g_{ab}. The action for the theory is given by

$$S = \int d^4x \sqrt{-g} R(g_{ab}) + \int d^4x \sqrt{-g} \mathcal{L}(\text{matter}), \qquad (7.1)$$

where g is the determinant of g_{ab}, $R(g_{ab})$ is the curvature scalar and we have also included a term to take into account possible couplings to matter, although we will largely concentrate on the vacuum case. The equations of motion are obtained by varying the action with respect to the spacetime metric,

$$\frac{\delta S}{\delta g^{ab}} = 0 \quad : \quad R_{ab} - \tfrac{1}{2} R g_{ab} = \frac{\delta S_{\text{matter}}}{\delta g^{ab}}, \qquad (7.2)$$

and are the well known Einstein equations. The action is invariant under diffeomorphisms of the spacetime manifold (which can also be viewed as invariance under coordinate transformations). We will see that this symmetry is intimately tied into the structure of the Einstein equations.

7.2.2 The split into space and time

The standard Hamiltonian formulation for general relativity was developed by Arnowitt, Deser and Misner (ADM) [124]. To cast the theory in a canonical form, we need to split spacetime into space and time. Without a notion of time, there is no notion of evolution and therefore no Hamiltonian in the traditional sense. This may seem odd at first; one of the main points of general relativity is to cast space and time on the same footing and this approach seems to separate them again. We will see that the issue is more subtle. Although the canonical formalism manifestly breaks the spacetime covariance of the theory by singling out a particular time direction, in the end the formalism itself will tell us that it really did not matter which direction of time we took to begin with. The covariance is restored by certain relations that appear in the canonical formulation and the time picked is a "fiducial" one for construction purposes only. This is a similar situation to that which one faces when one formulates the theory of a relativistic particle canonically. We will see more details of this immediately.

We consider a spacetime ^4M with metric g_{ab} that has topology $^3\Sigma \times R$ where $^3\Sigma$ is a space-like surface with respect to g_{ab}. We will assume $^3\Sigma$ is a Cauchy surface, i.e., a surface such that the light cones emanating from it span all the spacetime to the future of $^3\Sigma$. Associated with the foliation is a time-like, future directed, vector t^a and a function on spacetime t such that its level surfaces coincide with the leaves of the foliation $^3\Sigma_t$ and such that $t^a \partial_a t = 1$. This vector field can be interpreted as describing the "flow of time" among the leaves of the foliation, but it should be realized that it has been introduced fiducially and cannot be connected with the measurements of any clock until we have a metric appropriately determined by the Einstein equations. We introduce a unit vector field n^a normal to the foliation. In combination with the spacetime metric this defines a unique, positive-definite spatial metric on the three-dimensional slice,

$$q_{ab} := g_{ab} + n_a n_b. \tag{7.3}$$

Notice that since we have a spacetime metric all indices are raised and lowered with it. The vector field t^a can be decomposed in components normal and tangential to $^3\Sigma$ as

$$t^a = N n^a + N^a, \tag{7.4}$$

where the scalar N is known as the "lapse" and N^a is a vector on $^3\Sigma$ and is usually referred to as the "shift" vector. The decomposition can be seen in figure 7.1. It is clear that the quantities N and N^a contain information about the particular foliation rather than information intrinsic to spacetime.

From the information contained in q_{ab}, N^a and N one can reconstruct the spacetime metric,

$$g^{ab} = q^{ab} - n^a n^b, \tag{7.5}$$

where n^a can be easily constructed from N and N^a and q^{ab} is the inverse of q_{ab} in the tangent space to $^3\Sigma_t$ (see Wald [123]). In fact, one can explicitly choose coordinates (t, x^i) such that the metric reads

$$ds^2 = -N^2 dt^2 + q_{ij}(dx^i + N^i dt)(dx^j + N^j dt), \tag{7.6}$$

where q_{ij} and N^i are the coordinate components of q_{ab} and N^a. We therefore see that the lapse has the interpretation of the "time time"

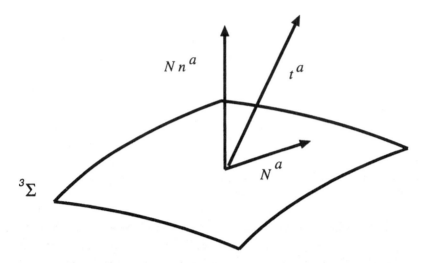

Fig. 7.1. The foliation introduced for the canonical formulation of general relativity.

component of the metric and the shift represents off-diagonal "time space" components.

An important quantity in the canonical description is the extrinsic curvature of the surface $^3\Sigma$. This is defined by

$$K_{ab} := q_a^c q_b^d \nabla_c n_d, \tag{7.7}$$

where ∇ is the torsion-free derivative compatible with g_{ab}. The extrinsic curvature measures the rate of change of the spatial metric along the congruence defined by n^a and therefore gives an idea of the "bending" of the spatial surfaces in spacetime. One can easily check that

$$K_{ab} = \tfrac{1}{2}\mathcal{L}_{\vec{n}} q_{ab}, \tag{7.8}$$

and also that

$$\dot{q}_{ab} := \mathcal{L}_{\vec{t}} q_{ab} = 2N K_{ab} + \mathcal{L}_{\vec{N}} \, q_{ab}. \tag{7.9}$$

That is, the extrinsic curvature allows us to give a measure of the variation of the three-dimensional metric with respect to the fiducial time introduced by the foliation, i.e., K_{ab} essentially contains the information about the "time derivative" of q_{ab}.

We have introduced up to now a series of quantities defined on the spatial surface in terms of which we can reconstruct the spacetime metric and its time derivatives. We now proceed to rewrite the Einstein action in terms of these variables (see reference [123]),

$$S = \int dt L, \tag{7.10}$$

$$L = \int d^3x N \sqrt{q}(^3R + K_{ab}K^{ab} - K^2), \tag{7.11}$$

where q is the determinant of q_{ab} in a basis adapted to $^3\Sigma$ such that $\sqrt{-g} = N\sqrt{q}$, ^3R is the (intrinsic) curvature of the spatial metric and $K := K_{ab}q^{ab}$. To achieve this particular form of the action surface terms have to be added appropriately. In this book we will always deal with compact three-surfaces (like those that arise in some cosmologies) and will therefore ignore these issues. If one wants to consider non-compact spatial slices (as is needed in asymptotically flat spacetimes like those that describe stars and black holes) one can achieve the same form of the action by imposing appropriate boundary conditions at infinity. This can be done in a straightforward manner (see reference [123]).

We now have the action of general relativity in a reasonable form to allow a canonical formulation. We have it expressed in terms of variables that are functions of "space" and that "evolve in time". This is the usual setup for constructing canonical formulations.

We pick as the canonical variable the three-metric q_{ab} and compute its conjugate momentum,

$$\tilde{\pi}_{ab} := \frac{\delta L}{\dot{q}_{ab}} = \sqrt{q}(K^{ab} - Kq^{ab}) \tag{7.12}$$

and we see that the conjugate momentum to the metric is essentially given by the extrinsic curvature ("time derivative").

The variables N and N^a have vanishing conjugate momenta, since the action (7.11) does not contain time derivatives of them. This implies the canonical formulation will have constraints.

We can now perform the Legendre transform and obtain the Hamiltonian of the theory

$$H(\tilde{\pi}, q) = \int d^3x(\tilde{\pi}^{ab}q_{ab} - \tilde{\mathcal{L}}), \tag{7.13}$$

where $\tilde{\mathcal{L}}$ is the Lagrangian density ($L = \int d^3x\tilde{\mathcal{L}}$). Replacing $\dot{q}$ in terms of $\tilde{\pi}$ one gets

$$H(\tilde{\pi}, q) = \int d^3x(N(-\sqrt{q}R + (\sqrt{q})^{-1}(\tilde{\pi}^{ab}\tilde{\pi}_{ab} - \tfrac{1}{2}\tilde{\tilde{\pi}}^2)) - 2N^b D_a \tilde{\pi}^a_b, \tag{7.14}$$

where $\tilde{\pi} = \tilde{\pi}^{ab}q_{ab}$ (and squared has double density weight) and D_a is the torsion-free covariant derivative compatible with q_{ab}.

The variables q_{ab} and $\tilde{\pi}^{ab}$ have the straightforward simplectic structure of conjugate pairs,

$$\{q_{ab}(x), \tilde{\pi}^{cd}(y)\} = \delta_a^c \delta_b^d \delta(x - y). \tag{7.15}$$

7.2.3 Constraints

Having cast the theory in a Hamiltonian form, let us step back a minute and analyze the formalism that we have built. We started from a four-dimensional metric g^{ab} and we now have in its place the three-dimensional q^{ab} and the "lapse" and "shift" functions N and N^a. We defined a conjugate momentum for q_{ab}. However, notice that nowhere in the formalism does a time derivative of the lapse or shift appear. That means their conjugate momenta are zero. That is, our theory has constraints. In fact, if we rewrite the action using the expression for the Hamiltonian given above, we get

$$S = \int dt \int d^3x \, ((\tilde{\pi}_{ab}\dot{q}^{ab} + \underset{\sim}{N}(-qR + (\tilde{\pi}^{ab}\tilde{\pi}_{ab} - \tfrac{1}{2}\tilde{\pi}^2))) - 2N^b D_a \tilde{\pi}_b^a), \tag{7.16}$$

where the inverse-densitized lapse $\underset{\sim}{N}$ is defined as $(\sqrt{q})^{-1}N$. If we vary the action with respect to $\underset{\sim}{N}$ and N^b in order to get their respective equations of motion, we get four expressions, functions of $\tilde{\pi}$ and q which should vanish identically, and are usually called $\tilde{C}^a$ and $\underset{\sim}{\tilde{\mathcal{H}}}$,

$$\tilde{C}_a(\pi, q) = 2D_b \tilde{\pi}_a^b, \tag{7.17}$$

$$\underset{\sim}{\tilde{\mathcal{H}}}(\pi, q) = -\tilde{q}R + (\tilde{\pi}^{ab}\tilde{\pi}_{ab} - \tfrac{1}{2}\tilde{\pi}^2). \tag{7.18}$$

For calculational simplicity, these equations are usually "smoothed out" with arbitrary test fields on the three-manifold, $C(\vec{N}) = \int d^3x N^a \tilde{C}_a$, $\mathcal{H}(\underset{\sim}{N}) = \int d^3x \underset{\sim}{N} \tilde{\mathcal{H}}$.

These equations are "instantaneous" laws, i.e., they must be satisfied *on each hypersurface*. They tell us that if we want to prescribe data for a gravitational field, not every pair of $\tilde{\pi}$ and q will do; equations (7.17), (7.18) should be satisfied. The counting of degrees of freedom is done in the following way: we have a 12-dimensional phase space. In that space we have four constraints and we can fix four gauge conditions. We are therefore left with a four-dimensional constraint-free phase space, which gives two degrees of freedom. (General relativity being a field theory the previous counting holds per each point of the spatial surface.)

These equations have the same character as the Gauss law has for electromagnetism, which tells us that not any vector field would necessarily work as an electric field, it must have vanishing divergence in vacuum. As is well known, the Gauss law appears as a consequence of the $U(1)$ invariance of the Maxwell equations. An analogous situation appears here.

To understand this, consider the Poisson bracket of any quantity with the constraint $C(\vec{N})$. It is straightforward to check that (exercise)

$$\{f(\tilde{\pi},q),C(\vec{N})\} = \mathcal{L}_{\vec{N}}f(\tilde{\pi},q). \qquad (7.19)$$

Therefore we see that the constraint $C(\vec{N})$ "Lie drags" the function $f(\pi,q)$ along the vector $\vec{N}$. Technically, it is the infinitesimal generator of diffeomorphisms of the three-manifold in phase space. As the Gauss law (in the canonical formulation of Maxwell's theory) is the infinitesimal generator of $U(1)$ gauge transformations, the constraint here is the infinitesimal generator of spatial diffeomorphisms. This clearly shows why we have this constraint in the theory: it is the canonical representation of the fact that the theory is invariant under spatial diffeomorphisms. The constraint $C(\underset{\sim}{N})$ is analogously associated with the invariance under spacetime diffeomorphisms of general relativity, it is related to the time reparametrization invariance of the theory.

We can now work out the equations of motion of the theory by either varying the action with respect to q^{ab} and $\tilde{\pi}_{ab}$ or taking the Poisson bracket of these quantities with the Hamiltonian constraint.

The above system of constraints is first class (for the definition of this see chapter 3). Computing the Poisson algebra one gets

$$\{C(\vec{N}),C(\vec{M})\} = C(\mathcal{L}_{\vec{M}}\vec{N}), \qquad (7.20)$$

$$\{C(\vec{N}),\mathcal{H}(\underset{\sim}{M})\} = \mathcal{H}(\mathcal{L}_{\vec{N}}\underset{\sim}{M}), \qquad (7.21)$$

$$\{\mathcal{H}(\underset{\sim}{N}),\mathcal{H}(\underset{\sim}{M})\} = C(\vec{K}), \qquad (7.22)$$

where the vector $\vec{K}$ is defined by $K^a = qq^{ab}(\underset{\sim}{N}\partial_b\underset{\sim}{M} - \underset{\sim}{M}\partial_a\underset{\sim}{N})$. The reader should notice, however, that the algebra is not a true Lie algebra, since one of the structure constants (the one defined by the last equation) is not a constant but depends on the fields q^{ab} (through the definition of the vector $\vec{K}$). At a quantum mechanical level this will imply that the fields should appear to the left of the constraint in the appropriate factor ordering to ensure consistency.

7.2.4 Quantization

Having cast the theory in a canonical form, we can now proceed to a canonical quantization, following the general quantization scheme outlined in chapter 3. One picks as canonical algebra the pair q_{ab} and $\tilde{\pi}^{ab}$, and represents them as quantum operators acting on a set of wavefunctionals $\Psi[q]$ in the obvious fashion: $\hat{q}_{ab}$ as a multiplicative operator and $\hat{\pi}^{ab} = -i\delta/\delta q_{ab}$. One wants the wavefunctions to be invariant under the symmetries of the theory. As we saw the symmetries are represented in this language as constraints. The requirement that the wavefunctions be annihilated by the constraints (promoted to operatorial equations) implements the symmetries at the quantum level. The wavefunctionals that are annihilated by the constraints are the physical states of the theory. Notice that we do not yet have a Hilbert space. One needs to introduce an inner product on the space of physical states in order to compute expectation values and make physical predictions. Only at this point does one have an actual Hilbert space. How to find this inner product is not prescribed by standard canonical quantization (we will discuss this in the next section). Under this inner product the physical states should be normalizable. The expectation values, by the way, only make sense for quantities that are invariant under the symmetries of the theory (quantities that classically have vanishing Poisson brackets with all the constraints). We call them physical observables. For the gravitational case *none* is known for compact spacetimes (we will return to this issue later). The observables of the theory should be self-adjoint operators with respect to the inner product in order to yield real expectation values.

It is at the level of the constraints that we run into trouble. We have to promote the constraints we discussed in the last subsection to quantum operators. This in itself is a troublesome issue, since general relativity being a field theory, issues of regularization and factor ordering appear. One can, — at least formally — find factor orderings in which the diffeomorphism constraint becomes the infinitesimal generator of diffeomorphisms on the wavefunctions. Therefore the requirement that a wavefunction be annihilated by it just translates itself in the fact that the wavefunction has to be invariant under diffeomorphisms. This is not difficult to accomplish (formally!). One simply requires that the wavefunctions be not actually functionals of the three-metric q^{ab}, but of the "three-geometry" (i.e., meaning the properties of the three-metric invariant under diffeomorphisms). Thus, what we are saying is just a restatement of the fact that the functional should be invariant under diffeomorphisms. One can come up with several examples of functionals that meet this requirement. The real trouble appears when we want the wavefunctions to be annihilated by the Hamiltonian constraint. This constraint does not have a simple geometrical interpretation in terms of three-dimensional quantities (remember that the idea that it represents "time evolution" does not help here, since we are always talking about equations that hold *on the three-surface* without any explicit reference to time). Therefore we are just forced to proceed directly: to promote the constraint to a wave equation, use some factor ordering (hopefully with some physical motivation), pick some regularization and try to solve the resulting equation (the Wheeler–DeWitt equation). It turns out that this task has never been accomplished in general (it has been in simplified mini-superspace examples). One of the difficulties encountered in this direction is the fact that the constraint is a non-polynomial function of the basic variables (remember it involves the scalar curvature, a non-polynomial function of the three-metric).

Therefore the program of canonical quantization stalls here. It could well be that the constraints do not admit a consistent factor ordering and the quantum theory may not exist. Having been unable to find the physical states of the theory we are in a bad position to introduce an inner product (since we do not know on what space of functionals to act) and actually make physical predictions. This issue is compounded by the fact that we do not know any observables for the system, which puts us in a more clueless situation with respect to the inner product. This state of affairs had already been reached in the work of DeWitt in the 1960s [125] and little improvement has been made until recently. We will see in the next section that the use of a new set of variables improves the situation with respect to the Hamiltonian constraint, giving hope of maybe allowing us to attack the problem of the inner product. Moreover, we will see that the new formulation allows a natural contact with the main ideas of this book.

7.3   The new Hamiltonian formulation

As we saw in the previous section, the traditional canonical approach to quantum general relativity faces serious obstructions at a very early stage. On the other hand, as we saw, the canonical quantization of Yang–Mills theories has been more successful. For many years efforts were directed towards casting general relativity in such a way that it resembled a Yang–Mills theory more with the hope that quantization techniques and ideas developed for the latter would become applicable to general relativity. This led to several attempts that started from a gauge theory approach with the aim of deriving a theory of gravity based on gauging a particular symmetry group. This, in general, led to new theories of gravity that involved higher order terms in the Hilbert action [126]. There is another possible approach: to keep the Einstein equations for the gravitational theory but reinterpret them as statements about a connection instead of a metric. The simplest way to achieve such a reformulation is to consider the Palatini variational principle. In this, one varies the metric and the spacetime connection as independent variables. One retrieves the Christoffel definition of the connection as one of the field equations. Attempts to formulate gravity in terms of connections in this way go back to Einstein and Dirac in the 1940s. In order to have a formalism as close as possible to a usual Yang–Mills formulation, one could take the Palatini principle based on tetrads and $SO(3,1)$ connections. This route was studied in some detail by Kijowski [127]. Unfortunately, the canonical theory based on such connections has second class constraints (in 3+1 dimensions). When one eliminates these, non-polynomialities are introduced and one is led back to the traditional Hamiltonian formulation [61]. It is remarkable that in 2+1 dimensions one actually can formulate the theory in terms of connections, although historically this was realized later and through a different construction. We will review the 2+1 case later.

In 3+1 dimensions, the only successful attempt to obtain a canonical theory in terms of a connection that yields first class constraints is that due to Ashtekar [51]. It is based on the use of self-dual connections. Not only do the constraints remain first class but they are relatively simple polynomial functions. The price to be paid is that the self-dual connections are complex. In the next subsections we will develop this formalism. The treatment will follow closely the book by Ashtekar [2], we direct the reader to it for extensive details.

7.3.1 Tetradic general relativity

To introduce the new variables, we first need to introduce the notion of tetrads. A tetrad is a vector basis in terms of which the metric of spacetime looks locally flat,

$$g_{ab} = e_a^I e_b^J \eta_{IJ}, \tag{7.23}$$

where $\eta_{IJ} = \mathrm{diag}(-1,1,1,1)$ is the Minkowski metric, and equation (7.23) simply expresses that g_{ab}, when written in terms of the basis e_a^I, is locally flat. If spacetime were truly flat, one could perform such a transformation globally, integrating the basis vectors into a coordinate transformation $e_a^I = \partial x^I / \partial x'^a$. In a curved spacetime these equations cannot be integrated and the transformation to a flat space only works locally, the flat space in question being the "tangent space". From equation (7.23) it is immediate to see that given a tetrad, one can reconstruct the metric of spacetime. One can also see that although g_{ab} has only ten independent components, the e_a^I have sixteen. This is due to the fact that equation (7.23) is invariant under Lorentz transformations on the indices $I, J \dots$. That is, these indices behave as if existing in flat space. In summary, tetrads have all the information needed to reconstruct the metric of spacetime but there are extra degrees of freedom in them, and this will have a reflection in the canonical formalism.

7.3.2 The Palatini action

We now write the Einstein action in terms of tetrads. We introduce a covariant derivative via $D_a K_I = \partial_a K_I + \omega_{aI}{}^J K_J$. Here $\omega_{aI}{}^J$ is a Lorentz connection (its associated covariant derivative annihilates the Minkowski metric). We define a curvature by $\Omega_{ab}{}^{IJ} = \partial_{[a}\omega_{b]}{}^{IJ} + [\omega_a, \omega_b]^{IJ}$, where $[\,,]$ is the commutator in the Lorentz Lie algebra. The Ricci scalar of this curvature can be expressed as $e^a_I e^b_J \Omega_{ab}^{IJ}$ (indices I, J are raised and lowered with the Minkowski metric). The Einstein action can be written as

$$S(e, \omega) = \int d^4x \; e \; e^a_I e^b_J \Omega_{ab}^{IJ}, \qquad (7.24)$$

where e is the determinant of the tetrad (equal to $\sqrt{-g}$).

We will now derive the Einstein equations by varying this action with respect to e and ω as independent quantities. To take the metric and connection as independent variables in the action principle was first considered by Palatini [128].

As a shortcut to performing the calculation (this derivation is taken from reference [2]), we introduce a (torsion-free) connection compatible with the tetrad via $\nabla_a e^b_I = 0$. The difference between the two connections we have introduced is a field $C_{aI}{}^J$, defined by $C_{aI}{}^J V_J = (D_a - \nabla_a)V_I$. We can compute the difference between the curvatures (R_{ab}^{IJ} is the curvature of ∇_a), $\Omega_{ab}{}^{IJ} - R_{ab}{}^{IJ} = \nabla_{[a}C_{b]}{}^{IJ} + C_{[a}{}^{IM}C_{b]M}{}^J$. The reason for performing this intermediate calculation is that it is easier to compute the variation by reexpressing the action in terms of ∇ and $C_a{}^{IJ}$ and then noting that the variation with respect to $\omega_a{}^{IJ}$ is the same as the variation with respect to C_a^{IJ}. The action therefore is

$$S = \int d^4x \; e \; e^a_I e^b_J (R_{ab}{}^{IJ} + \nabla_{[a}C_{b]}{}^{IJ} + C_{[a}{}^{IM}C_{b]M}{}^J). \qquad (7.25)$$

The variation of this action with respect to $C_a{}^{IJ}$ is easy to compute: the first term simply does not contain $C_a{}^{IJ}$ so it does not contribute. The second term is a total divergence (notice that ∇ is defined so that it annihilates the tetrad), the last term yields $e^{[a}_M e^{b]}_N \delta^M_{[I} \delta^K_{J]} C_{bK}{}^N$. It is easy to check that the prefactor in this expression is non-degenerate and therefore the vanishing of this expression is equivalent to the vanishing of $C_{bK}{}^N$. So this equation basically tells us that ∇ coincides with D when acting on objects with only internal indices. Thus the connection D is completely determined by the tetrad and Ω coincides with R (some authors refer to this fact as the vanishing of the torsion of the connection). We now compute the second equation, straightforwardly varying with respect to the tetrad. We get (after substituting $\Omega_{ab}{}^{IJ}$ by $R_{ab}{}^{IJ}$ as given by the previous equation of motion)

$$e_I^c R_{cb}{}^{IJ} - \tfrac{1}{2} R_{cd}{}^{MN} e_M^c e_N^d e_b^J = 0, \qquad (7.26)$$

which, after multiplication by e_{Ja} just tells us that the Einstein tensor $R_{ab} - \tfrac{1}{2} R g_{ab}$ of the metric defined by the tetrads vanishes. We have therefore proved that the Palatini variation of the action in tetradic form yields the usual Einstein equations.

There is a difference between the first order (Palatini) tetradic form of the theory and the usual one. One sees that a solution to the Einstein equations we presented above is simply $e_J^b = 0$. This solution would correspond to a vanishing metric and is therefore forbidden in the traditional formulation since quantities, such as the Ricci or Riemann tensor are not defined for a vanishing metric. However, the first order action and equation of motion are well defined for vanishing triads. We therefore see that strictly speaking the first order tetradic formulation is a "generalization" of general relativity that contains the traditional theory in the case of non-degenerate triads. We will see this subtlety playing a role in subsequent chapters. It should be noticed that the potential of allowing vanishing metrics in general relativity offers new possibilities for some old questions, since one could envisage the formalism "going through", say, the formation of singularities. It also allows for topology change [129].

Is there any advantage in this formulation over the traditional one? The answer is no. If one performs a canonical decomposition of the first order tetradic action, one finds that the momentum canonically conjugate to the connection is quadratic in the tetrads. The factorizability of the momenta leads to new constraints in the theory that turn out to be second class. If one eliminates them through the Dirac procedure one returns to the traditional formulation [61].

7.3.3 *The self-dual action*

Up to now the treatment has been totally traditional. We will now take a conceptual step that allows the introduction of the Ashtekar variables. We will reconstruct the tetradic formalism of the previous subsection but we will introduce a change. Instead of considering the connection $\omega_a{}^{IJ}$ we will consider its self-dual part with respect to the internal indices and we will call it $A_a{}^{IJ}$, i.e., $iA_a{}^{IJ} = \tfrac{1}{2}\epsilon_{MN}{}^{IJ} A_a{}^{MN}$. Now, to really be able to do this, the connection must be complex (or one should work in an Euclidean signature). Therefore for the time being we will consider *complex general relativity* and we will then specify appropriately how to recover the traditional real theory. The connection now takes values in the (complex) self-dual subalgebra of the Lie algebra of the Lorentz group.

We will propose as action,

$$S(e, A) = \int d^4x \, e \, e_J^a e_K^b F_{ab}{}^{JK}, \qquad (7.27)$$

where $F_{ab}{}^{JK}$ is the curvature of the self-dual connection and it can be checked that it corresponds to the self-dual part of the curvature of the usual connection.

We can now repeat the calculations of the previous subsection for the self-dual case. When one varies the self-dual action with respect to the connection $A_a{}^{IJ}$ one obtains that this connection is the self-dual part of a torsion-free connection that annihilates the triad (if one repeated step by step the previous subsection argument, the self-dual part of $C_a{}^{IJ}$ would vanish). The variation with respect to the tetrad follows along very similar lines except that $\Omega_{ab}{}^{IJ}$ is everywhere replaced by $F_{ab}{}^{IJ}$. The final equation one arrives at again tells us that the Ricci tensor vanishes. Remarkably, the self-dual action leads to the (complex) Einstein equations. This essentially can be explained by the fact that the two actions differ by terms that on-shell are a pure divergence. This implies that the imaginary part of the equations of motion identically vanishes. If one works it out explicitly one finds that this corresponds to the Bianchi identities.

7.3.4 The new canonical variables

As we said before, if one takes the Palatini action principle in terms of tetrads and performs a canonical decomposition, second class constraints appear and one is led back to the traditional formulation. A quite different thing happens if one decomposes the self-dual action. Let us therefore proceed to do the 3+1 split. As we did before, we introduce a vector $t^a = Nn^a + N^a$. Taking the action

$$S(e, A) = \int d^4x \, e \, e_I^a e_J^b F_{ab}{}^{IJ} \qquad (7.28)$$

and defining the vector fields $E_I^a = q_b^a e_I^b$ (where $q_b^a = \delta_b^a + n^a n_b$ is the projector on the three-surface), which are orthogonal to n^a, we have

$$S(e, A) = \int d^4x \, (e \, E_I^a E_J^b F_{ab}{}^{IJ} - 2 \, e \, E_I^a e_J^d n_d n^b F_{ab}{}^{IJ}). \qquad (7.29)$$

We now define $\tilde{E}_I^a = \sqrt{q} E_I^a$, which is a density on the three-manifold. The determinant of the triad can be written as $e = N\sqrt{q}$. We also introduce the vector in the "internal space" induced by n^a, defined by $n_I = e_I^d n_d$. With these definitions, and exploiting the self-duality of $F_{ab}{}^{IJ}$ to write $F_{ab}{}^{IJ} = -i\frac{1}{2}\epsilon^{IJ}{}_{MN} F_{ab}{}^{MN}$, we get

$$S(e, A) = \int d^4x \, (-\tfrac{i}{2} N \tilde{E}_I^a \tilde{E}_J^b \epsilon^{IJ}{}_{MN} F_{ab}{}^{MN} - 2Nn^b \tilde{E}_I^a n_J F_{ab}{}^{IJ}). \qquad (7.30)$$

The action is now written in canonical form and the conjugate variables can be read off directly. The configuration variable is the self-dual connection A_a. The conjugate momentum is the self-dual part of $-i\tilde{E}^a_J \epsilon^J_{MN}$,

$$\tilde{\pi}^a_{MN} = \tilde{E}^a_{[M} n_{N]} - \frac{i}{2}\tilde{E}^a_I \epsilon^I_{MN}. \tag{7.31}$$

Now, in terms of the canonical variables the Lagrangian takes the form

$$\int_\Sigma d^3x \mathrm{Tr}(-\tilde{\pi}^a \mathcal{L}_t A_a + N^a \tilde{\pi}^b F_{ab} - (A \cdot t) D_a \tilde{\pi}^a - \underset{\sim}{N}\tilde{\pi}^a \tilde{\pi}^b F_{ab}), \tag{7.32}$$

where all references to the internal vector n^I have disappeared. The projection of the spacetime connection on the time-like direction $(A \cdot t)$ is arbitrary and acts as a Lagrange multiplier.

Since n_I is not a dynamical variable it can be gauge fixed. We fix $n^I = (1, 0, 0, 0)$ and therefore $\epsilon^{IJKL} n_L = \epsilon^{IJK0}$. Since A^{IJ}_a and $\tilde{\pi}^a_{IJ}$ are self-dual, they can be determined by their $0I$ components. We may therefore define

$$A^i_a = iA^{0I}_a, \quad \tilde{E}^a_i = \tilde{\pi}^a_{0I}, \tag{7.33}$$

where internal indices i, j refer to the $SO(3)$ Lie algebra. In fact, as is well known the self-dual Lorentz Lie algebra is isomorphic to the (complexified) $SO(3)$ algebra

The new variables satisfy the Poisson bracket relations

$$\{A^i_a(x), \tilde{E}^b_j(y)\} = +i\delta^b_a \delta^i_j \delta^3(x-y). \tag{7.34}$$

The constraints may be read off from the Lagrangian (7.32) and take the form

$$\tilde{\mathcal{G}}^i = D_a \tilde{E}^{ai}, \tag{7.35}$$

$$\tilde{C}_a = \tilde{E}^b_i F^i_{ab}, \tag{7.36}$$

$$\underset{\sim}{\tilde{\mathcal{H}}} = \epsilon^{ij}_k \tilde{E}^a_i \tilde{E}^b_j F^k_{ab}, \tag{7.37}$$

and the Hamiltonian is again a linear combination of the constraints.

The last four equations correspond to the usual diffeomorphism and Hamiltonian constraints of canonical general relativity. The first three equations are extra constraints that stem from our use of triads as fundamental variables. These equations, which have exactly the same form as a Gauss law of an $SU(2)$ Yang–Mills theory, are the generators of infinitesimal $SU(2)$ transformations. They tell us that the formalism is invariant under triad rotations, as it should be.

Notice that a dramatic simplification of the constraint equations has occurred. In particular the Hamiltonian constraint is a polynomial function of the canonical variables, of quadratic order in each variable. Moreover, the canonical variables, and the phase space of the theory are exactly those of a (complex) $SU(2)$ Yang–Mills theory. The reduced phase space is actually a subspace of the reduced phase space of a (complex) Yang–Mills theory (the phase space modulo the Gauss law), since general relativity has four more constraints that further reduce its phase space. This resemblance of the formalism to that of a Yang–Mills theory will be the starting point of all the results we will introduce in the rest of the book.

In terms of the new variables, the structure of the constraints is simple enough for the reader to be able to compute the constraint algebra without great effort (this computation can also be carried out with the traditional variables and the results are the same). We only summarize the results here. To express them in a simpler form (and to avoid confusing manipulations of distributions while performing the computations), it is again convenient to smooth out the constraints with arbitrary test fields and to perform some recombinations. We denote

$$\mathcal{G}(N_i) = \int d^3x N_i (\mathcal{D}_a \tilde{E}^a)^i, \tag{7.38}$$

$$C(\vec{N}) = \int d^3x N^b \tilde{E}^a_i F^i_{ab} - \mathcal{G}(N^a A^i_a), \tag{7.39}$$

$$\mathcal{H}(\underset{\sim}{N}) = \int d^3x \underset{\sim}{N} \epsilon^{ij}{}_k \tilde{E}^a_i \tilde{E}^b_j F^k_{ab}, \tag{7.40}$$

and as before the notation is unambiguous. The constraint algebra then reads

$$\{\mathcal{G}(N_i), \mathcal{G}(N_j)\} = \mathcal{G}([N_i, N_j]), \tag{7.41}$$

$$\{C(\vec{N}), C(\vec{M})\} = C(\mathcal{L}_{\vec{M}} \vec{N}), \tag{7.42}$$

$$\{C(\vec{N}), \mathcal{G}(N_i)\} = \mathcal{G}(\mathcal{L}_{\vec{N}} N_i), \tag{7.43}$$

$$\{C(\vec{N}), \mathcal{H}(\underset{\sim}{M})\} = \mathcal{H}(\mathcal{L}_{\vec{N}} \underset{\sim}{M}), \tag{7.44}$$

$$\{\mathcal{G}(N_i), \mathcal{H}(\underset{\sim}{N})\} = 0, \tag{7.45}$$

$$\{\mathcal{H}(\underset{\sim}{N}), \mathcal{H}(\underset{\sim}{M})\} = C(\vec{K}) - \mathcal{G}(A^i_a K^a), \tag{7.46}$$

where the vector $\vec{K}$ is defined by $K^a = 2\tilde{E}^a_i \tilde{E}^b_i (\underset{\sim}{N} \partial_a \underset{\sim}{M} - \underset{\sim}{M} \partial_a \underset{\sim}{N})$. Here we clearly see that the constraints are first class. The reader should notice, however, that the algebra is not a true Lie algebra, since one of the structure constants (the one defined by the last equation) is not a constant but depends on the fields $\tilde{E}^a_i$ (through the definition of the vector $\vec{K}$).

The new variables are simply related to the traditional Hamiltonian variables:

$$A^i_a = \Gamma^i_a - iK^i_a, \qquad qq^{ab} = \tilde{E}^a_i \tilde{E}^b_i, \tag{7.47}$$

where $K_a^i = K_{ab}E^{bi}$ and Γ_a^i is the spin connection compatible with the triad.

The evolution equations for the canonical variables are obtained taking the Poisson brackets of the variables with the Hamiltonian,

$$\dot{A}_a^i = -i\epsilon^{ijk} \underset{\sim}{N} \tilde{E}_j^b F_{abk} - N^b F_{ab}^i, \tag{7.48}$$

$$\dot{\tilde{E}}_i^a = i\epsilon_i^{jk} D_b(\underset{\sim}{N} \tilde{E}_j^a \tilde{E}_k^b) - 2D_b(N^{[a} \tilde{E}^{b]i}). \tag{7.49}$$

A similar simplification to that introduced in the constraints is evident in the equations of motion.

As we mentioned above, because of the self-duality used in the definition of the canonical variables, these are in general complex. The situation is totally analogous to that introduced when we discussed the harmonic oscillator and Maxwell theory in the Bargmann representation in section 4.5. If we want to recover the classical theory we must take a "section" of the phase space that corresponds to the dynamics of real relativity. This can be done. One gives data on the initial surface that correspond to a real spacetime and the evolution equations will keep these data real through the evolution. Now, strictly speaking, this procedure is not really canonical, since we are imposing these conditions by hand at the end. That does not mean it is not useful*. In fact, one can eliminate the reality conditions and have a canonical theory. However, much of the beauty of the new formulation is lost, in particular the structure of the resulting constraints is basically that of the traditional formalism.

The issue of the reality conditions acquires a different dimension at the quantum level. A point of view that is strongly advocated, and may turn out to be correct, is the following. Start by considering the complex theory and apply the usual steps towards canonical quantization After the space of physical states has been found, when one looks for an inner product, the reality conditions are used in order to choose an inner product that implements them. That is, the reality conditions can be a guideline to finding the appropriate inner product of the theory. One simply requires that the quantities that have to be real according to the reality conditions of the classical theory become self-adjoint operators under the chosen inner product. This solves two difficulties at once, since it allows us to recover the real quantum theory and the appropriate inner product at the same time. This point of view is strictly speaking a deviation from standard Dirac quantization, and works successfully for several model problems [130]. The success or failure in quantum gravity of this approach

---

* A non-trivial example where it can be worked to the end is the Bianchi II cosmology [132].

is yet to be tested and is one of the most intriguing and attractive features of the formalism. (For a critical viewpoint, see reference [131].)

In terms of the basic variables, the reality conditions are

$$(\tilde{E}_i^a \tilde{E}^{bi})^* = \tilde{E}_i^a \tilde{E}^{bi},$$ (7.50)

$$(\epsilon^{ijk}\tilde{E}_i^{(a} D_c(\tilde{E}_k^{b)}\tilde{E}_j^c))^* = (\epsilon^{ijk}\tilde{E}_i^{(a} D_c(\tilde{E}_k^{b)}\tilde{E}_j^c)).$$ (7.51)

This particular form of the reality conditions may be useful to select real initial data for classical evolutions. However, if one wants to impose the conditions as adjointness relations of operators with respect to a quantum inner product, it is clear that one would need to recast the conditions in terms of physical observables, since these are the only quantities defined in the space of physical states. In particular equations (7.50),(7.51) are not well defined in that space.

Up to now we have discussed the theory in vacuum. There is no difficulty in incorporating matter fields in the new variable formulation. The constraints can be made polynomial in a natural fashion for coupling to scalar fields, Yang–Mills fields, and fermions. It is remarkable that Dirac fermions can be introduced only coupled to the self-dual part of the connection. A complete discussion can be found in references [133, 2].

It is immediate to include a cosmological constant in the framework. In the Einstein action the cosmological constant appears as $\int d^4x\sqrt{-g}\Lambda$. This action can be immediately canonically decomposed as

$$S_\Lambda = \int dt \int d^3x \underset{\sim}{N} q\Lambda,$$ (7.52)

and this can be written in terms of the new variables noting that the determinant of the three-metric is given by

$$q = \frac{1}{6}\eta_{abc}e^{ijk}\tilde{E}_i^a \tilde{E}_j^b \tilde{E}_k^c.$$ (7.53)

The only change introduced in the canonical theory is that the Hamiltonian constraint gains an extra term,

$$\mathcal{H}(\underset{\sim}{N}) = \int d^3x \underset{\sim}{N} \epsilon^{ij}{}_k \tilde{E}_i^a \tilde{E}_j^b F_{ab}^k + \frac{\Lambda}{6}\int d^3x \underset{\sim}{N} \eta_{abc}e^{ijk}\tilde{E}_i^a \tilde{E}_j^b \tilde{E}_k^c.$$ (7.54)

And again, is a polynomial expression. There is no modification to the other constraints, since the entire term in the action is proportional to $\underset{\sim}{N}$.

7.4 Quantum gravity in terms of connections

7.4.1 Formulation

The casting of general relativity as a theory of a connection has important implications at the quantum mechanical level. One can now proceed to quantize the theory exactly like we did in chapter 5, picking a polarization in which wavefunctions are functionals of a connection

$$\Psi[A]. \tag{7.55}$$

The Gauss law will immediately require that these be gauge invariant functions, i.e., functionals in the space of connections modulo gauge transformations. Notice that this is a significant departure from the traditional picture where one considered functionals of a three-metric, or if one imposed the diffeomorphism constraint, of a three-geometry.

As in the Yang–Mills case a representation for the Poisson algebra of the canonical variables considered can be simply achieved by representing the connection as a multiplicative operator and the triad as a functional derivative:

$$\hat{A}_a^i \Psi(A) = A_a^i \Psi(A), \tag{7.56}$$

$$\hat{\tilde{E}}_i^a \Psi(A) = \frac{\delta}{\delta A_a^i} \Psi(A). \tag{7.57}$$

It should be emphasized that a difference with the Yang–Mills case arises since the connection is complex. The wavefunctions considered are holomorphic functions of the connection and the functional derivative treats as independent the connection and its complex conjugate.

We would now like to use this choice in the representation of the canonical algebra to promote the constraint equations to operatorial equations. Since the constraint equations involve operator products, a regularization is needed. This is a fundamental point. Most of the issues one faces when promoting the constraints to wave equations do not have a unique answer unless one has a precise regularization. There is not a complete regularized picture of the theory at present. We will introduce some of the issues in this chapter and will return to them in chapters 8 and 11 as we develop the quantum theory and some of its consequences.

Ignoring for the time being the regularization issue, one can promote the constraints formally to operator equations if one picks a factor ordering. Two factor orderings have been explored: with the triads either to the right or the left of the connections.

7.4.2 Triads to the right and the Wilson loop

If one orders the triads to the right, the constraints become

$$\hat{\tilde{G}}^i = D_a \frac{\delta}{\delta A_a^i}, \tag{7.58}$$

$$\hat{\tilde{C}}_a = F_{ab}^i \frac{\delta}{\delta A_b^i}, \tag{7.59}$$

$$\hat{\tilde{\mathcal{H}}} = \epsilon^{ijk} F_{ab}^i \frac{\delta}{\delta A_a^j} \frac{\delta}{\delta A_b^k}. \tag{7.60}$$

This ordering was first considered by Jacobson and Smolin [134] because the Gauss law and the diffeomorphism constraint formally (without a regularization) generate gauge transformations and diffeomorphisms on the wavefuctions.

There is a potential problem when one considers the algebra of constraints. Remember that it is not a true algebra, but as we discussed, the commutator of two Hamiltonians has a structure "constant" that depends on one of the canonical variables, the triad. This means that in this ordering such a "constant" would have to appear to the right of the resulting commutator, which is not expected. In fact, an explicit calculation of the formal commutator shows the triads appear to the right. Therefore, it is not immediate that acting on a solution the commutator of two Hamiltonians vanishes and it has to be checked explicitly.

The simplest solution to the constraints in this representation is

$$\Psi[A] = \text{constant}. \tag{7.61}$$

This state is annihilated by all the constraints formally and it is easy to check that it is also annihilated with simple point-splitting regularizations. This state is less trivial than one may imagine. It has been explored in the context of Bianchi models and it has a quite non-trivial form if transformed into the traditional variables [135].

Jacobson and Smolin set out to find less obvious solutions to the constraint equations in this formalism. If one starts by considering the Gauss law, one would like the wavefunctionals to be invariant under $SU(2)$ gauge transformations. An example of such functionals is the Wilson loop,

$$W(A, \gamma) = \text{Tr}\left(\text{Pexp} \oint_\gamma dy^a A_a(y) \right). \tag{7.62}$$

In fact, as we have seen *any* gauge invariant function of a connection can be expressed as a combination of Wilson loops. In view of this, one can consider Wilson loops as an infinite family of wavefunctions in the connection representation parametrized by a loop $\Psi_\gamma(A) = W(\gamma, A)$ that forms an (overcomplete) basis of solutions to the quantum Gauss law constraint.

What happens to the diffeomorphism constraint? Evidently Wilson loops are not solutions. When a diffeomorphism acts on a Wilson loop, it gives as a result a Wilson loop with the loop displaced by the diffeomorphism performed. Therefore they are not annihilated by the diffeomorphism constraint and cannot become candidates for physical states of quantum gravity. In spite of that, they are worth exploring a bit more.

Remember they form an overcomplete basis in terms of which any physical state should be expandable (since any physical state has to be gauge invariant). We will therefore explore what happens when we act with the Hamiltonian constraint on them. To perform this calculation we only need the formula for the action of a triad on a holonomy along an open path $\gamma_o^{o'}$,

$$\tilde{\underline{E}}_i^a(x)U(\gamma_o^{o'}) = \frac{\delta}{\delta A_a^i(x)}U(\gamma_o^{o'}) = \oint_\gamma dy^a \delta^3(x-y)U(\gamma_o^y)\tau^i U(\gamma_y^{o'}), \quad (7.63)$$

where τ^i are $-i\sqrt{2}/2$ times the Pauli matrices.

The reason why we are considering an open path is to avoid ambiguities when we act with the second derivative. The expression for the action on the Wilson loop we are interested in is obtained in the limit in which o and o' coincide. We now act with a second triad,

$$\frac{\delta}{\delta A_a^i(x)}\frac{\delta}{\delta A_b^j(x)}U(\gamma_o^{o'}) =$$

$$\oint_\gamma dy^b \oint_{\gamma_o^y} dz^a \delta(x-y)\delta(x-z)U(\gamma_o^z)\tau^i U(\gamma_z^y)\tau^j U(\gamma_y^{o'})$$

$$+ \oint_\gamma dy^b \oint_{\gamma_y^{o'}} dz^a \delta(x-y)\delta(x-z)U(\gamma_o^y)\tau^j U(\gamma_y^z)\tau^i U(\gamma_z^{o'}). \quad (7.64)$$

We now take the trace and obtain the action of the Hamiltonian,

$$H(x)\Psi_\gamma[A] =$$

$$F_{ab}^k(x)\epsilon_{ijk}\left[\oint_\gamma dy^b \oint_{\gamma_o^y} dz^a \delta(x-y)\delta(x-z)\mathrm{Tr}(\tau^i U(\gamma_z^y)\tau^j U(\gamma_{yo}^z))\right.$$

$$\left. + \oint_\gamma dy^b \oint_{\gamma_y^{o'}} dz^a \delta(x-y)\delta(x-z)\mathrm{Tr}(\tau^j U(\gamma_y^z)\tau^i U(\gamma_{zo}^y))\right], \quad (7.65)$$

where the notation $U(\gamma_{yo}^z)$ denotes the portion of the loop going from y to z through the basepoint o.

If the loop has no kinks or intersections, the portion γ_z^y shrinks to a point due to the presence of the Dirac delta functions and the action of the Hamiltonian can be written as

$$\hat{\mathcal{H}}(x)\Psi_\gamma[A] =$$

$$F_{ab}^k(x)\epsilon_{ijk}\left[\oint_\gamma dy^b \oint_\gamma dz^a \delta(x-y)\delta(x-z)\mathrm{Tr}(\tau^i\tau^j U(\gamma_{yo}^z))\right.$$

$$+ \oint_\gamma dy^b \oint_\gamma dz^a \delta(x-y)\delta(x-z)\mathrm{Tr}(\tau^j\tau^i U(\gamma_{zo}^y)) =$$

$$\left. \oint_\gamma dy^b \oint_\gamma dz^a \delta(x-y)\delta(x-z)\mathrm{Tr}(\delta^{ij}U(\gamma_{zo}^y))\right], \quad (7.66)$$

where we have extended the second integral along the whole loop, since no additional contributions are added due to the fact that the loop is smooth.

Notice that we have a quantity $F^i_{ab}\epsilon_{ijk}$ which is antisymmetric in both a, b and i, j contracted with an expression that is symmetric in both a, b and i, j. Therefore, the expression vanishes! We have just proved that a Wilson loop formed with the Ashtekar connection is a (formal) solution of the Hamiltonian constraint of quantum gravity. This is a remarkable fact. Notice that up to this discovery *no* solution of this constraint was known in a general case (without making mini-superspace approximations). Historically, this discovery fostered the interest for loops in this context and led to the use of the loop representation.

A key to this result was the consideration of smooth non-intersecting loops. If the loops have intersections or kinks, the proof we presented above does not work. Moreover, it should be stressed that the result is *formal*. The expressions considered involve one-dimensional integrals of three-dimensional Dirac delta functions. In a particular coordinate system they are proportional to $\delta^2(0)$. Therefore we are canceling divergent terms.

To see if this result holds beyond the formal level, a regularization is needed. Two different regularizations were considered by Jacobson and Smolin [134]. The first one is based on "flux tubes", a process in which the loops are thickened out. The main drawback of this method is that it is not gauge invariant. Under this regularization, smooth loops solve the constraint with suitable prescriptions for limiting procedures. The second regularization method is based on a point-splitting of the two functional derivatives of the Hamiltonian constraint. Although point-splitting in general breaks gauge invariance (since point-split quantities exist at different points of the manifold and transform with different transformation matrices) one can restore gauge invariance connecting the point-split quantities with holonomies along paths connecting the split points. Unfortunately, under this procedure smooth Wilson loops *fail* to satisfy the constraint. An anomaly appears that is proportional to terms that depend on the curvature of the loops ("acceleration terms") and is non-vanishing. We will see that the role of the acceleration terms is different in the loop representation and there is a sense in which smooth loops correspond to solutions of the constraints. We will return to these and other regularization issues later.

Even ignoring the regularization issues of the Hamiltonian constraint, there are two main drawbacks to these solutions: they do not solve the diffeomorphism constraint and they fail to solve the Hamiltonian constraint if the loops have intersections.

Why care about loops with intersections? Why not just restrict our-

selves to smooth loops? The problem appears when we try to get some
sort of understanding of what these wavefunctionals are. The first question that comes to mind is what is the metric for such a state. This, in
principle, is a meaningless question, since the metric is not an observable
in the sense of Dirac, but let us ask it anyway to see where it leads. The
metric acting on one of these states gives

$$\hat{\tilde{q}}^{ab}(x)\Psi_\gamma(A) = \frac{\delta}{\delta A_a^i}\frac{\delta}{\delta A_b^i}\Psi_\gamma(A) = X^{ax}X^{bx}\Psi_\gamma(A). \qquad (7.67)$$

Again, this expression needs to be regularized. At a formal level we
see that Wilson loops are eigenstates of the metric operator if the loops
considered are smooth. Notice that the metric only has support distri-
butionally along the direction of the tangent to the loop. Moreover, the
metric has only one non-vanishing component, the one along the loop.
Therefore it is a degenerate metric. Now, this statement is still meaningless in a diffeomorphism invariant context, but it actually can be given
a rigorous meaning with a little elaboration. Consider the Hamiltonian
constraint for general relativity with a cosmological constant, given by
expression (7.54). The only difference with the vacuum constraint is the
term involving the determinant of the spatial metric. This term can be
promoted to the connection representation with similar regularization difficulties as the rest of the constraint. It is easy to see that the additional
term formally annihilates a Wilson loop based on a smooth loop. Therefore the determinant of the three-metric vanishes for these states, as is
expected for a degenerate metric. Since these states are annihilated by the
vacuum Hamiltonian constraint *and* the determinant of the three-metric,
this means they are states for an arbitrary value of the cosmological constant! That spells serious trouble. General relativity with and general
relativity without a cosmological constant are *very* different theories, and
one does not expect them to share a common set of states, except for
special situations, such as for degenerate metrics.

It turns out, one can improve the situation a little using intersections.
One can find some solutions to the Hamiltonian constraint for the intersecting case by considering linear combinations of holonomies in such
a way that the contributions at the intersections cancel [134, 136, 26].
However, unexpectedly, this is not enough to construct non-degenerate
solutions. All the solutions constructed in this fashion, if they satisfy the
Hamiltonian constraint, are also annihilated by the determinant of the
metric [26]. This, plus the fact that they do not satisfy the diffeomorphism constraint, shows that these solutions are of little physical use in
this context. They were, however, very important historically as motiva-
tional objects for the study of loops. We will show later how, when one
works in the loop representation, it is possible to generate solutions to

all the constraints that, although still based on loops, do not have this degeneracy problem.

7.4.3 Triads to the left and the Chern–Simons form

If one orders the constraints with the triads to the left, there is potential for a problem: as we have said, apparently in this factor ordering the diffeomorphism constraint fails to generate diffeomorphisms on the wavefunctions. This would be a reason to abandon this ordering altogether. However, by considering a very generic regularized calculation one can prove that the diffeomorphism constraint actually generates diffeomorphisms, so this is not a problem [137]. Besides, there is the advantage that when one considers the constraint algebra, one obtains (these are only formal unregulated results) the correct closure [51].

Let us see how the regularized version of the constraint in this factor ordering generates diffeomorphisms. We consider a point-split version of the diffeomorphism constraint,

$$\hat{C}(\vec{N}) = \lim_{\epsilon \to 0} \int d^3x d^3y N^a(x) f_\epsilon(x-y) \frac{\delta}{\delta A_b^i(x)} F_{ab}^i(y), \qquad (7.68)$$

where $\lim_{\epsilon \to 0} f_\epsilon(x-y) = \delta(x-y)$. This expression differs from that in the factor ordering with triads to the right by the term in which the functional derivative acts on F_{ab}^i, $\int d^3y f_\epsilon(x-y)\delta F_{ab}^i(y)/\delta A_a^i(x)$. When the functional derivative acts on the portion of F_{ab}^i linear in A_a^i one gets a contribution of the form $\int d^3x N^a(x) \int d^3y \partial_b \delta(x-y) f_\epsilon(x-y)$. If one considers a regulator that is symmetric in x, y, $f_\epsilon(x-y) = f_\epsilon(y-x)$, this contribution vanishes. The action of the functional derivative on the term quadratic in the connections vanishes due to the antisymmetry of the structure constants ϵ^{ijk} of $SU(2)$. We have therefore proved that the expression for the constraint with the triads to the left coincides, if one considers symmetric regulators, with the expression with the triads to the left. Since the former generates diffeomorphisms on the wavefunctions the latter does so as well. Therefore the diffeomorphism constraint regains its natural geometric interpretation and can be solved by considering wavefunctionals of the connection $\Psi[A]$ that are invariant under diffeomorphisms.

In this ordering, Wilson loops do not solve the Hamiltonian constraint. However, there is a very interesting and rich solution one can construct. Consider the following state, a function of the Chern–Simons form built with the Ashtekar connection,

$$\Psi_\Lambda[A] = \exp\left(-\frac{6}{\Lambda} \int \tilde{\epsilon}^{abc} \mathrm{Tr}[A_a \partial_b A_c + \tfrac{2}{3} A_a A_b A_c]\right). \qquad (7.69)$$

This functional has the property that the triad equals the magnetic field constructed from the Ashtekar connection (in the language of Yang–Mills theory, the electric field equals the magnetic field),

$$\frac{\delta}{\delta A_a^i} \Psi_\Lambda[A] = \frac{3}{\Lambda} \tilde{\epsilon}^{abc} F_{bc}^i \Psi_\Lambda[A]. \tag{7.70}$$

Moreover, it is well known that this functional is invariant under (small) gauge transformations and diffeomorphisms. One can check that it is annihilated by the corresponding constraints (with the proviso of the symmetric regulator in the diffeomorphism constraint introduced above). What may come as a surprise is that it is actually annihilated by the Hamiltonian constraint with a cosmological constant. This is easy to see. Consider the constraint

$$\hat{\mathcal{H}} = \epsilon_{ijk} \frac{\delta}{\delta A_a^i} \frac{\delta}{\delta A_b^j} F_{ab}^k - \frac{\Lambda}{6} \epsilon_{ijk} \tilde{\epsilon}_{abc} \frac{\delta}{\delta A_a^i} \frac{\delta}{\delta A_b^j} \frac{\delta}{\delta A_c^k} \tag{7.71}$$

and notice that the rightmost derivative of the determinant of the metric reproduces the term on the left when acting on the wavefunction. Notice that the result holds without even considering the action of the other derivatives, and therefore is very robust *vis a vis* regularization. This result was noticed independently by Ashtekar [53] and Kodama [54]. A nice feature of this result is that the metric is non-degenerate in the sense that we discussed in the previous section. The metric is just given by the trace of the product of two magnetic fields built with the Ashtekar connection. Such a property holds classically for spaces of constant curvature. This has led some authors to suggest that this wavefunction is associated with the DeSitter geometry [55].

The reader may question the relevance of the Chern–Simons state. First of all, it is only one state. Moreover, a similar state is present in Yang–Mills theory (this is easy to see, since the Hamiltonian is $E^2 + B^2$ and adjusting constants one gets for the corresponding state $E = iB$) and is known to be non-physical since it is not normalizable. This is true, but it is also true that the nature of a theory defined on a fixed background as a Yang–Mills theory is expected to be radically different from that of a theory invariant under diffeomorphisms, such as general relativity. Therefore normalizability under the inner product of one theory does not necessarily imply or rule out normalizability under the inner product of the other. The non-normalizability in the Yang–Mills context is under the Fock inner product, and it is expected that inner products of that kind will not have any relevance in the context of general relativity. At the moment, however, the normalizability or not of any state in general relativity cannot be decided, since we lack an inner product for the theory.

It is remarkable that the Chern–Simons form, which is playing such a prominent role in particle physics nowadays, should have such a singular role in general relativity. It is the only non-trivial state in the connection representation that we know that may have something to do with a non-degenerate geometry. We will also see in chapters 10 and 11 that the state plays a prominent role in the progress made in finding states in the loop representation and has opened up new connections between general relativity, topological field theories and knot theory.

There are more things one could say about the connection representation. There is the compelling work of Ashtekar, Balachandran and Jo [61] concerning the CP violation problem and the partial success (in the linearized theory) of Ashtekar [56] in addressing the issue of time. We do not have space here to do justice to these pieces of work and we refer the reader to the relevant literature. In particular, a good summary of these topics appears in the book by Ashtekar [2].

7.5 Conclusions

We have formulated gravity canonically and discussed the general features of its canonical quantization. We have discussed the difficulties associated with the traditional metric variables and introduced a new set of variables that allows some progress in the definition of the quantum constraint equations and their solutions. We have discussed some of the factor ordering and regularization issues and set the stage for the introduction of a loop representation, which we will do in the following chapter.

8

The loop representation
of quantum gravity

8.1 Introduction

Having cast general relativity as a Hamiltonian theory of a connection, we are now in a position to apply the same techniques we used to construct a loop representation of Yang–Mills theories to the gravitational case. We should recall that we are dealing with a complex $SU(2)$ connection. However, we can use exactly the same formulae that we developed in chapter 5 since few of them depend on the reality of the connections. Whenever the presence of a complex connection introduces changes, we will discuss this explicitly .

As we have seen, we can introduce a loop representation either through a transform or through the quantization of a non-canonical algebra. The initial steps are exactly the same as those in the $SU(2)$ Yang–Mills case. The differences arise when we want to write the constraint equations. In the Yang–Mills case the only constraint was the Gauss law and one had to represent the Hamiltonian in terms of loops. In the case of gravity one has to impose the diffeomorphism and Hamiltonian constraints in terms of loops. In order to do so one can either use the transform or write them as suitable limits of the operators in the T algebra. We will outline both derivations for the sake of comparison. As we argued in the Yang–Mills case both derivations are formal and in a sense equivalent, although the difficulties are highlighted in slightly different ways in the two derivations.

The space of states of an $SU(2)$ theory in terms of the loop representation has been discussed in detail in chapter 3. It is formed by wavefunctions with support on the group of loops,

$$\Psi(\gamma), \tag{8.1}$$

that satisfy the basic Mandelstam identities,

$$\Psi(\gamma) = \Psi(\gamma^{-1}), \tag{8.2}$$

$$\Psi(\gamma \circ \eta) = \Psi(\eta \circ \gamma), \tag{8.3}$$

$$\Psi(\gamma \circ \eta \circ \beta) + \Psi(\gamma \circ \eta \circ \beta^{-1}) = \Psi(\eta \circ \gamma \circ \eta) + \Psi(\eta \circ \gamma \circ \beta^{-1}), \tag{8.4}$$

and by combination of these identities one can find an infinite number of linear relations among the wavefunctions.

In many papers on the subject, multiloops have been used to build the loop representation. As we discussed in chapter 3, for the $SU(2)$ case all expressions in terms of multiloops can be rewritten as single-loop expressions via Mandelstam identities. We will therefore restrict ourselves here to single-loop wavefunctions.

The outline of this chapter is as follows. In the next two sections we will derive the expression of the constraints of quantum gravity in the loop representation both as a limit of the T algebra and via the loop transform. We will then discuss the regularization of the Hamiltonian in terms of loops and briefly discuss the solution space. We will return to the issue of solution to the constraints in chapters 10 and 11.

8.2 Constraints in terms of the T algebra

We need to write the classical diffeomorphism and Hamiltonian constraints in terms of the T operators. It is quite simple to write the diffeomorphism constraint as a limit of a T^1 operator. Consider a one-parameter family $\gamma_{\hat{a}\hat{b}}^{\delta}(x)$ of closed curves in the $\hat{a}\hat{b}$ coordinate plane basepointed at the point x such that in the limit $\delta \to 0$ the loops shrink to a point. The area element of the loop is given by

$$\sigma^{cd}(\gamma_{\hat{a}\hat{b}}^{\delta}) = \delta^2 \delta_{\hat{a}}^{[c} \delta_{\hat{b}}^{d]}. \tag{8.5}$$

The diffeomorphism constraint is given by the limit

$$C(\vec{N}) = \lim_{\delta \to 0} \frac{1}{\delta^2} \int d^3x N^{\hat{a}}(x) T^{\hat{b}}(\gamma_{\hat{a}\hat{b}}^{\delta}(x)). \tag{8.6}$$

To prove this, notice that in this limit the holonomy is given by

$$\lim_{\delta \to 0} \mathbf{H}(\gamma_{ab}^{\delta}(x)) = 1 + \tfrac{1}{2} \sigma^{cd}(\gamma_{\hat{a}\hat{b}}(x)) \mathbf{F}_{cd}(x). \tag{8.7}$$

When one takes the trace to construct the T^1, the contribution from the identity drops out because of the tracelessness of the triad and the leading contributions is $\mathrm{Tr}(\tilde{\mathbf{E}}^a(x)\mathbf{F}_{ab}(x))$, which corresponds with the usual expression of the vector constraint.

A remarkable fact is that the constraint algebra is consistently reproduced in the limit

$$C^\delta(\vec{N}) \quad \rightarrow \quad \{C^\delta(\vec{N}), C^\delta(\vec{M})\}$$
$$\downarrow \qquad\qquad\qquad \downarrow \qquad\qquad (8.8)$$
$$C(\vec{N}) \quad \rightarrow \quad \{C(\vec{N}), C(\vec{M})\}$$

i.e., computing the Poisson bracket of two T^1s and shrinking the loops yields the same results as shrinking the loops and computing the Poisson algebra of the constraints [139].

To obtain the Hamiltonian constraint we will introduce a double limiting procedure, which in what follows will be useful as a regularization procedure for the quantum calculation. We will consider the point-split classical Hamiltonian,

$$C(\underset{\sim}{N}) = \lim_{\epsilon \to 0} C^\epsilon(\underset{\sim}{N}) = \lim_{\epsilon \to 0} \int d^3x \underset{\sim}{N}(x) \int d^3y f_\epsilon(x-y)$$
$$\times \mathrm{Tr}(\tilde{\mathbf{E}}^a(y)\mathbf{H}(\mu_y^x)\tilde{\mathbf{E}}^b(x)\mathbf{F}_{ab}(x)\mathbf{H}(\mu_x^y)), \qquad (8.9)$$

where we have introduced an arbitrary infinitesimal path μ_y^x. The introduction of this path is needed in order to have a gauge invariant point-split Hamiltonian. Since the T variables are gauge invariant it would be impossible to retrieve a non-invariant quantity from them. The contribution from the holonomy $\mathbf{H}(\mu_y^x)$ reduces to the identity in the limit.

We will present a shrinking loop procedure that will yield the split constraint $C^\epsilon(\underset{\sim}{N})$, and from there one recovers the usual constraint in the limit $\epsilon \to 0$. We introduce a one-parameter family of shrinking loops as before $\gamma_{\hat{a}\hat{b}}^\delta(x)$. The Hamiltonian constraint is given by

$$C^\epsilon(\underset{\sim}{N}) = \lim_{\delta \to 0} \frac{1}{\delta^2} \int d^3x \underset{\sim}{N}(x) \int d^3y f_\epsilon(x-y) T^{[\hat{a}\hat{b}]}(\mu_x^y, \mu_y^x \circ \gamma_{\hat{a}\hat{b}}^\delta(x)). \quad (8.10)$$

The proof follows similar lines as before: in the shrinking limit the holonomy yields two contributions; the one proportional to the identity vanishes due to the antisymmetrization in the $\hat{a}\hat{b}$ indices (if not one would get the metric $\mathrm{Tr}(\tilde{\mathbf{E}}^a\tilde{\mathbf{E}}^b)$ as leading contribution) and the term proportional to $\mathbf{F}_{ab}$ yields the constraint.

We therefore have classical expressions relating the constraints and the T operators. This allows us to find expressions for the constraints as quantum mechanical operators by promoting their definitions in terms of the T quantities to quantum mechanical operators. The quantum mechanical expressions for the T operators were introduced in chapter 5, choosing a factor ordering with the triads to the right. We recall here their expression

$$\hat{T}^0(\eta)\Psi(\gamma) \equiv \Psi(\gamma \circ \eta) + \Psi(\gamma \circ \eta^{-1}), \qquad (8.11)$$

$$\hat{T}^a(\eta_x^x)\Psi(\gamma) \equiv \sum_{\epsilon=-1}^{1} \epsilon \oint dy^a \delta(x-y)\Psi(\gamma \circ \eta^\epsilon), \qquad (8.12)$$

$$\hat{T}^{ab}(\eta_x^y, \eta_y^x)\Psi(\gamma) = X^{ax}(\gamma)X^{by}(\gamma)[\Psi(\gamma_x^y \circ \bar{\eta}_y^x, \gamma_y^x \circ \bar{\eta}_x^y)$$
$$+\Psi(\gamma_x^y \circ \eta_y^x, \gamma_y^x \circ \eta_x^y) + \Psi(\gamma_x^y \circ \bar{\eta}_y^x \circ \bar{\gamma}_y^y \circ \eta_y^x)$$
$$+\Psi(\gamma_y^x \circ \bar{\eta}_x^y \circ \bar{\gamma}_y^y \circ \eta_x^y)]. \qquad (8.13)$$

We now promote the relation (8.6) to an operatorial equation,

$$\hat{C}(\vec{N})\Psi(\gamma) = \lim_{\delta\to 0} \frac{1}{\delta^2} \int d^3x N^{\hat{a}}(x)\hat{T}^{\hat{b}}(\gamma_{\hat{a}\hat{b}}^\delta(x))\Psi(\gamma)$$

$$= \lim_{\delta\to 0} \frac{-1}{2\delta^2} \int d^3x N^{\hat{a}}(x) \sum_{\epsilon=-1}^{1} \epsilon \oint dy^{\hat{b}}\delta(x-y)\Psi(\gamma \circ (\gamma_{\hat{a}\hat{b}}^\delta(x))^\epsilon),$$

$$(8.14)$$

and we notice that the introduction of the infinitesimal loop $\gamma_{\hat{a}\hat{b}}^\delta(x)$ with the two possible orientations given by the power ϵ corresponds to the action of the loop derivative. Since the loop derivative along the reversed loop introduces a minus sign the two contributions $\epsilon = \pm 1$ add up to give

$$\hat{C}(\vec{N})\Psi(\gamma) = -\int d^3x N^a(x) \oint dy^b \delta(x-y)\Delta_{ab}(\gamma_o^x)\Psi(\gamma) \qquad (8.15)$$

and we see that the diffeomorphism constraint in the loop representation can be obtained in the limit of shrinking loops from the T^1 operator. As the derivation shows, the loop derivative arises because the action of the T^1 operator corresponds to the introduction of a small loop of precisely the same form as in the loop derivative.

The Hamiltonian constraint can be obtained through manipulations that are very similar to those of the diffeomorphism constraint. Since the final expression coincides exactly with the one we will obtain in the next section via the loop transform we do not give the explicit calculation. For details see reference [139]. We will just outline the first steps of the calculation to facilitate the comparison with the expression that we derive in the next secion. We need to compute

$$T^{ab}(\mu_x^y, \mu_y^x \circ \gamma^\delta{}_{ab}(x))\Psi(\gamma) = X^{ax}(\gamma)X^{bx}(\gamma)[\Psi(\gamma_x^y \circ \mu_y^x, \bar{\gamma}_{ab}^\delta(x) \circ \mu_x^y \circ \gamma_y^x)$$
$$+\Psi(\gamma_x^y \circ \mu_y^x\gamma^\delta{}_{ab}(x), \mu_x^y\gamma_y^x) + \Psi(\gamma_x^y \circ \mu_y^x \circ \bar{\gamma}_x^y \circ \mu_y^x \circ \gamma^\delta{}_{ab}(x))$$
$$+\Psi(\gamma_y^x \circ \bar{\gamma}_{ab}^\delta(x) \circ \mu_x^y \circ \bar{\gamma}_y^x \circ \mu_x^y)] \qquad (8.16)$$

and using the Mandelstam identities and recalling that we are only interested in the antisymmetric part of T^{ab} we get

$$\hat{C}^\epsilon(\underset{\sim}{N}) = \lim_{\delta\to 0} \frac{1}{\delta^2} \int d^3x \underset{\sim}{N}(x) \int d^3y f_\epsilon(x-y) X^{[a|x}(\gamma) X^{b]y}(\gamma)$$

$$\times \left[2\Psi(\gamma^\delta{}_{ab}(x) \circ \gamma^y_x \circ \mu^x_y \circ \bar{\gamma}^y_x \circ \mu^x_y) \right.$$

$$\left. + 2\Psi(\gamma^\delta{}_{ab}(x) \circ \bar{\gamma}x^y \circ \mu^x_y \circ \gamma^y_x \circ \mu^x_y) \right] \tag{8.17}$$

which, taking into account the definition of the loop derivative, yields a regularized expression for the Hamiltonian constraint that we will present in an explicit fashion in the next section.

8.3   Constraints via the loop transform

To obtain the quantum version of the constraints via the loop transform, we proceed in the same way as we did for an $SU(2)$ Yang–Mills theory in chapter 6. There is a difference, however, due to the fact that the connection in the general relativity case is complex. In principle, its complex conjugate is a complicated expression given by the reality conditions. Therefore we cannot quite write for the transform as we did in chapter 6,

$$\Psi(\gamma) = \int dA W_\gamma[A]^* \Psi[A] \tag{8.18}$$

since the expression for $W_\gamma[A]^*$ would, in principle, be a complicated non-polynomial expression in terms of A. Moreover, as we argued before, it is not clear that one wants to implement the reality conditions at this level. One may want to impose them later as relations among observables of the theory.

In order to be able to proceed we will assume in the following manipulations that A is real. This is not unjustified, since the manipulations in terms of real As yield operator expressions in the loop representation that have exactly the same commutation relations as their counterparts in the connection representation. In this sense the loop transform is a very useful heuristic device for finding appropriate loop counterparts to operators in the connection representation. The reader should be aware that the following calculations are heuristic and not meant to be precise derivations. It is remarkable that through this procedure one can recover exactly the same expression for the constraints as we did in the previous section. This suggests that a measure may exist such that the manipulations can be made rigorous taking into account the complex nature of the connections.

We therefore define

$$\hat{O}\Psi(\gamma) \equiv \int dA W_\gamma[A] \hat{O}\Psi[A] = \int dA \hat{O}^\dagger W_\gamma[A] \Psi[A], \tag{8.19}$$

where by $\hat{O}^\dagger$ we mean the operator $\hat{O}$ but with a reverse factor ordering. Therefore the practical calculation of transforming an operator consists in evaluating its action on a Wilson loop as if it were a calculation in the connection representation and rearranging the result as a manipulation purely in terms of loops. One should remember that when considering the action on the Wilson loop one should choose for the operator one wishes to transform the *opposite* factor ordering to the one chosen for its action on wavefunctions $\Psi[A]$.

We start with the vector constraint. Its action on a Wilson loop is given by

$$F_{ab}^i(x)\frac{\delta}{\delta A_a^i(x)}W_\gamma[A] = F_{ab}^i(x)\oint_\gamma dy^a\delta(y-x)\mathrm{Tr}(\mathbf{H}(\gamma_o^y)\tau^i\mathbf{H}(\gamma_y^o))$$

$$= \oint_\gamma dy^a\delta(y-x)\mathrm{Tr}(\mathbf{H}(\gamma_o^x)\mathbf{F}_{ab}\mathbf{H}(\gamma_x^o)). \qquad (8.20)$$

Recalling the action of a loop derivative on a Wilson loop introduced in chapter 1 we get

$$F_{ab}^i(x)\frac{\delta}{\delta A_a^i(x)}W_\gamma[A] = \oint_\gamma dy^a\delta(y-x)\Delta_{ab}(\gamma_o^y)W_\gamma[A], \qquad (8.21)$$

and therefore we can write for the diffeomorphism constraint in the loop representation

$$\hat{C}(\vec{N}) = \int d^3x N^b(x)\oint_\gamma dy^a\delta(x-y)\Delta_{ab}(\gamma_o^y). \qquad (8.22)$$

This is exactly the expression we introduced in the first chapter as the generator of diffeomorphisms on functions of the group of loops and we checked in that chapter that it satisfied the correct algebra of diffeomorphisms,

$$[\hat{C}(\vec{N}),\hat{C}(\vec{M})] = \hat{C}(\mathcal{L}_{\vec{N}}\vec{M}). \qquad (8.23)$$

Sometimes one may use the shorthand notation

$$\hat{C}(\vec{N}) = \int d^3x N^b(x)X^{ax}(\gamma)\Delta_{ab}(\gamma_o^x), \qquad (8.24)$$

where $X^{ax}(\gamma)$ is the first order multitangent to the loop, but care should be exercised if the loop has multiple points (intersections).

The reader may appreciate the remarkable fact that a formalism so heuristic in nature manages to yield the expected result. We started with the action of the diffeomorphism constraint in the connection representation and by the most direct and obvious manipulation we end up with an expression with the desired geometric action in terms of loops. Encouraged by this result we will follow the same procedure for the Hamiltonian constraint.

The calculations for the Hamiltonian constraint are of the same nature, the only care to be taken is the presence of a second functional derivative, which requires a regularization. We will perform here only a formal calculation in order to simplify the presentation, we postpone the discussion of regularization issues to the next section. In fact, at the formal level we have already performed the required calculation in the previous chapter,

$$\hat{\mathcal{H}}(x)W_\gamma[A] = \epsilon^{ijk}F^i_{ab}(x)\frac{\delta}{\delta A^j_a}\frac{\delta}{\delta A^k_b}$$

$$= F^k_{ab}(x)\epsilon_{ijk}\oint_\gamma dy^b \oint_{\gamma^y_o} dz^a \delta(x-y)\delta(x-z)\text{Tr}(\tau^i\mathbf{H}(\gamma^y_z)\tau^j\mathbf{H}(\gamma^z_{y\,o}))$$

$$+ F^k_{ab}(x)\epsilon_{ijk}\oint_\gamma dy^b \oint_{\gamma^{o'}_y} dz^a \delta(x-y)\delta(x-z)\text{Tr}(\tau^j\mathbf{H}(\gamma^z_y)\tau^i\mathbf{H}(\gamma^y_{z\,o})).$$

$$(8.25)$$

We now rearrange this expression using the identity,

$$i\epsilon^{lmn}\text{Tr}(\tau^m\mathbf{A}\tau^n\mathbf{B}) = \text{Tr}(\tau^l\mathbf{A})\text{Tr}(\mathbf{B}) - \text{Tr}(\mathbf{A})\text{Tr}(\tau^l\mathbf{B}), \qquad (8.26)$$

where $\mathbf{A}, \mathbf{B}$ are $SU(2)$ matrices. The integrands can then be rewritten as

$$\epsilon^{ijk}\text{Tr}(\tau^i\mathbf{H}(\gamma^y_z)\tau^j\mathbf{H}(\gamma^z_{y\,o})) = \text{Tr}(\tau^k\mathbf{H}(\gamma^y_z))\text{Tr}(\mathbf{H}(\gamma^z_{y\,o}))$$
$$-\text{Tr}(\mathbf{H}(\gamma^y_z))\text{Tr}(\tau^k\mathbf{H}(\gamma^z_{y\,o})), \quad (8.27)$$

$$\epsilon^{ijk}\text{Tr}(\tau^j\mathbf{H}(\gamma^z_y)\tau^i\mathbf{H}(\gamma^y_{z\,o})) = \text{Tr}(\tau^k\mathbf{H}(\gamma^y_{z\,o}))\text{Tr}(\mathbf{H}(\gamma^z_y))$$
$$-\text{Tr}(\mathbf{H}(\gamma^y_{z\,o}))\text{Tr}(\tau^k\mathbf{H}(\gamma^z_y)), \quad (8.28)$$

and noticing that

$$\text{Tr}(\mathbf{H}(\gamma^z_{y\,o})) = \text{Tr}(\mathbf{H}(\gamma^y_{z\,o})), \qquad (8.29)$$

$$\text{Tr}(\tau^k\mathbf{H}(\gamma^y_z)) = -\text{Tr}(\tau^k\mathbf{H}(\gamma^z_y)), \qquad (8.30)$$

we get for the action of the Hamiltonian,

$$\hat{\mathcal{H}}(x)W_\gamma[A] = F^k_{ab}(x)(\oint_\gamma dy^b \oint_{\gamma^y_o} dz^a + \oint_\gamma dy^b \oint_{\gamma^o_y} dz^a)\delta(x-y)\delta(x-z)$$

$$\times\text{Tr}(\tau^k\mathbf{H}(\gamma^y_{z\,o}))\text{Tr}(\mathbf{H}^y_z) - \text{Tr}(\mathbf{H}^y_{z\,o})\text{Tr}(\tau^k\mathbf{H}(\gamma^z_y)). \quad (8.31)$$

The two sets of integrals can be combined into a single one, and inserting the F^i_{ab} in the holonomies we get,

$$\hat{\mathcal{H}}(x)W_\gamma[A] = \oint_\gamma dy^b \oint_\gamma dz^a \delta(x-y)\delta(x-z)$$

$$\times\text{Tr}(\mathbf{F}_{ab}(x)\mathbf{H}(\gamma^y_{z\,o}))\text{Tr}(\mathbf{H}(\gamma^z_y) - \text{Tr}(\mathbf{H}^y_{z\,o})\text{Tr}(\mathbf{F}_{ab}(x)\mathbf{H}(\gamma^z_y)). \quad (8.32)$$

We now rearrange the products of holonomies into a single one using the generalization of the Mandelstam identities when elements of the algebra are involved. One could have left the expression as it was and then the action of the Hamiltonian constraint on a wavefunction of a single loop would be a function of a multiloop. This has been the approach taken in some papers [138]. Here, as we said before, we reexpress everything in terms of single loops. The identity needed is

$$\text{Tr}(\tau^i \mathbf{A})\text{Tr}(\mathbf{B}) = \text{Tr}(\tau^i(\mathbf{AB} + \mathbf{AB}^{-1})) = \text{Tr}(\tau^i(\mathbf{BA} + \mathbf{B}^{-1}\mathbf{A})), \quad (8.33)$$

where $\mathbf{A}, \mathbf{B}$ are again elements of the $SU(2)$ group. Rearranging terms with this identity, we get

$$\hat{\mathcal{H}}(x)W_\gamma[A] = \oint_\gamma dy^{[b} \oint_\gamma dz^{a]}\delta(x-y)\delta(x-z)$$
$$\times \text{Tr}(\mathbf{F}_{ab}(x)[\mathbf{H}(\gamma_y^z)\mathbf{H}(\gamma_{yo}^z) + \mathbf{H}(\gamma_{yo}^z)\mathbf{H}(\gamma_y^z)]), \quad (8.34)$$

We can rearrange this expression in terms of loop derivatives,

$$\hat{\mathcal{H}}(x)W_\gamma[A] = \oint_\gamma dy^{[b} \oint_\gamma dz^{a]}\delta(x-y)\delta(x-z)$$
$$\times \Delta_{ab}(\gamma_o^x)\text{Tr}([\mathbf{H}(\gamma_y^z \circ \gamma_{yo}^z) + \mathbf{H}(\gamma_{yo}^z \circ \gamma_y^z)]), \quad (8.35)$$

from which we can read off the expression of the Hamiltonian constraint in the loop representation,

$$\hat{\mathcal{H}}(\underset{\sim}{N})\Psi(\gamma) = \int d^3x \underset{\sim}{N}(x) \oint_\gamma dy^{[b} \oint_\gamma dz^{a]}\delta(x-y)\delta(x-z)$$
$$\times \Delta_{ab}(\gamma_o^x)[\Psi(\gamma_y^z \circ \gamma_{yo}^z) + \Psi(\gamma_{yo}^z \circ \gamma_y^z)]. \quad (8.36)$$

It should be pointed out that the notation in the above two expressions for the loop derivative precisely means

$$\Delta_{ab}(\gamma_o^x)\Psi(\gamma_y^z \circ \gamma_{yo}^z) \equiv \Delta_{ab}(\beta_o^x)\Psi(\beta)|_{\beta=\gamma_y^z \circ \gamma_{yo}^z}, \quad (8.37)$$

and similarly for the action of the loop derivative on the holonomy. From now on we will use this notation whenever the Hamiltonian constraint is involved. Again, this expression coincides with the one introduced in the previous section directly obtained as a limit of the T operators. We see that the two approaches yield the same constraints.

One can perform another rearrangement that simplifies the expression of the Hamiltonian constraint even further. Going back to the expression in terms of F_{ab}^i (8.34), there are two terms in the expression of the Hamiltonian. Each of them is a trace of an element of the algebra times elements of the group. Such traces are equal to minus the trace of the inverse argument. If one replaces the argument of the second trace by its inverse, one obtains exactly the same expression as the argument of the first trace, with y and z exchanged. One can relabel y and z in the second term (one gains an additional minus sign from the antisymmetrization in $dy^{[a}dz^{b]}$) and one gets back (in the limit in which the regulator is removed) the same term as the first one. Continuing with the derivation as presented above one gets for the final action of the Hamiltonian

$$H(\underset{\sim}{N})\Psi(\gamma) = 2 \int d^3x \underset{\sim}{N}(x) \oint_\gamma dy^{[b} \oint_\gamma dz^{a]} \delta(x-y)\delta(x-z)$$
$$\times \Delta_{ab}(\gamma_o^x)\Psi(\gamma_y^z \circ \gamma_{y\,o}). \tag{8.38}$$

Because the equality presented only holds in the limit in which the regulator is removed, the above expression can be thought of as a different regularization of the Hamiltonian constraint introduced before.

It is remarkable that such a compact expression embodies all the information of the time evolution of the Einstein equations in the language of loops.

The constraint algebra involving the Hamiltonian constraint that we derived above has been computed at the formal level in reference [141] and it reproduces the classical algebra. Care should be exercised when computing the constraint algebra, since the problem necessarily requires a regularization, as has been emphasized in the papers by Tsamis and Woodard[142] and Friedman and Jack [143]. The formal computation of the constraints is useful, however, to illustrate a series of computational techniques in loop space and to clarify the meaning of the expressions of the constraints in the loop representation.

8.4　Physical states and regularization

In the previous section we found expressions for the Hamiltonian and diffeomorphism constraints of quantum gravity in the loop representation. In this section we will discuss the construction of solutions to these constraints. We will start with the diffeomorphism constraint and then we will analyze the Hamiltonian. We will elaborate further on the Hamiltonian constraint in chapters 9, 10 and 11. In order to operate properly with the quantum constraints on wavefunctions we will be required to study the regularization of the constraints.

8.4.1　Diffeomorphism constraint

Let us start with the diffeomorphism constraint. In section 1.3.4 we showed that the diffeomorphism constraint acts on functions of loops by infinitesimally deforming the loop argument along a vector $\vec{N}$. The deformation is the same that the loop would suffer if it existed in a spatial manifold on which a diffeomorphism is performed along a vector $\vec{N}$. Therefore if a wavefunction $\Psi(\gamma)$ in the loop representation is to be annihilated by the diffeomorphism constraint it should be invariant under deformations of the loop argument. Such functions are known as knot invariants. Another way of putting this is to say that the function only depends on the knot class of the loop. The knot class of a loop is given by the orbit of the diffeomorphism group in loop space that contains the given loop.

Therefore by considering such functions of loops one immediately solves the diffeomorphism constraint. The diffeomorphism invariance of general relativity therefore is very elegantly coded into knot invariance in the loop representation. There is an abundant literature on the study of knot invariants, and we will return in more detail to issues of knot theory in the next chapter. Notice that the situation is qualitatively different from that in the traditional variables for quantum gravity. There one considered functionals of a spatial metric $\Psi[q]$. The invariance under diffeomorphisms implied that one was dealing with functionals of the "geometry" (or more precisely its diffeomorphism invariant properties) rather than functionals of a metric. The situation is also qualitatively different from that in the connection representation that we discussed in the previous chapter. Again, there one had to consider functions of a connection that were invariant under diffeomorphisms $\Psi[A]$. Although some isolated examples of these can be given, it is quite evident that one can construct many more examples of functions of loops invariant under diffeomorphisms. For instance, functions that depend on the number of intersections of a loop or the number of corners or kinks in the loops are examples of functions that are invariant under diffeomorphisms. So are the "characteristic functions" in loop space: functions that give 1 if the argument is in a certain knot class and zero otherwise. Although we have seen that the use of loops played a role in the connection representation, we see that the shift in point of view offered by the loop representation is very important in the task of finding the physical states that are annihilated by the constraints. We will find many solutions to the constraints in the loop representation of which the counterpart in terms of connections is either not known and is expected to be quite complicated or ill defined. Knot theory captures in a natural way the non-local, topological properties of a theory invariant under diffeomorphisms. The connection between knot theory and quantum gravity was first noticed by Rovelli and Smolin [38].

8.4.2 Hamiltonian constraint: formal calculations

In order to discuss the solutions to the Hamiltonian constraint one needs to introduce a regularization. The issue of the regularization of the Hamiltonian constraint is the subject of intense investigations at present. Basically the problem is that all known regularization procedures are difficult to make compatible with diffeomorphism invariance and typically introduce conflicts or ambiguities in the resulting regularized theory. We will first introduce a point-splitting regularization in loop space and discuss the action of the Hamiltonian constraint on a generic function of loop $\Psi(\gamma)$. We will *not* at the moment assume that the function is invariant under deformations of the loops, i.e., the state will not, in general, be annihilated by the diffeomorphism constraint. This is the most natural thing to do, since the Hamiltonian constraint is an operator that is not invariant under diffeomorphisms and therefore its action is not well defined on the space of knot invariants. In general the action of the Hamiltonian on a knot invariant will produce as a result a function of a loop that is not invariant under diffeomorphisms.

There is a second motivation for considering the action of the Hamiltonian on all function of loops, related to the details of the definitions we give for the constraints. This is due to the fact that the loop derivative that we defined in chapter 1 is not, in general, well defined on functions that are invariant under diffeomorphisms. This can be readily seen. The notion of a loop derivative involves, in general, a change of topology in the loop. Therefore in its definition,

$$\Psi(\pi_o^x \circ \delta\gamma \circ \pi_x^o \circ \gamma) = (1 + \tfrac{1}{2}\sigma^{ab}(x)\Delta_{ab}(\pi_o^x))\Psi(\gamma), \qquad (8.39)$$

it could happen that the loop argument of Ψ in the left-hand side is in a different knot class that that of the right-hand side. The addition of the infinitesimal loop would therefore not amount to a small change in the loop function and the limit involved in the derivative is not well defined. The situation is the loop analog of the derivative of the Heaviside theta function at the origin in elementary calculus. The usual way to deal with this problem (that leads to the calculus of distributions) is to consider the Heaviside function as a limit of a set of differentiable functions. Similarly here we would like to regard the functions invariant under diffeomorphisms as suitable limits of non-invariant functions that are loop differentiable. The action of the Hamiltonian constraint on a diffeomorphism invariant function will also be defined in a limiting process.

There have been several proposals for the Hamiltonian constraint in the loop representation [39, 138, 139, 16, 140]. Some of them do not involve the use of loop derivatives or use derivatives that are different from the one we introduce in this book. All of them, however, are based on the idea of appending an infinitesimal loop to the knot and therefore do not have a clear and unambiguous topological action in terms of knots.

We consider the Hamiltonian introduced in the last section

$$\hat{\mathcal{H}}(\underset{\sim}{N})\Psi(\gamma) = \int d^3x \underset{\sim}{N}(x) \oint_\gamma dy^{[b} \oint_\gamma dz^{a]} \delta(x-y)\delta(x-z)$$
$$\times \Delta_{ab}(\gamma_o^x)\Psi(\gamma_y^z \circ \gamma_{yo}^z). \qquad (8.40)$$

As we pointed out before, the above expression is formal and a regularization is needed for its proper definition. Before discussing the regularization let us qualitatively study the action of the formal constraint on a function of a loop. Taking the results from the connection representation as a guide, we know that the action of the Hamiltonian constraint is different if loops with and without intersections are involved. In the loop representation wavefunctions must take values for *all* piecewise differentiable loops. We will therefore study separately the action of the Hamiltonian constraint on a generic loop function Ψ assuming that the argument is a smooth loop, a loop with a kink or a loop with intersections.

The action of the formal Hamiltonian on a function of a loop $\Psi(\gamma)$ is very simple in the case in which the argument is a smooth non-intersecting loop at the point where the Hamiltonian acts. In that case, in the formal expression of the Hamiltonian there is a single contribution per point x belonging to the loop γ. The contribution is proportional (through a divergent factor) to the double contraction of the tangent to the loop at the point with the loop derivative $\dot{\gamma}^a \dot{\gamma}^b \Delta_{ab} \Psi(\gamma)$ (where $\dot{\gamma}^a$ is the tangent vector to the loop in a certain parametrization). Since one is contracting a symmetric tensor with an antisymmetric one the result vanishes. This is the counterpart in the loop representation of the same result that we found in the connection representation at the formal level: non-intersecting smooth loops yield solutions of the Hamiltonian constraint. In general, the action of the Hamiltonian involves a splitting and rerouting of the argument of the wavefunction. For the case of non-intersecting loops or kinks, the contribution gives back the same loop as the original one since $\gamma_y^z \to \gamma$ and $\gamma_{y\,o}^z \to \iota$ in the limit (or vice-versa depending on the order of y and z along the loop). The rerouting is non-trivial only at intersections. At the formal level of this discussion, the Hamiltonian has a non-vanishing contribution at intersections and kinks but not at points where the loops are smooth.

The fact that the Hamiltonian constraint has a (formally) vanishing action at points where loops are smooth and non-intersecting led [38, 39] to the construction of a historically very important set of "physical states" of quantum gravity by simply considering wavefunctions $\Psi(\gamma)$ with support only on smooth non-intersecting loops, i.e.,

$$\Psi(\gamma) = \begin{cases} \Psi_0(\gamma) & \text{if } \gamma \text{ is smooth and non-intersecting,} \\ 0 & \text{otherwise,} \end{cases} \tag{8.41}$$

where $\Psi_0(\gamma)$ is any knot invariant. Formally the Hamiltonian has vanishing action on this state since it gives no contribution if the loop γ is either smooth (for the reasons explained above) or intersecting (since the state vanishes for such loops). This state has the appearance of a "step function" in loop space. The reader may question the applicability of a

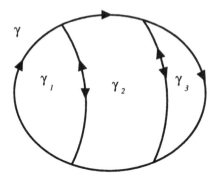

Fig. 8.1. The loop used in the Mandelstam identity that is not satisfied by the naive states

differential operator in loop space to such a state. In principle, the action could be well defined since the Hamiltonian in this case does not change the number of intersections of the loop and therefore has a separate action in the two regions into which the definition of the state partitions the loop space.

Unfortunately, there is a serious objection to these kinds of naive states. This was noticed by Rovelli and Smolin ([39] page 135). The problem is that, as we emphasized at the beginning of this chapter and throughout this book, a state in the loop representation is not any function of a loop, but has to satisfy several properties, among them the Mandelstam identities. The Mandelstam identities imply relations among the values that a wavefunction takes when evaluated on loops with and without intersection. It is easy to check that the above proposed wavefunctions do *not* satisfy the appropriate relations. For instance, consider a non-intersecting loop γ obtained by the composition of loops γ_1, γ_2 and γ_3 as shown in the figure 8.1, and apply the Mandelstam identity

$$\Psi(\gamma_1 \circ \gamma_2 \circ \gamma_3) + \Psi(\gamma_1 \circ \gamma_2 \circ \gamma_3^{-1}) = \Psi(\gamma_2 \circ \gamma_1 \circ \gamma_3) + \Psi(\gamma_2 \circ \gamma_1 \circ \gamma_3^{-1}). \quad (8.42)$$

The first term in the left-hand side is $\Psi(\gamma)$ and all the other terms involve intersections (and multiple lines) between the different components. Therefore the state has vanishing value on all the terms in the expression except on the first where it is $\Psi_0(\gamma)$ and one is led to the contradiction: $\Psi_0(\gamma) = 0$.

One could think of constructing a set of states motivated by the non-intersecting ones by assigning proper values to loops with intersections via the Mandelstam identity. This was suggested in reference[39]. Very recently, the introduction of the spin-network [146] ideas gave a concrete meaning to this construction. There is rapid development at present in trying to exploit these states for physical purposes [144].

There is another way in which states based on non-intersecting loops can be thought of as generating genuine solutions to the Hamiltonian constraint, using the notions of bras and kets. Consider the space of kets $|\Psi>$ and let us assume that we know an inner product in loop space such that the Hamiltonian is a self-adjoint operator (notice that the inner product is not on the physical space but on all states). We define the bra $<\alpha|$ by

$$\Psi(\alpha) \equiv < \alpha|\Psi > . \tag{8.43}$$

Notice that the bras, from their definition, satisfy the Mandelstam identities, for instance $<\alpha| =< \alpha^{-1}|$, etc.

By definition, the action of the Hamiltonian on $\Psi(\alpha)$ is

$$\hat{H}\Psi(\alpha) \equiv < \alpha|H|\Psi >, \tag{8.44}$$

from which one can immediately read off the action of the Hamiltonian on a bra $< \alpha|$, being given by the usual expression in the loop representation. If one now considers a bra $< \alpha|$ with α a smooth loop then $< \alpha|\hat{H}| = 0$. Making use of the assumption that the Hamiltonian is a self-adjoint operator one has that $\hat{H}|\alpha >= 0$ and therefore

$$< \gamma|\hat{H}|\alpha >= \hat{H}\Psi_\alpha(\gamma) = 0. \tag{8.45}$$

That is, if one knows the inner product in the space of loops under which the Hamiltonian is a self-adjoint operator, one can construct a family of functions of loops $\Psi_\alpha(\gamma)$ (where the smooth non-intersecting loop α plays the role of a parameter) that are annihilated by the Hamiltonian constraint simply by taking the inner product $< \gamma|\alpha >$. These states satisfy the Mandelstam constraint. Notice that the wavefunctions depend on a loop γ that can have arbitrary intersections and kinks. Though this construction constitutes an interesting observation, the fact that it relies on the introduction of an inner product in loop space under which the Hamiltonian is self-adjoint makes it of little use in practice.

There is a chance that one could modify the definition of the naive states in order make them compatible with the Mandelstam constraints. In particular, Smolin[145] has a proposal based on the use of an area operator; however, it is not clear whether under the proposed modification one still manages to solve the Hamiltonian constraint.

Let us now discuss the regularized action of the Hamiltonian constraint.

8.4.3 Hamiltonian constraint: regularized calculations

We again consider the Hamiltonian introduced in the last section,

$$\hat{\mathcal{H}}(\underset{\sim}{N})\Psi(\gamma) = \int d^3x \underset{\sim}{N}(x) \oint_\gamma dy^{[b} \oint_\gamma dz^{a]}\delta(x - y)\delta(x - z)$$

$$\times \Delta_{ab}(\gamma_o^x)\Psi(\gamma_y^z \circ \gamma_{y\,o}^z), \qquad (8.46)$$

but we point split one of the Dirac delta functions,

$$\hat{\mathcal{H}}_\epsilon(\underset{\sim}{N})\Psi(\gamma) = \int d^3x \underset{\sim}{N}(x) \oint_\gamma dy^{[b} \oint_\gamma dz^{a]} \delta(x-y)f_\epsilon(y-z)$$
$$\times \Delta_{ab}(\gamma_o^y)\Psi(\gamma_y^z \circ \gamma_{y\,o}^z), \qquad (8.47)$$

where $f_\epsilon(y-z)$ is a usual symmetric regulator. For the sake of concreteness, we can consider a family of Gaussians,

$$f_\epsilon(x-y) = (\pi\epsilon)^{-3/2} \exp\left(\frac{-|x-y|^2}{\epsilon}\right). \qquad (8.48)$$

One can consider other families of regulators, like families of Heaviside functions $f_\epsilon(x,y) = \Theta_\epsilon(x,y)/\epsilon^3$ where $\Theta_\epsilon(x,y) = 3/4\pi$ if $|x-y| < \epsilon$ and zero otherwise. The background metric enters in all cases since one has to compute the distance between x and y.

Notice that there are several possibilities to regularize and the regularized expressions will, in general, be different and coincide only in the limit. For instance, we could have split the other delta function that appears in the definition of the Hamiltonian.

The introduction of the point-splitting implies that the paths that appear in the expression of the regularized constraint do not close a loop. This is equivalent to the introduction of a non-gauge invariant point-splitting in the connection representation, the breaking of gauge invariance being manifest in the loop representation in the appearance of open paths. When the regulators are removed, the open ends of paths coincide and one recovers closed loops and gauge invariance. One could simply choose to work in a regularized framework with open loops and recover gauge invariance only as a limit after regularization. Another procedure is to close the loops by adding arbitrary small paths and restore gauge invariance in the regularized expressions. In the limit, the contributions from the added paths drop out. In the connection representation one does not have any privileged paths to restore gauge invariance in the point-splitting. In the loop representation one can always choose to close the loops through their original trajectory before reroutings and splittings, as was done in references [138, 139], or through other prescribed paths [16, 39]. Notice that these constructions hide implicit assumptions about the behavior of the wavefunctions of loops $\Psi(\gamma)$. It is not true that for all functions the contributions of the infinitesimal added paths drop out in the limit. These kinds of statements imply a certain notion of continuity of the functions in loop space that at the moment is not well understood.

Let us now redo the calculation of the action of the Hamiltonian constraint acting on a function of loops in the case in which it acts on a point of the loop that has no kinks nor intersections. To make the calculation as explicit as possible we introduce a parametrization for the loop $\gamma(s)^a$ with $s \in [0, 1]$ and we rewrite the Hamiltonian (8.47),

$$\hat{\mathcal{H}}_\epsilon(\underset{\sim}{N})\Psi(\gamma) = \int_0^1 ds \int_0^1 dt \dot{\gamma}^{[b}(s)\dot{\gamma}^{a]}(t)\underset{\sim}{N}(\gamma(s))$$
$$\times f_\epsilon(\gamma(s) - \gamma(t))\Delta_{ab}(\gamma_o^s)\Psi(\gamma_s^t \circ \gamma_{s\,o}^t). \qquad (8.49)$$

We now split the integral in t,

$$\hat{\mathcal{H}}_\epsilon(\underset{\sim}{N})\Psi(\gamma) = \left(\int_0^1 ds \int_s^1 dt + \int_0^1 ds \int_0^s dt \right) \dot{\gamma}^{[b}(s)\dot{\gamma}^{a]}(t)$$
$$\times \underset{\sim}{N}(\gamma(s)) f_\epsilon(\gamma(s) - \gamma(t))\Delta_{ab}(\gamma_0^s)\Psi(\gamma_s^t \circ \gamma_{s\,o}^t). \quad (8.50)$$

The above expression involves open loops, as we discussed. One needs to close them appending infinitesimal loops going from s to t in one of the terms and from t to s in the other. Since we assume the point of action is smooth, there is no ambiguity in the closing process and one gets $\gamma_s^t \circ \gamma_{s\,o}^t \to \gamma^{-1}$ when $t > s$ and $\gamma_s^t \circ \gamma_{s\,o}^t \to \gamma$ when $t < s$.

If we now replace, in the limit $\epsilon \to 0$, $\dot{\gamma}^a(t) \to \dot{\gamma}^a(s) + \ddot{\gamma}^a(s)(t - s)$ and $\gamma^a(s) - \gamma^a(t) \to \dot{\gamma}^a(s)(s - t)$, the terms involving two tangent vectors cancel out, exactly as they did in the formal calculation. Introducing the variable u, defined as $t - s$ for the first integral and $s - t$ for the second, one is left with

$$\hat{\mathcal{H}}_\epsilon(\underset{\sim}{N})\Psi(\gamma) = 2 \int_0^1 \int_0^1 ds\,du\, u\dot{\gamma}^{[b}(s)\ddot{\gamma}^{a]}(s)\underset{\sim}{N}(\gamma(s)) f_\epsilon(u\dot{\gamma}(s))\Delta_{ab}(\gamma_0^s)\Psi(\gamma)$$
$$(8.51)$$

and noticing that with the Gaussian regulator

$$\frac{\epsilon}{2|\dot{\gamma}|^2}\partial_u f_\epsilon(u\dot{\gamma}(s)) = -u f_\epsilon(u\dot{\gamma}(s)), \qquad (8.52)$$

we get for the leading action of the Hamiltonian,

$$\hat{\mathcal{H}}_\epsilon(\underset{\sim}{N})\Psi(\gamma) = -\frac{1}{\pi^{3/2}\epsilon^{1/2}} \int_0^1 ds \frac{\dot{\gamma}^{[b}(s)\ddot{\gamma}^{a]}(s)}{|\dot{\gamma}(s)|^2}\underset{\sim}{N}(\gamma(s))\Delta_{ab}(\gamma_0^s)\Psi(\gamma). \quad (8.53)$$

We see that the action of the Hamiltonian is divergent. This will be the case for all kinds of loops and points in the loop and we will be forced to define a renormalized Hamiltonian as the regulated operator that has a finite limit for $\epsilon \to 0$, i.e.,

$$\hat{H} = \lim_{\epsilon \to 0} \sqrt{\epsilon}\hat{H}^\epsilon. \qquad (8.54)$$

We see that the action of the Hamiltonian constraint on a smooth point of a loop, after the constraint is appropriately regularized and renormalized, is non-vanishing, contrary to what the naive calculation suggested. The resulting terms depend on higher derivatives of the loop and are usually referred to as "acceleration terms" [134]. The result (8.53) is invariant under reparametrization of the loops but depends explicitly on a background metric through $|\dot{\gamma}(s)|^2$, reflecting the fact that the regulator we took is not invariant under diffeomorphisms.

Notice that the expression (8.53) can be reinterpreted as the action on a loop state of a diffeomorphism along the vector field $\dot{\gamma}^{[b}(s)\ddot{\gamma}^{a]}(s)/|\dot{\gamma}(s)|^2$. This is not a standard diffeomorphism along a fixed external vector field, but the vector field is defined by the loop. If the loop has intersections, then this vector field is not well defined. If the loop is smooth, however, one could construct smooth vector fields $\vec{N}$ on the manifold such that on the loop take the same value as $\dot{\gamma}^{[b}(s)\ddot{\gamma}^{a]}(s)/|\dot{\gamma}(s)|^2$ and the wavefunction should be annihilated by them (if it is invariant under diffeomorphisms). Therefore we see that the contribution from the acceleration terms vanishes if one considers wavefunctions of smooth loops that are invariant under diffeomorphisms and one can solve the Hamiltonian constraint. This is an improvement on the situation in the connection representation. As we pointed out in the previous chapter, there one also finds acceleration terms when one regulates using point-splitting and that means that the Wilson loops do not satisfy the Hamiltonian constraint. In the loop representation, since we can deal with diffeomorphism invariant states, one can make the contributions from the acceleration terms vanish. Therefore we see that — ignoring the objections already stated concerning the Mandelstam constraints — the naive states based on loops without intersections also solve the constraints when a proper regularization is taken into account.

Let us now consider the action of the Hamiltonian at a point where the loop has a kink [138], i.e., a discontinuity in the tangent vector to the curve, but there is only one line going in and out of the point, i.e., there are no intersections. Such a situation is illustrated in the figure 8.2. In the expression of the Hamiltonian there is now a contribution of lower order than in the previous case, stemming from the fact that at the point of the kink s_0 there are two possible values for the tangent to the loop which we denote $\dot{\gamma}_+^a$ and $\dot{\gamma}_-^a$. Therefore, in the formal computation one gains a term $\dot{\gamma}_+^a\dot{\gamma}_-^a\Delta_{ab}$ that does not vanish. The regularized calculation gives as result

$$\hat{\mathcal{H}}_\epsilon(\underset{\sim}{N})\Psi(\gamma) = 2\frac{\dot{\gamma}_+^{[b}\dot{\gamma}_-^{a]}\underset{\sim}{N}(x_i)}{(\pi\epsilon)^{3/2}}\int_0^1\int_0^1 ds\,dt$$

$$\times \exp\left(-\frac{s^2+t^2+2st\vec{\dot{\gamma}}_+\cdot\vec{\dot{\gamma}}_-}{\epsilon}\right)\Delta_{ab}(\gamma_o^{x_i})\Psi(\gamma),\quad (8.55)$$

where x_i is the point at which the kink lies. If there were more than one kink in the loop, the expression would be the same for each of them and a

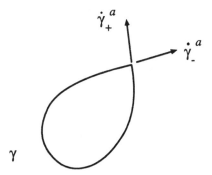

Fig. 8.2. A loop with a kink. Notice the convention for the tangent vectors $\dot\gamma^a_\pm$.

discrete sum along all the kinks should be introduced. In this expression we have assumed that a parametrization was chosen such that $|\dot\gamma_\pm|^2 = 1$.

The integral can be explicitly computed, giving

$$\hat{\mathcal{H}}_\epsilon(\underset{\sim}{N})\Psi(\gamma) = 2\frac{\dot\gamma^{[b}_+\dot\gamma^{a]}_-}{\sqrt{1-(\vec{\dot\gamma}_+\cdot\vec{\dot\gamma}_-)^2}}\frac{\underset{\sim}{N}(x_i)}{(\pi\epsilon)^{1/2}}$$
$$\times\left(\frac{1}{4}-\frac{\arcsin(\vec{\dot\gamma}_+\cdot\vec{\dot\gamma}_-)}{2\pi}\right)\Delta_{ab}(\gamma^{x_i}_o)\Psi(\gamma). \qquad (8.56)$$

Again, we see this contribution from the Hamiltonian has to be renormalized with a factor $\sqrt{\epsilon}$ to obtain a finite contribution. We also see that the expression is background dependent through the angle that the two tangents to the loop at the kink form measured with the background metric. The expression of the action of the Hamiltonian on a kink can be rewritten in terms of a quantity called the normalized area element,

$$\sigma^{ab}_N(\gamma) = \frac{\dot\gamma^{[b}_+\dot\gamma^{a]}_-}{\sqrt{1-(\vec{\dot\gamma}_+\cdot\vec{\dot\gamma}_-)^2}}. \qquad (8.57)$$

The word normalized is used in the sense that the norm of the vector dual to the area element is independent of the angle of the tangent vectors of the loop and therefore is independent of the background metric introduced for the regularization. The normalized area element is ill defined when the two tangent vectors coincide. However the product that appears in the action of the Hamiltonian on a kink,

$$\sigma^{ab}_N(\gamma)\left(\frac{1}{4}-\frac{\arcsin(\vec{\dot\gamma}_+\cdot\vec{\dot\gamma}_-)}{2\pi}\right), \qquad (8.58)$$

is well defined. It vanishes in the limit in which the two tangent vectors are the same and therefore the loop is smooth. This agrees with the result that we derived before in which the tangent–tangent contribution to the Hamiltonian at smooth points vanished, the leading order being given by the acceleration terms. We will notice a different behavior in the case of intersections.

It is remarkable that much like in the case of the acceleration terms, the action of the Hamiltonian on a kink can be reduced to a diffeomorphism. Consider the usual expression for the diffeomorphism constraint,

$$\hat{C}(\vec{N})\Psi(\gamma) = \int d^3x N^a(x) \oint_\gamma dy^b \delta(x - y)\Delta_{ab}(\gamma_o^y)\Psi(\gamma) \qquad (8.59)$$

and consider the particular vector field

$$N_\epsilon^a(x) = \underset{\sim}{M}(x) \oint_\gamma dz^a \frac{1}{\sqrt{\pi\epsilon}} \exp\left(-\frac{|z - x|^2}{\epsilon}\right). \qquad (8.60)$$

It is immediate to see that,

$$\hat{\mathcal{H}}(\underset{\sim}{M})\Psi(\gamma) = \lim_{\epsilon \to 0} \frac{1}{\pi\sqrt{\epsilon}}\hat{C}(\vec{N}_\epsilon)\Psi(\gamma). \qquad (8.61)$$

Therefore we see that the action of this particular diffeomorphism on the loop state is exactly the same as that of the Hamiltonian in the regularized limit if the loop is smooth with at most a finite number of kinks and no intersections. We therefore see another difference with the connection representation, where Wilson loops with kinks simply failed to solve the Hamiltonian constraint. In the loop representation, if one considers states that have support on loops with kinks and are diffeomorphism invariant, they automatically solve the Hamiltonian constraint (again there can be a conflict with the Mandelstam identities that prevents us from considering such functions as true states of the gravitational field).

We finally discuss the case of a loop with intersections. We will focus our attention on double intersections but higher order ones are a straightforward generalization. The calculation is very similar to the one we performed for the case of kinks, except that now there are four possible contributions coming from taking the four lines adjacent to the intersection in groups of two. The contribution per pair is exactly the same as that of a single kink (8.56) with the difference that the argument of the wavefunction is not the loop γ in the regularized limit but a rerouting of the loop at the intersection takes place. The vectors $\dot{\gamma}_\pm^a$ in this case correspond to the two tangent vectors in the particular pair of lines considered. An orientation convention has to be determined *a priori* as was done in the case of the kinks in figure 8.2.

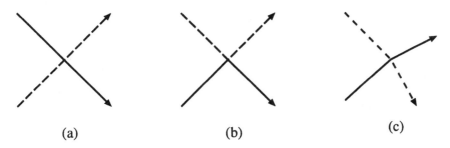

(a)   (b)   (c)

Fig. 8.3. Three different possibilities at a double intersection: (a) a straight-through intersection; (b) intersection with a kink; (c) intersection with more than two tangent vectors. Cases (b) and (c) are usually referred to as cases with "kinks at the intersection"

At a double intersection there are several different possibilities, illustrated in figure 8.3. The case of a straight-through intersection gives a qualitatively different result than the cases with kinks at the intersection. In the former case, the four contributions coming from taking the lines in pairs add up in such a way that the $\arcsin(\vec{\gamma}_+ \cdot \vec{\gamma}_-)$ terms in (8.56) all drop out and we get

$$\hat{\mathcal{H}}_\epsilon(\underset{\sim}{N})\Psi(\gamma) = 2\sigma_N^{ab}(\gamma)\frac{N(x_i)}{(\pi\epsilon)^{1/2}}\Delta_{ab}(\gamma_o^{x_i})\Psi(\gamma_{x_i}^{x_i} \circ \gamma_{x_i o}^{x_i}). \qquad (8.62)$$

It is remarkable that the expression depends on the tangent vectors only through the normalized area element and therefore it is independent of the background metric used for the regularization. This result was first noticed by Rovelli and Smolin [140]. Unfortunately, the resulting expression is ill defined in the limit in which the tangent vectors coincide, as opposed to the case of a single kink.

If there are kinks at the intersection, the above cancellation of the $\arcsin(\vec{\gamma}_+ \cdot \vec{\gamma}_-)$ terms does not happen and one is left with a background dependent result. Several terms appear, some having the same rerouting effect as in the straight-through intersection but others having as the argument of the wavefunction the loop γ, as happened at a kink.

The action of the Hamiltonian on an intersection cannot be rewritten as a genuine diffeomorphism as was the case of the action on a kink or the acceleration terms. Attempts have been made to interpret the Hamiltonian at intersections in this way ("shift operator") [39, 139] but they all amount to a reinterpretation of the terms we have derived, without a genuine connection with diffeomorphisms. These reinterpretations may help to visualize the action of the Hamiltonian at intersections. At a smooth point in the loop the action of the Hamiltonian can be viewed as a diffeomorphism along the tangent to the loop.

As can be concluded from this section, the action of the regularized Hamiltonian in loop space is only non-trivial at points where the loops have intersections. The resulting action of the Hamiltonian at such points is relatively simple, it amounts to the sum of terms consisting of a straight-forward rerouting of the argument of the wavefunction acted upon by a loop derivative contracted with the normalized area element of the loop at the intersection point.

At this point it is worthwhile pondering whether the point-splitting procedure introduced has been enough to produce well defined expressions for the constraints in the loop representation. The answer is positive if one makes certain assumptions about the wavefunctions considered. A strong assumption is the existence of a loop derivative of the wavefunctions. As was mentioned above, the loop derivative is ill defined for wavefunctions that are diffeomorphism invariant. In general, the action of appending an infinitesimal loop does not preserve the knot class of a given loop. Moreover, the particular way in which the infinitesimal loop is added can influence the final result. The way in which this conflict may be resolved is through the use of suitable limiting procedures for the definition of the wavefunctions, such that they are diffeomorphism invariant in the limit. Outside the limit, the loop derivative is well defined. A practical implementation of this proposal is the use of extended loops, to which we will return in chapter 11. Another proposal is to take the limits involved in a different way such that loop derivatives do not explicitly appear. We refer the reader to reference [140] for more details.

8.5   Conclusions

We have applied the loop representation ideas to the quantization of general relativity based on the Ashtekar new variables formulation. We introduced explicit expressions for the constraint equations at a formal and regularized level. We discussed some general issues concerning the space of states of the theory. In the following chapters we will discuss applications of these ideas. In the next chapter we will discuss the inclusion of matter and the use of approximations. In chapter 10 we will elaborate on the connections with knot theory. In chapter 11 we will discuss a regularization that gives rise to a new representation in terms of extended loops.

9
Loop representation:
further developments

9.1 Introduction

In the previous chapter we discussed the basics of the loop representation for quantum gravity. We obtained expressions for the constraints at both a formal and a regularized level and discussed generalities about the physical states of the theory. In this chapter we would like to discuss several developments that are based on the loop representation. We will first discuss the coupling of fields of various kinds: fermions using an open path formalism, Maxwell fields in a unified fashion and antisymmetric fields with the introduction of surfaces. These examples illustrate the various possibilities that matter couplings offer in terms of loops. We then present a discussion of various ideas for extracting approximate physical predictions from the loop representation of quantum gravity. We discuss the semi-classical approximation in terms of weaves and the introduction of a time variable using matter fields and the resulting perturbation theory. We end with a discussion of the loop representation of $2 + 1$ gravity as a toy model for several issues in the $3 + 1$ theory.

9.2 Inclusion of matter: Weyl fermions

As we did for the Yang–Mills case, we now show that the loop representation for quantum gravity naturally accommodates the inclusion of matter. In the Yang–Mills case, in order to accommodate particles with Yang–Mills charge one needed to couple the theory to four-component Dirac spinors. A Dirac spinor is composed of two two-component spinors that transform under inequivalent representations of the group. This made the addition of matter complicated and one had to resort to the staggered fermion techniques. Although one could couple Dirac fermions to gravity, in the gravitational case the simplest and most natural kind

209

of matter to couple would be uncharged spinning particles. These are described by two-component Weyl fermions. From a particle point of view we will be studying the coupling of neutrinos to general relativity, which is described by the Einstein–Weyl fermion theory. The first discussion of this system in terms of the new variable formulation is due to Morales-Técotl and Rovelli [149].

In order to describe Weyl fermions we need to use two-component spinors. We recall some basic definitions of these mathematical objects, a more complete treatment is in the appendix of reference [2]. Consider a two-dimensional complex vector space. Consider a two-form in that space ϵ_{AB} and its inverse ϵ^{AB}, defined by $\epsilon_{AB}\epsilon^{BC} = \delta_A{}^C$. The linear mappings $L_A{}^C$ which preserve the two-form ϵ_{AB} must have unit determinant, i.e., they are elements of $SL(2,C)$. The two-forms ϵ provide an isomorphism between the two-dimensional vector space and its dual, which we can denote by raising and lowering of indices with the following conventions (care should be exercised because of the antisymmetry of ϵ_{AB}): $\eta^A = \epsilon^{AB}\eta_B$ and $\eta_B = \eta^A\epsilon_{AB}$. Since the elements of this vector space are complex, a natural notion arises of the vector space of the complex conjugate elements and its dual. A vector in the complex conjugate space is denoted by a prime in its index $\eta^{A'}$ or $\eta_{A'}$ if it is in its dual. Primed indices are raised and lowered with the matrix $\bar{\epsilon}^{A'B'}$ and $\bar{\epsilon}_{A'B'}$.

In terms of two spinors one can define a vector space V of objects of the form $\beta^{AA'}$ such that $\bar{\beta}^{AA'} = -\beta^{AA'}$. It is straightforward to check that this has the structure of a four-dimensional real vector space equipped with a natural metric $\epsilon_{AB}\epsilon_{A'B'}$ of signature $(-,+,+,+)$. Consider now a four-dimensional spacetime and a fiber bundle over it with fibers isomorphic to the two-dimensional vector space introduced above. It is natural to identify the tangent space at each point of the spacetime with V,

$$\sigma^a_{AA'}\beta^{AA'} \equiv \beta^a \tag{9.1}$$

in such a way that the metric of V is mapped to the metric of spacetime, $g^{ab} = \sigma^a_{AA'}\sigma^b_{BB'}\epsilon^{AB}\bar{\epsilon}^{A'B'}$. If σ exists globally on the spacetime we say that it admits an $SL(2,C)$ spinor structure. Objects of the form η^A are called unprimed spinors and those of the form $\eta^{A'}$ primed spinors; the matrices σ are called soldering forms. The role of the soldering forms is the analogue in spinor language of the role of the tetrad fields in tetradic language. Both entities carry enough information to reconstruct the spacetime metric and they are determined by the metric up to local transformations ($SO(3,1)$ in the case of the tetrads, $SL(2,C)$ in the case of the soldering forms). Their relation can be explicitly written $\sigma^a_{AB'} = e^a_I\sigma^I_{0\,AB'}$ where the $\sigma^I_{0\,AB'}$ matrices are constant $SL(2,C)$ matrices. A basis of such matrices is given by $\sigma^I = (\mathbf{1}, \tau^i)$, where τ^i are $i\sqrt{2}/2$ times the Pauli matrices [148].

The Lagrangian for general relativity coupled to Weyl fermions in terms of self-dual first order variables was independently introduced by Jacobson [147] and Ashtekar, Romano and Tate [133]. It is given by

$$S(e, A, \bar{\psi}, \psi) = \int d^4x \left[e\, e^a_J e^b_K F^{JK}_{ab} + \sqrt{2}\, e\, e^a_I \sigma^I_{0\,AA'} \bar{\psi}^{A'} D_a \psi^A \right], \quad (9.2)$$

where the notation is the same as in section 7.3. The fields ψ^A and $\bar{\psi}^{A'}$ are Grassmann-valued (anticommuting) $SL(2,C)$ spinors. The covariant derivative on spinors is defined as $D_a \psi^A \equiv \partial_a \psi^A + A_a{}^A{}_B \psi^B$ and the self-dual connection is defined in terms of the connection defined in section 7.3.3 by $A_a{}^A{}_B \equiv A_{aIJ} \sigma_0^{I}{}^A{}_{C'} \sigma_0^{J}{}^{C'}{}_B$.

One can perform a canonical decomposition of this action along the same lines as that performed in chapter 7 for pure gravity. We will not give the details here (they are discussed in reference [2]). The main point is that the introduction of the unit normal n^a introduces an isomorphism in the spinor space that casts the formalism in terms of $SU(2)$ spinors.

$SU(2)$ spinors are defined in the same fashion as $SL(2,C)$ spinors but one introduces an additional structure, a Hermitian inner product among spinors defined by $< \psi | \phi > \equiv \bar{\psi}^{A'} G_{A'A} \phi^A$ with $\bar{G}_{AA'} = G_{AA'}$. It follows that the transformations that leave invariant *both* ϵ_{AB} and $G_{AA'}$ are $SU(2)$ transformations. The metric G defines an operation "$\dagger$" relating the primed and unprimed spinors $(\beta^A)^\dagger \equiv -\epsilon^{AB} G_{BA'} \bar{\beta}^{A'}$. If one now considers the space H of objects $\beta^A{}_B$ such that $\beta^A{}_A = 0$ and $(\beta^\dagger)^A{}_B = \beta^A{}_B$, it turns out that it has the structure of a three-dimensional vector space equipped with a positive definite metric $(\beta, \eta) = -\beta^A{}_B \eta^B{}_A$. It can then be made isomorphic to the tangent space of a curved three-manifold with metric $q^{ab} \equiv -\sigma^{aA}{}_B \sigma^{bB}{}_A$, the matrices σ again are called soldering forms and are related to the (undensitized triads) by $\sigma^{aA}{}_B = E^a_i \tau^i$ with τ^i as defined above.

Continuing with the discussion of the canonical decomposition of the action, the introduction of the unit normal n^a and its associated $SL(2,C)$ spinor $n^{AA'} = i\sqrt{2} n^a \sigma_a^{AA'}$ gives the matrix $G^{AA'}$ that implements the Hermitian inner product that introduces the $SU(2)$ spinors in the formalism. The three- and four- dimensional soldering forms are related by $\sigma_a{}^A{}_B \equiv q^c_a \sigma_c^{AA'} G_{A'B}$. The canonical variables end up being A^i_a and ψ^A and the corresponding canonically conjugate momenta are $\tilde{E}^a_i$ and $\tilde{\pi}_A = -ie(\psi_A)^\dagger$. The theory has the same constraints as usual general relativity (the theory is invariant under the same symmetries) but the constraints are appropriately modified to generate the corresponding transformations in the fermionic variables. The constraints are given by

$$\mathcal{G}^i = D_a \tilde{E}^{ai} - \frac{i}{\sqrt{2}} \tilde{\pi}_A \psi_B \sigma_0^{iAB}, \quad (9.3)$$

$$C_a = \tilde{E}_i^b F_{ab}^i + \frac{i}{\sqrt{2}} \tilde{\pi}_A D_a \psi^A, \tag{9.4}$$

$$\mathcal{H} = \epsilon_{ijk} \tilde{E}_i^a \tilde{E}_j^b F_{ab}^k - 2\tilde{E}_i^a \tilde{\pi}_A D_a \psi_B (\sigma_0)^{iAB}. \tag{9.5}$$

The Weyl field is Grassmann-valued, so the canonical Poisson bracket between $\tilde{\pi}_A$ and ψ^B has a $+$ sign.

We now define an algebra of gauge invariant operators for the theory. Apart from the usual T variables constructed with the connection A_a^i one can define the following variables, based on open paths such as the ones defined in chapter 1:

$$Z(\eta_x^y) = \psi^A(x) U_A{}^B(\eta_x^y)\psi_B(y), \tag{9.6}$$

$$Y(\eta_x^y) = \tilde{\pi}^A(x) U_A{}^B(\eta_x^y)\psi_B(y). \tag{9.7}$$

These objects form a closed algebra under Poisson brackets with the T variables. One could define two other variables, one by considering ψ and $\tilde{\pi}$ in the reverse order in Y and another with two $\tilde{\pi}$s at the ends. Although one does not need these variables to write the Hamiltonian they are needed to write other gauge invariant quantities.

The open path variables satisfy a series of identities; first of all notice that the dependence is on a path, in the sense of chapter 1, so retraced portions do not contribute. Moreover, they satisfy the relations

$$Z(\eta_y^x) = Z(\eta_x^y), \tag{9.8}$$

$$Z(\eta_x^y \circ \beta_y^z)T^0(\gamma) = Z(\eta_x^y \circ \gamma \circ \beta_y^z) + Z(\eta_x^y \circ \gamma^{-1} \circ \beta_y^z), \tag{9.9}$$

$$Z(\eta_x^{y_1})Z(\eta_x^{y_2})Z(\eta_x^{y_3}) = 0. \tag{9.10}$$

The first identity (retracing) stems from the fact that the spinor fields are Grassmanian and as a consequence $U(\gamma)^A{}_B = U(\gamma^{-1})_B{}^A$ and also $U(\gamma)_{AB} = U(\gamma^{-1})_{AB}$. The second identity is the Mandelstam identity for open paths. In that identity the loop γ is connected by a tree to the point y to connect with the open path. The third identity, which is also valid for three open paths ending at the same point comes from the fact that the spinor fields are Grassmanian and being two-component objects one cannot have more than two at a given point. These identities are the same as those we found in chapter 6 for Yang–Mills theories coupled to fermions, with the exception of the retracing identity, which was absent in that case.

The algebra of these quantities is

$$\{Z(\eta_x^y), Z(\gamma_z^w)\} = 0, \tag{9.11}$$

$$\{Y(\eta_x^y), Z(\gamma_z^w)\} = \delta(x - w)Z(\gamma_z^w \circ \eta_x^y) + \delta(x - z)Z(\gamma_w^z \circ \eta_x^y), \tag{9.12}$$

$$\{Y(\eta_x^y), Y(\gamma_z^w)\} = \delta(x - w)Y(\gamma_z^w \circ \eta_x^y) + \delta(y - z)Y(\eta_x^y \circ \gamma_w^z), \tag{9.13}$$

and their commutators with the T variables can be seen in reference [149] but are similar to the commutators of the T variables with T^0 since the fermionic parts do not contribute. The algebra of the Z and Y operators can be viewed, as in the Yang–Mills case, as a set of rules of fusion and splitting of paths.

The diffeomorphism and Hamiltonian constraint can be written purely in terms of the T variables and the Y variable but we will not present a detailed derivation here.

A quantum representation of this algebra can be obtained in terms of operators that act on a space of wavefunctions of loops and open paths. One can reduce the loop dependence, using Mandelstam identities, to a single loop and a series of open paths $\beta_{x_1}^{y_1}, \ldots, \beta_{x_n}^{y_n}$. The quantum representation is given by

$$\hat{Z}(\pi_x^y)\Psi(\beta_{x_1}^{y_1}, \ldots, \beta_{x_n}^{y_n}, \gamma) = \Psi(\pi_x^y, \beta, \gamma), \tag{9.14}$$

$$\hat{Y}(\pi_x^y)\Psi(\beta, \gamma) = i \sum_{k=1}^{n} \left[\delta(x - x_k)\Psi(\ldots, \beta_{y_k}^{x_k} \circ \pi_x^y, \ldots, \gamma) \right.$$
$$\left. + \delta(x - y_k)\Psi(\ldots, \beta_{x_k}^{y_k} \circ \pi_x^y, \ldots, \gamma) \right]. \tag{9.15}$$

The operator Z simply appends the open path it has as argument to the wavefunction. The operator Y appends its open path argument at the beginning and at the end of each of the open paths on which the wavefunction depends.

The quantum constraints can be written in a straightforward fashion. We will not discuss in detail the realization of the diffeomorphism constraint. The effect is the expected geometric one: the loops and paths are deformed along the diffeomorphism flow. The additional terms in the constraint take care of moving the end points of the open paths. It is immediate to construct the solution space to that constraint in a geometric fashion, much in the same spirit as in the purely gravitational case. The solution space is composed of wavefunctions of the generalized knot classes, the sets of knots and open paths that are related by the orbits of the diffeomorphism group. The concept of knotting when open paths are involved is non-trivial due almost only to the possible presence of intersections. If no intersections (or self-intersections) are present, all open paths are equivalent under diffeomorphisms. The "almost only" accounts for the fact that configurations with non-intersecting paths can be diffeomorphism inequivalent if the number of paths is different.

We will not present in an exhaustive fashion the general action of the Hamiltonian on a wavefunction of a multipath with arbitrary intersections and self-intersections, since it resembles very closely the case of pure gravity when written in terms of multiloops [138]. The action of the operator can be found by writing it in terms of the algebra of gauge invariant operators that we introduced above or in terms of the loop transform,

$$\Psi(\beta_{x_1}^{y_1}, \ldots, \beta_{x_n}^{y_n}, \gamma) = \int dA \int d\psi \Psi[A, \psi] Z(\beta_{x_1}^{y_1}) \times \cdots \times Z(\beta_{x_n}^{y_n}) T^0(\gamma). \tag{9.16}$$

We would, however, like to illustrate the new contributions that arise due to the fermionic parts by considering the action of the Hamiltonian on a state dependent on a single, possibly self-intersecting, path. In order to compute this, we consider the action of the fermionic part of the Hamiltonian on one of the Zs that appear in the loop transform,

$$\hat{H}_{\text{Weyl}}(z) Z(\pi_x^y) = -2 \hat{\tilde{E}}_i^a(z) \hat{\tilde{\pi}}_A(z) D_a \hat{\psi}_B(z) (\sigma_0)^{iAB} Z(\pi_x^y)$$

$$= -2 D_a \psi_B(z) (\sigma_0)^{iAB} \frac{\delta}{\delta A_a^i(z)} \frac{\delta}{\delta \psi_A(z)} Z(\pi_x^y). \tag{9.17}$$

The result is

$$\hat{H}_{\text{Weyl}}(z) Z(\pi_x^y) = 2(\delta(z-x) X^{ax}(\pi) D_a^x - \delta(z-y) X^{ay}(\pi) D_a^y) Z(\pi_x^y), \tag{9.18}$$

where D_a^y is the Mandelstam covariant derivative we introduced in chapter 1, with the generalization that it acts not only at the end point of the open path but also at the beginning. From this result we can read off the action of the Weyl part of the Hamiltonian on a state dependent on a single open path,

$$\hat{H}_{\text{Weyl}}(z) \Psi(\pi_x^y, \gamma) = 2(\delta(z-x) X^{ax}(\pi) D_a^x - \delta(z-y) X^{ay}(\pi) D_a^y) \Psi(\pi_x^y, \gamma). \tag{9.19}$$

The geometric meaning of the Weyl part of the Hamiltonian is to translate the ends of the open paths in the direction of the tangent vector at those points. It is remarkable that the action of the purely gravitational Hamiltonian we discussed in the previous section on non-intersecting loops has a rather similar effect, in the sense that it can be interpreted as a diffeomorphism along the loop. In this sense, if one considers the purely gravitational Hamiltonian in terms of loops and extends naturally its action to open paths one is automatically left with the Einstein–Weyl fermion theory, without the need to input details about the Weyl Hamiltonian. In this sense the loop representation of quantum gravity naturally "predicts" the Dirac equation for fermions [149].

As in the previous section, one must regularize and renormalize the operators; the techniques involved are similar so we omit a detailed discussion. Morales Técotl and Rovelli [149] study the issue in detail using regularization ideas that we will discuss in section 9.5.1 in the context of pure gravity using a matter clock.

9.3   Inclusion of matter: Einstein–Maxwell and unification

Once a theory is cast in terms of a connection we can build a loop representation for it. We have done so for Yang–Mills theories and also for general relativity. What happens when one couples such theories? The obvious answer is to consider a mixed loop representation with some loops associated with the connection of a certain theory and others to the other. Such an approach can be pursued for all gauge fields that are coupled in gauge invariant fashion, as are all Yang–Mills fields coupled to gravity. In essence, the resulting description is faithful to the spirit of this book in which each gauge field has been treated as quantizable in its own right. For many years, however, the trend in particle physics has been towards viewing the different gauge fields as different low energy manifestations of a single unified theory that is apparent only at high energies. The question therefore arises: are loop descriptions estranged from unification ideas or can they be made compatible to a certain extent? Such a subject is largely unexplored at present. What we would like to show in this section is that the seeds for a unified description of gauge fields in terms of loops may be present. We will illustrate the idea with the simplest possible example, that of Einstein–Maxwell theory. However, we will see that the idea goes through largely unchanged if one replaces Maxwell theory with a Yang–Mills field.

The Einstein–Maxwell theory in the canonical formulation based on Ashtekar's new variables is described in terms of the usual variables for the gravitational part plus a $U(1)$ vector potential a_a, its associated field tensor $f_{ab} = \partial_{[a}a_{b]}$ and the electric field $\tilde{e}^a$. The constraint equations are

$$\partial_a \tilde{e}^a = 0, \tag{9.20}$$

$$D_a \tilde{E}^{ai} = 0, \tag{9.21}$$

$$\tilde{E}^a_i F^i_{ab} + i\frac{\sqrt{2}}{4}\tilde{e}^a f_{ab} = 0, \tag{9.22}$$

plus a Hamiltonian constraint. The first equation is the $U(1)$ Gauss law of Maxwell theory, the second set is the gravitational Gauss law and the third set is the diffeomorphism constraint. Due to the fact that the Gauss laws for both gauge groups appear separately one could build, as argued above, a loop representation based on two separate sets of loops, one associated with the $U(1)$ invariance and other with the $SU(2)$ invariance. In that loop representation, each set of loops would operate independently and be subject to separate Mandelstam identities.

We now show that the above gauge symmetries can be cast in a unified fashion, suitable for the introduction of a loop representation based on a *single* kind of loop that still captures the information of the two interacting theories. Let us introduce a $U(2)$ connection $\mathcal{A}_a$ in the following way,

$$\mathcal{A}_a \equiv A_a^i \tau_i + i a_a \mathbf{1}, \tag{9.23}$$

where τ_i are the Pauli matrices with our usual conventions and $\mathbf{1}$ is the identity matrix in two dimensions. One can similarly introduce a $U(2)$ electric field $\tilde{\mathcal{E}}^a$ and from the $U(2)$ connection build a field tensor $\mathcal{F}_{ab}$ and a covariant derivative $\mathcal{D}_a$.

The remarkable fact is that the constraints we wrote above can now be written

$$\mathcal{D}_a \tilde{\mathcal{E}}^a = 0, \tag{9.24}$$

$$\text{Tr}(\tilde{\mathcal{E}}^a \mathcal{F}_{ab}) = 0, \tag{9.25}$$

and the Hamiltonian constraint can also be written in terms of these variables, though we will not need its particular expression here. We refer the reader to reference [97] for more details. The point is that at the kinematical level, the theory looks *exactly the same as vacuum general relativity* but with an enlarged gauge group, $U(2)$ instead of $SU(2)$. This construction can also be carried out for general relativity coupled to a Yang–Mills field with gauge group G^{YM}, the resulting group is $SU(2) \times G^{YM}$ [150].

Therefore one can now construct a loop representation based on a single kind of loop for the $U(2)$ symmetry. In such a representation the unified Gauss law (9.24) is automatically solved. The wavefunctions are functions of multiloops subject to the $U(2)$ Mandelstam constraints,

$$\Psi(\gamma_1 \circ \gamma_2) = \Psi(\gamma_2 \circ \gamma_1), \tag{9.26}$$

$$\Psi(\gamma_1, \gamma_2, \gamma_3) = \Psi(\gamma_1 \circ \gamma_2, \gamma_3) + \Psi(\gamma_2 \circ \gamma_3, \gamma_1) + \Psi(\gamma_3 \circ \gamma_1, \gamma_2)$$
$$- \Psi(\gamma_1 \circ \gamma_2 \circ \gamma_3) - \Psi(\gamma_1 \circ \gamma_3 \circ \gamma_2). \tag{9.27}$$

Two comments are in order. First notice that there is no retracing identity, $\Psi(\gamma) \neq \Psi(\gamma^{-1})$. Second, notice that the second Mandelstam identity is considerably different from that of $SU(2)$. In the $SU(2)$ case the second Mandelstam identity allowed us to express a wavefunction of n loops as a combination of wavefunctions of $n-1$ loops and could be used recursively to reduce any wavefunction of a multiloop to a single-loop wavefunction. In the present case, the identity allows us to reduce a wavefunction of n loops to a combination of wavefunctions of $n-1$ and $n-2$ loops. This implies in particular that one can only reduce a wavefunction of an arbitrary multiloop to a wavefunction of *two* loops.

Remarkable we therefore come to the conclusion that wavefunctions in the unified loop representation depend on two loops, exactly as if we had built two independent representations for gravity and electromagnetism. There is an important difference: in the unified case there is no distinction between the two loops and the Mandelstam identities for both of them are the same. We therefore see that a unified setting arises as a consequence of going to the language of loops. Similar considerations hold for the case in which the group is not $U(2)$ but $SU(2) \times G^{YM}$ though the minimum number of loops involved is higher.

There are several aspects of this unification that are interesting enough to merit investigation. We will only briefly discuss them here since the subject is largely unexplored. By inspection we can tell what the diffeomorphism constraint of the unified theory implies in the loop representation. Since it has exactly the same form as the usual diffeomorphism constraint of general relativity, we know it will require that the wavefunctions be invariant under smooth deformations of the loops. Therefore we know how to solve that constraint: we just need to consider functions of two loops that are invariant under deformations of the loops. Notice that if one had pursued a loop representation based on separate loops for both gauge invariances, the action of this constraint would be considerably less geometrical and more involved. Some of the results we introduced in the connection representation for gravity in the previous chapter apply to the unified model. For instance, if one constructs a state based on the exponential of the Chern–Simons form of the unified connection, such a state solves all the constraints of the theory with cosmological constant. As we will see in chapters 10 and 11, such a state has importance in the loop representation, being related to the Jones polynomial. The same relationship appears for the unified model. If one considers the inclusion of fermions in the unified model, one would have to proceed as in the previous section by introducing open paths. However, in the unified model, opening the loops implies introducing not only a charge at the level of the gravitational Gauss law (spin) but also one for the Maxwell Gauss law (electric charge). This means that the most natural form of matter in the unified model has an electric charge if it has spin.

Finally, what happens with the Hamiltonian? The Hamiltonian of Einstein–Maxwell can be written in terms of the unified variables but its form is slightly cumbersome (it still is polynomial) and differs from that of vacuum gravity. There is nothing to prevent us from realizing it in the loop representation and studying its solutions, though this issue is as yet unexplored. In an interesting development Chakraborty and Peldán [150] have noticed that one can write a Hamiltonian in terms of the unified variables that looks quite similar to that of vacuum gravity. The resulting theory is not Einstein–Maxwell but reduces to it in the weak field limit. The loop representation of such a model could appear quite naturally and lead to new insights of the unified theory.

9.4   Kalb–Ramond fields and surfaces

Kalb–Ramond fields [171] are antisymmetric second rank tensor gauge fields. They found physical application in the field theories that arise as low energy limits of string theory [151] and as models of dark cosmological matter[152]. We will discuss Abelian Kalb–Ramond fields here because they couple very simply to gravity and because their gauge symmetry makes them associated with surfaces which are a natural higher dimensional generalization of loops. They are suited to a geometric quantum formulation completely analogous to the loop representation for usual gauge symmetries, but based on surfaces instead of loops. These surfaces can later be used in quantum gravity to measure properties of the metric, as was argued by Smolin [172] and which we will see in section 9.5.1. The first analysis of antisymmetric tensor fields in terms of surfaces was discussed by Arias, Di Bartolo, Fustero, Gambini and Trias [155]. Although non-Abelian antisymmetric tensor fields have been considered [154], it has not been possible to give them a geometric formulation.

Let us start with a brief discussion of the properties of the Abelian group of surfaces and then relate it to the loop representation of an Abelian Kalb–Ramond field.

9.4.1   The Abelian group of surfaces

Consider the set S of closed two-dimensional oriented surfaces in $\mathbf{R}^3$. For each surface s we denote by $\bar{s}$ the reverse-oriented surface. We define the following product,

$$s_1 \circ s_2 \equiv s_1 \cup s_2, \qquad s_1, s_2 \in S, \tag{9.28}$$

which is associative and commutative, but lacks an inverse element. In order to define this we introduce, in the same spirit as for loops, the notion of a tree. We define as trees all elements of S such that the integral of all scalar functions on them is zero. We introduce an equivalence relation in S by identifying two elements if their composition is a tree. The quotient set is an Abelian group. We denote it by Σ and its elements by σ_i. This structure is easily generalizable to the set Λ of surfaces with boundary. One can naturally view the group of closed surfaces without boundary as the group of deformations of surfaces with boundaries, very much as loops can be viewed as deformations of paths. We will not discuss open surfaces here, details can be seen in references [155, 172]. We will see that from these group structures we can recover all the kinematical content of a Kalb–Ramond theory, in the same sense as the group of loops contained all kinematical information of usual gauge theories. In order to unravel this connection we proceed as we did for gauge theories, by introducing the infinitesimal generators of the group.

Consider the following two infinitesimal elements in S. The first one which we call $\delta\sigma(x, \vec{u}, \vec{v}, \vec{w})$ is an infinitesimal three-dimensional parallelepiped with vertex at the point x and sides along the three vectors $\vec{u}, \vec{v}, \vec{w}$. The second element which we call $\delta\sigma(\pi_o^x, \vec{u}, \vec{v})$ is defined by an infinitesimal parallelogram similar to the one we used to define the loop derivative in chapter 1 attached to a path going from a basepoint to the point x.

Consider now a representation of the Abelian group that associates to each element σ_i a complex number $U(\sigma_i)$, acting on a space of functions on the group of surfaces $\Psi(\sigma_i)$,

$$U(\sigma)\Psi(\sigma') = \Psi(\sigma \circ \sigma'), \tag{9.29}$$

$$U(\sigma_1 \circ \sigma_2) = U(\sigma_1)U(\sigma_2), \tag{9.30}$$

and we fix $U(\sigma_0) = 1$ with σ_0 the identity element in Σ.

To find the infinitesimal generators of the group we assume the representation is differentiable in the sense that the following expansions exist:

$$\Psi(\delta\sigma(z, \vec{u}, \vec{v}, \vec{w}) \circ \sigma) = (1 + u^a v^b w^c \Delta(z)_{abc})\Psi(\sigma), \tag{9.31}$$

$$\Psi(\delta\sigma(\pi_o^z, \vec{u}, \vec{v}) \circ \sigma) = (1 - u^a v^b \delta(\pi_o^x)_{ab})\Psi(\sigma). \tag{9.32}$$

As in the case of loops, we now have differential operators and we would like to find relations among them. Exactly like when we proved the Bianchi identity for loop derivatives, we start from an identity in the space of surfaces,

$$\delta\sigma(x, \vec{u}, \vec{v}, \vec{w}) = \delta\sigma(\pi_o^x, \vec{u}, \vec{v}) \circ \delta\sigma(\pi_o^x, \vec{u}, \vec{w}) \circ \delta\sigma(\pi_o^x, \vec{v}, \vec{w})$$
$$\circ \delta\sigma(\pi_o^{x+u}, \vec{w}, \vec{v}) \circ \delta\sigma(\pi_o^{x+v}, \vec{u}, \vec{w}) \circ \delta\sigma(\pi_o^{x+w}, \vec{v}, \vec{u}), \tag{9.33}$$

simply stating that the parallelepiped can be obtained by joining together six parallelograms. Introducing the parallel derivative acting on the path dependence of functions,

$$(u^a \partial_a)\delta(\pi_o^x) = \delta(\pi_o^{x+u}) - \delta(\pi_o^x), \tag{9.34}$$

the geometric identity (9.33) implies

$$\Delta_{abc}(z) = \partial_a \delta_{bc}(\pi_o^z) + \partial_b \delta_{ca}(\pi_o^z) + \partial_c \delta_{ab}(\pi_o^z). \tag{9.35}$$

This relation is the analogue for surfaces of the identity between the connection and the loop derivatives that we proved in chapter 1. As in that case, we have an identity that relates path dependent objects with path independent ones. If one seeks a description that is path independent, one must associate with each point x a prescription of a fiducial path π_o^x, with which we can identify $\delta(x) \to \delta(\pi_o^x)$. Modifications of the fiducial path prescription amount to gauge transformations.

Representations of the Abelian group of surfaces in $U(1)$ yield as a result the kinematical structure of the Kalb–Ramond fields. The two derivatives Δ_{abc} and δ_{ab} become the field tensor and antisymmetric tensor potential. Let us now discuss the usual formulation of a Kalb–Ramond field to make explicit contact with the ideas that we have introduced in this subsection.

9.4.2 Kalb–Ramond fields and surface representation

An Abelian Kalb–Ramond field is a two-form $A_{ab} = -A_{ba}$. Its field strength is a three-form $F_{abc} = \partial_{[a} A_{bc]}$. The action for these fields is defined in analogy with the Maxwell action,

$$S_{KR} = \int d^4x \sqrt{-g} F_{abc} F^{abc}, \tag{9.36}$$

where indices are raised with a spacetime metric g^{ab} that defines the coupling to gravity. The action is invariant under gauge transformations

$$A_{ab} \to A'_{ab} = A_{ab} + \partial_{[a} \Lambda_{b]}. \tag{9.37}$$

It is easy to introduce a Hamiltonian formulation of the theory. The canonical variables are the pull-back to the three-surface of the Kalb–Ramond field A_{ab} and its canonically conjugate momentum $\tilde{\pi}^{ab}$. The theory has a Gauss law constraint associated with the gauge symmetry

$$\partial_c \tilde{\pi}^{cd} = 0, \tag{9.38}$$

and the coupling to gravity is achieved by adding the following terms to the usual Hamiltonian and diffeomorphism constraints of vacuum general relativity:

$$\tilde{C}_a = \underset{\sim}{\epsilon}_{abc} \tilde{\pi}^{bc} \tilde{\epsilon}^{cde} F_{cde}, \tag{9.39}$$

$$\mathcal{H} = \tfrac{1}{2} (\tilde{\epsilon}^{abc} F_{abc})^2 + \tfrac{1}{2} \tilde{\pi}^{ab} \tilde{\pi}^{cd} \underset{\sim}{\epsilon}_{abe} \underset{\sim}{\epsilon}_{cdf} \tilde{q}^{ef}. \tag{9.40}$$

One can build a quantum representation for the joint Einstein–Kalb–Ramond system coordinatized by loops and surfaces. We start by constructing a non-canonical algebra of quantities associated with the Kalb–Ramond field, to supplement the usual T algebra for gravity. To each surface we associate the gauge invariant quantity,

$$U(\sigma) = \exp\left(i \int_\sigma d^2 S^{ab} A_{ab} \right), \tag{9.41}$$

which is the analogue of the Wilson loop for Kalb–Ramond fields and materializes the representation of the group of surfaces we discussed in the previous subsection. This object forms an algebra with the gauge invariant quantity $\tilde{\pi}^{ab}$,

$$\{U(\sigma), \tilde{\pi}^{ab}\} = i \int d^2 S^{ab}(y)\delta(x-y)U(\sigma). \tag{9.42}$$

One can construct a representation of this non-canonical algebra in terms of functions of the group of surfaces $\Psi(\sigma)$; it is given by

$$\hat{W}(\sigma')\Psi(\sigma) = \Psi(\sigma' \circ \sigma), \tag{9.43}$$

$$\hat{\tilde{\pi}}^{ab}\Psi(\sigma) = \int d^2 S^{ab}(y)\delta(x-y)\Psi(\sigma). \tag{9.44}$$

A joint representation for gravity and Kalb–Ramond fields can be obtained by considering states that are functions of loops and surfaces $\Psi(\gamma,\sigma)$. We will see in the next section how to build diffeomorphism invariant quantum observables for such a system. Finally, the formalism involving open surfaces is useful for representing the coupling of Kalb–Ramond fields to matter, in particular to Abelian one-form fields. Details can be seen in references [155, 172, 156]. It is a complete analogue of what happened with usual gauge theories and coupling to fermions, which was achieved through the introduction of open paths on which loops acted as deformations.

9.5 Physical operators and weaves

In canonical quantization, as we outlined in chapter 3, after finding the space of physical states one needs to introduce an inner product under which the observables are self-adjoint operators. One can then compute expectation values and make measurable physical predictions.

Several difficulties prevent us from completing these steps for general relativity. Although we discussed some possible solutions to the constraints and in the next two chapters we will introduce further ones, one is far from knowing at present the space of all solutions to the constraints. On the other hand, for general relativity on compact spatial manifolds, we do not know at present a single observable in the Dirac sense. As a consequence, we are far from knowing a suitable inner product. In spite of these drawbacks, one would like to know if the structures that we have developed in these chapters have any connection, even at a kinematical or formal level, with possible physical ideas. An example of this was the argument presented in chapter 7 concerning the value of the determinant of the metric operator on the space of loops without intersections. This was not an argument based on a Dirac observable and yet it allowed us to draw conclusions about the space of states discussed. A further motivation for this kind of studies is that by discussing the action of operators in loop space one gains knowledge that may be useful whenever operators that commute with the constraints are found. Apart from this, one is, in general, interested in making a connection between the quantum description of gravity in terms of loops and the classical picture of general relativity as a theory of a metric.

All these considerations lead us to discuss in some detail some operators in the space of loop states. These operators will not commute with the constraints and therefore are not observables. We will see, however, that some of them could be related to observables (or at least commute with some of the constraints) if matter sources are introduced. They will allow us to give a notion of classical geometry associated with a certain set of loop states that play the role of "semi-classical states". We will see the crucial role that diffeomorphism invariance plays in the regularization of these operators.

9.5.1 Measuring the geometry of space in terms of loops

Consider the metric in terms of the new variables

$$\tilde{\tilde{q}}^{ab} = \tilde{E}^a_i \tilde{E}^b_i. \qquad (9.45)$$

It would be easy using the technology introduced in chapters 7 and 8 to promote this to an operator in loop space. One can regularize and renormalize the expression but it will depend on the particular details of the background metric introduced in order to regularize it. This is a general result. One regularizes expressions that involve products of Dirac delta functions, as the metric operator does. The renormalization procedure always amounts to replacing the product of two delta functions by a single one. The problem is that the Dirac delta function $\delta(x - y)$ is not only a distribution but also a density*. Therefore any procedure that converts the product of two deltas into one has to supply a factor with appropriate density weight. Since there are no natural scalar densities defined in a manifold without a metric, one is forced to introduce a density weight constructed with an external metric structure. Therefore renormalized expressions will always depend on a background metric.

There is a way out of this general objection that is based on the definition of operators that, unlike the metric, are well defined via a regularization procedure but *without a renormalization*. Let us give an example of such an operator, first introduced by Ashtekar, Rovelli and Smolin [167].

---

* In reference [153] this is emphasized by writing it as $\delta(x, y)$ since the expression $\delta(x - y)$ only makes sense when a background metric is defined. Then $\delta(x, y)$ is a density in *one* of its arguments.

Consider the classical expression

$$Q(\omega) = \int d^3x \sqrt{\tilde{E}_i^a(x)\tilde{E}_i^b(x)\omega_a(x)\omega_b(x)}, \qquad (9.46)$$

where ω is an arbitrary smooth one-form. This quantity associates a real number to any one-form and it evidently contains information about the spatial metric in the sense that one can reconstruct the spatial metric from knowledge of $Q(\omega)\ \forall\omega$. To construct the quantum version of this operator we regulate the classical expression, $Q(\omega) = \lim_{\epsilon\to 0} Q^\epsilon(\omega)$, where

$$Q^\epsilon(\omega) = \int d^3x \sqrt{\int d^3y \int d^3z f_\epsilon(x,y) f_\epsilon(x,z)\tilde{E}_i^a(y)\tilde{E}_i^b(z)\omega_a(y)\omega_b(z)}. \qquad (9.47)$$

We now promote the quantity under the square root to an operator in the loop representation. This is accomplished in a straightforward fashion using the calculational techniques of the previous chapter. The result is

$$\int d^3y \int d^3z f_\epsilon(x,y) f_\epsilon(x,z)\hat{\tilde{E}}_i^a(y)\hat{\tilde{E}}_i^b(z)\omega_a(y)\omega_b(z)\Psi(\gamma) =$$

$$2\int d^3y \int d^3z f_\epsilon(x,y) f_\epsilon(x,z)\omega_a(y)\omega_b(z)$$

$$\times \oint_\gamma dv^a \oint_\gamma dw^b \delta(z-v)\delta(y-w)\left[2\Psi(\gamma_z^w \circ \gamma_{zo}^w) + \Psi(\gamma)\right]. \qquad (9.48)$$

Notice that this operator, when acting on a state $\Psi(\gamma)$ with support on loops without intersections like the ones we discussed in the previous chapter, returns a contribution proportional to $\Psi(\gamma)$. Therefore the square root is well defined, as we shall see. If the wavefunction has support on loops with intersections, the definition of the square root is more involved. It turns out that one can ignore the intersections in the definition of the operator. To see this, notice that the argument of the square root is a function of x that coincides with its value on loops without intersections at all xs except at a finite number of x_is, the intersection points. Therefore if one assumes that the value of the square root at those points is finite, one can ignore their contribution to the integral defining the operator $\hat{Q}(\omega)$.

Therefore at points without intersections one can explicitly compute the square root that appears in the definition of $Q(\omega)$. The result is, in the limit in which $\epsilon \to 0$,

$$\left|\sqrt{6}\oint_\gamma dy^a f_\epsilon(x,y)\omega_a(y)\right|\Psi(\gamma) + O(\epsilon), \qquad (9.49)$$

which gives for the operator $Q(\omega)$, after noting that it is the integral of a positive quantity with support along the loop,

$$\hat{Q}(\omega)\Psi(\gamma) = \sqrt{6} \oint_{\gamma} |dx^a \omega_a(x)| \Psi(\gamma). \tag{9.50}$$

The operator so defined is finite and independent of the background structure that was used to define the regulator. No renormalization was needed in its definition.

Let us now define another quantity in the same spirit as the previous one, also suggested by Ashtekar, Rovelli and Smolin [167]. This quantity is associated with a surface S with normal vector n^a. Its infinitesimal element of area is given by $dA = d^2 S^{ab} \sqrt{h} n^c \epsilon_{abc}$ where $h = \tilde{q}^{ab} n_a n_b$. From here we can give a polynomial expression in terms of the new variables for the square of the infinitesimal area element,

$$d\mathcal{A}(S) = dS^{ab} dS^{cd} \epsilon_{abe} \epsilon_{cdf} \tilde{E}_i^e \tilde{E}_i^f. \tag{9.51}$$

Additional comments on this formula for the area can be seen in reference [166]. From the above expression one can compute the area of the surface partitioning it into a countable number N of small area elements and writing

$$\mathcal{A}(S) = \lim_{N \to \infty} \sum_{i=1}^{N} \sqrt{\mathcal{A}_{\text{approx}}^2(S_i)}, \tag{9.52}$$

where the quantity

$$\mathcal{A}_{\text{approx}}^2(S_i) = \int_{S_i} d^2 S_i^{ab}(x) \int_{S_i} d^2 S_i^{cd}(y) \epsilon_{abe} \epsilon_{cdf} T^{ef}(\eta_x^y, \eta_y^x) \tag{9.53}$$

approximates the infinitesimal element of area in the limit in which S_i shrinks to a point and therefore the points x and y coincide and η (a loop contained in S_i that passes through x and y) shrinks to a point as well. Recall that when the loop η shrinks to a point $T^{ab}(\eta_x^y, \eta_y^x) \to \tilde{E}_i^a(x)\tilde{E}_i^b(x)$. We could have also proceeded as for the $\hat{Q}$ operator and introduced the limit in a gauge non-invariant fashion. We do it here in terms of the T^2 for illustration purposes.

The above expression for $\mathcal{A}_{\text{approx}}^2(S_i)$ is immediately promoted to a quantum operator in the loop representation replacing the T by its corresponding quantum operator, which we introduced in formula (8.13). Assuming we act on a wavefunction $\Psi(\gamma)$ of a loop γ without intersections on the surface S_i the action of $\hat{T}^2$ gives four terms that in the limit in which S_i shrinks to a point are, after Mandelstam rearrangements, identical,

$$\lim_{\eta \to \iota} \hat{T}^{ab}(\eta_x^y, \eta_y^x)\Psi(\gamma) = 6X^{ax}(\gamma)X^{bx}(\gamma)\Psi(\gamma). \tag{9.54}$$

and in the limit $\eta \to \iota$ the point $x \to y$.

On the other hand, the integral $\int_{S_i} d^2 S^{ab}(x)\epsilon_{abc}X^{ax}(\gamma)$ gives as a result the intersection number of the surface S_i and the loop γ, $I[S_i, \gamma]$, meaning that it counts the (oriented) number of times the loop γ pierces the surface S_i. As a result, we can write for the approximate area element in the limit in which it shrinks to a point,

$$\hat{A}^2_{\text{approx}}(S_i)\Psi(\gamma) = 6I[S_i, \gamma]^2\Psi(\gamma). \tag{9.55}$$

We can therefore write as a result the expression for the area operator,

$$\hat{A}(S)\Psi(\gamma) = \sqrt{6}I^+[S, \gamma]\Psi(\gamma), \tag{9.56}$$

where $I^+[S, \gamma]$ is the unoriented intersection number of the loop γ and the surface S. That is, the area operator counts the (unoriented) number of times the loop pierces the surface S.

In contrast to the case of the $\hat{Q}$ operator, one cannot simply ignore intersections in the definition of the $\hat{A}$ operator. If the wavefunction on which it acts has support on intersecting loops, it is no longer an eigenstate of $\hat{A}^2$ and the square root cannot be taken. Smolin [145] proposed an extension of the $\mathcal{A}(S)$ operator to the intersecting case using a constructive procedure. This is based on the application of the Mandelstam identity to infer the value of $\mathcal{A}(S)$ on loops with intersections. Some simple examples are given in reference [145] but a complete definition can only be introduced through the use of spin network states [146].

Rovelli and Smolin[146] have introduced a regularized definition of the volume element along the same lines as the operators we have introduced here. In that case the relevant states on which the operator is non-vanishing have intersections with three independent tangents. The attractive feature of this is that as a consequence its eigenstates can be naturally described by spin networks. Spin networks also seem to simplify the definition of $\mathcal{A}(S)$ for intersecting loops. Further discussion of the $\hat{Q}$ and $\hat{A}$ operators can be found in the Les Houches lectures of Ashtekar [3] as well as in the previously quoted references.

In both the $\hat{Q}$ and $\hat{A}$ operators we saw that in order to define the square roots it was simpler if the states on which they operated had support on loops without intersections. As we discussed in chapter 8 it is apparently inconsistent with the Mandelstam identities to consider such states. Therefore it is useful to introduce the notation of bras, in terms of which it is well defined to say that

$$< \alpha|\hat{Q}(\omega) = \sqrt{6}\oint_\alpha |dx^a\omega_a(x)| < \alpha|, \tag{9.57}$$

$$< \alpha|\hat{A}(S) = I^+[S, \alpha] < \alpha|, \tag{9.58}$$

if the loop α is smooth and non-intersecting.

A few comments are in order about the above expressions. The integrals that appear in the right-hand side of these equations are, *prima facie*, not well defined as functions of loops, in the sense of chapter 1. If one adds a tree to the loop γ their value changes. In the first integral this is due to the absolute value and in the second one it is due to the fact that we are considering the unoriented intersection number (which can be arbitrarily changed adding trees that pierce the surface). In both cases the problem can be traced back to the definition of an operator as the square root of a square, which introduces a sign ambiguity. This difficulty can be remedied by defining the operators in the following way. Consider the action of the operator on a loop γ. Then define a curve p obtained by stripping all trees from a representative curve of γ and compute the integrals in the right-hand side of the definitions of the operators using the curve p. The result is a function of loops.

As we mentioned above the definition introduced for $\hat{A}(S)$ is only valid for smooth non-intersecting loops and a more elaborate one is needed for intersections. It seemed that the definition of $\hat{Q}$ was free of these kinds of complications since intersections only constituted a set of measure zero in the integral on the loop that appeared in the definition of the operator. This, however, assumes that intersections only appear at isolated points. If one considers loops with lines traversed several times each of them would contribute a non-negligible amount to the integral and in those cases the operator that appears inside the square root is no longer in an eigenstate. It is possible that this complication can be handled in the same fashion as for the $\hat{A}$ operator with the use of spin networks. Loops with multiple lines are inevitable in any formulation that preserves the Mandelstam identities as we discussed in chapter 8, figure 8.1.

The idea of considering operators that are naturally densities of order one through the introduction of a square root in order to define them without renormalization is a general one. It could be applied to any operator with those characteristics. An interesting point would be to apply this idea to the Hamiltonian constraint as a means to overcome the regularization ambiguities and background dependences discussed in chapter 8. The Hamiltonian constraint is naturally a density of weight two. Its square root is a density of weight one and its integral on the three-manifold is likely to be well defined without the introduction of additional background structures. The trouble is that the canonical formulation of vacuum general relativity requires that the constraint vanish point by point and not only as an integral.

There is a context, however, in which the integral of the square root of the Hamiltonian constraint arises naturally. This has been explored by Rovelli and Smolin [140]. Suppose one couples general relativity to a massless minimally coupled scalar field $T(x)$. In the canonical formulation a new canonical pair arises: $T(x)$ and $\tilde{P}_T$. The constraints are

$$C_a(x) = C_a^0(x) + P_T(x)\partial_a T(x) = 0, \tag{9.59}$$

$$H(x) = H^0(x) + \tilde{P}_T^2 + \tfrac{1}{4}\tilde{E}_i^a \tilde{E}_i^b \partial_a T(x)\partial_b T(x) = 0, \tag{9.60}$$

where we have used a zero superscript to denote the usual diffeomorphism and Hamiltonian constraints of general relativity.

One can now use the scalar field to keep track of time in this problem by considering a foliation of spacetime defined by leaves in which $T(x)$ is constant. Such a foliation, in general, only exists locally but the approach is to explore the resulting quantum theory as long as such a foliation is acceptable. In that case the terms involving gradients of the fields drop out from the constraints and one can solve the Hamiltonian constraint for P_T classically,

$$P_T = \int d^3x \sqrt{-H(x)}, \tag{9.61}$$

where we have kept the same notation for P_T though it is now a constant due to the gauge fixing for the field T.

We therefore see that the square root of the integral of the Hamiltonian appears naturally in this context. One can now construct a quantum theory in which states are functions of loops parametrized by the "time" T, $\Psi(\gamma, T)$, and satisfying a Schrödinger-like equation,

$$i\frac{\partial}{\partial T}\Psi(\gamma, T) = \int d^3x \sqrt{-\hat{H}(x)}\Psi(\gamma, T), \tag{9.62}$$

in which the integral of the square root of the usual Hamiltonian constraint in the loop representation which we discussed in chapter 8 arises naturally. Rovelli and Smolin [140] studied this operator and found that it can be defined without the introduction of a background metric for smooth and intersecting loops without kinks at the intersections or elsewhere. The resulting Hamiltonian for the theory, being background independent, can be purely formulated as a set of topological operations on the space of knots. This requires setting a prescription for how one adds the infinitesimal loop that arises in the definition of the usual Hamiltonian constraint. There is an infinite-fold ambiguity on how to add the infinitesimal loop in the space of knots with intersections. Also, the definition of the addition of an infinitesimal loop is problematic in the space of knots, since the addition of the infinitesimal loop changes the knot character and therefore may lead to discontinuities unless one requires special properties of the wavefunctions. It is not even clear that there exists, in general, a diffeomorphism invariant assignment of a small loop. Moreover, viewing the Hamiltonian as the square root of a matrix may be problematic since the basis of knots with intersections is, in general, not a countable basis[†].

This approach has recently led to an important amount of activity with the introduction of topological Feynman rules for the interaction of gravity and matter, when combined with the formulation of fermions of Morales-Técotl and Rovelli that we discussed at the beginning of this chapter.

9.5.2   Semi-classical states: the weave

Usually, in a quantum theory, in order to extract physical information it is necessary to make some kind of correspondence with classical physics. This is accomplished through a semi-classical limit. This implies picking a preferred set of states that "approximate" the classical behavior with "small" quantum corrections. Typically what is meant by "approximate" and "small" refers to quantum fluctuations, which implies considering expectation values of observable quantities. Evidently, we cannot take this route here since we do not have an inner product or observables to compute such expectation values.

Ashtekar, Rovelli and Smolin [167] have suggested a different strategy. They consider a set of loop states based on collections of loops characterized by a certain (macroscopic) length scale L and a (microscopic) length scale l, associated with a classical geometry h. We will consider the operators introduced in the previous subsection and study their eigenvalues. We will see that if one considers the functions that parametrize the operators (w_a and the surface S) as smoothly varying with respect to L there is a unique value of the parameter l for which the eigenvalues coincide with the classical value of the quantities Q and $\mathcal{A}$ as calculated in the classical geometry determined by h.

The states that we will consider are constructed in the following way. Given the three-geometry h to which one wants to associate a state, sprinkle in a region of it a randomly distributed number of points N with a density $n = N/L^3$, where L is a certain (macroscopic) length scale. At each point draw a circle of radius $l = (1/n)^{1/3}$ with a random orientation. This results in a set of curves which we can consider a representative element of a multiloop $< \Delta|$. For the moment we consider multiloops without intersections, for simplicity, though the circles involved can be linked. One then considers, for instance, the action of the $\hat{Q}(w)$ operator on such a state with w a fixed one-form that is smooth on the scales de-

---

[†] An intuitive way of seeing that it is not countable is to consider quintuple intersections. In the extension of the braid group to intersections one needs to consider a two-parameter family of new elements with quintuple intersections, parametrized by the angles one of the tangents forms with the other three.

termined by L and smaller. Applying the results of the previous section we get

$$< \Delta | Q(\omega) = \sqrt{6} \sum_{i=1}^{N} \oint_{\alpha_i} |dy^a w_a| < \Delta |, \qquad (9.63)$$

where α_i is the circle located at the ith point.

Since w_a is slowly varying the integral at each circle can be computed, and averaging over all possible directions, we get

$$< \Delta | \hat{Q}(\omega) = \left(\sqrt{6} \pi \left(\frac{1}{l^2} \right) Q(w)[h] + O \left(\frac{l}{L} \right) \right) < \Delta | \qquad (9.64)$$

where $Q(w)[h]$ is the value of the quantity Q evaluated for the one-form w_a on the classical three-geometry h, which we have assumed to be flat for the particular form of the result presented. We therefore see that for the eigenvalue of the operator Q to approximate its classical value on the three geometry that we want to associate with the state $< \Delta |$ to order l/L we need the separation of the loops to be fixed to a value of order unity ($l^2 = \sqrt{6}\pi$), in natural units. That is, the loops need to be separated by a length approximately equal to the Planck length.

This result is remarkable in the sense that we have derived a natural cutoff scale in quantum gravity in order to recover classical physics. The *a priori* feeling would have been that the weave would have approximated the classical geometry better as the separation of the loops was smaller. The detailed calculation shows that this is true until one reaches scales of the order of the Planck scale. Then, beyond a certain value, the approximation is worse the finer the weave is. The appearance of a natural cutoff in a detailed calculation of quantum gravity opens up the possibility that the theory could be made finite by its own dynamics and shows a significant departure from the usual behavior of quantum field theories. It is also remarkable that the Planck length appears naturally as the cutoff, especially since the theory is a diffeomorphism invariant one and one has the expectation that no scales would be privileged in such theories. Natural cutoffs appear in various contexts in quantum gravity. For a discussion see the paper by Garay [173].

To consider that we have approximated a classical geometry because a single quantity — which does not commute with the constraints — approximates its classical value is clearly insufficient. One can repeat the above construction for the area operator, but it is clear that these kinds of calculations are at the moment only indicative of the kind of physics one should expect when one is able to perform similar calculations with genuine observables of the theory.

It is possible that this kind of calculation could be performed for general relativity coupled to matter, where one can construct quantities that commute with (at least some of) the constraints. For instance, Smolin [172] has considered general relativity coupled to Kalb–Ramond fields in a surface formulation similar to the one we discussed in section 9.4.2. In that context he can use the surfaces that characterize the Kalb–Ramond field to construct with the gravitational variables an operator similar to $\mathcal{A}(S)$. The difference now is that the surface S is determined by the fields of the theory, i.e., it is dynamical. As a consequence the operator defined is invariant under diffeomorphisms. A similar construction of diffeomorphism invariant quantities was performed for the Maxwell field by Husain [169], and for topological field theories by Rovelli [170].

A very important result is that of Iwasaki and Rovelli [168] who have carried this analysis one step further. They studied in detail the correspondence between the theory offered by the weave and its perturbations with the quantized theory of gravitational perturbations of a classical background (the usual linearized quantum gravity). They found a correspondence between sectors of the space of states of both theories such that now gravitons can be viewed as a particular family of perturbations of the weave.

9.6   2+1 gravity

General relativity in two spatial and one temporal dimensions offers a remarkable laboratory to test ideas of loop quantization. Because in three dimensions the Ricci tensor completely determines the Riemann tensor, the vacuum Einstein equations in three dimensions,

$$R_{\mu\nu} = 0, \tag{9.65}$$

imply that spacetime is locally flat. The only non-triviality of the Einstein theory in three spacetime dimensions comes from the topology of spacetime. The theory therefore does not have any local degrees of freedom. It has a finite number of topological degrees of freedom. Therefore the theory is exempt from the difficulties associated with the infinite number of degrees of freedom of field theories and yet it shares several features with the $3 + 1$ theory, foremost among which is the invariance under diffeomorphisms.

The Einstein action in $2 + 1$ dimensions is, written in first order form [157],

$$S(e, A) = \int d^3x \, \tilde{\epsilon}^{abc} e_{ai} F^i_{bc}, \tag{9.66}$$

where e_{ai} is a set of triads, A^i_a is the spin connection compatible with the triad and F^i_{bc} is the curvature. The indices i, j and k are $SO(2,1)$ indices in the tangent space to the three-dimensional spacetime. Notice that this is just the Palatini action for the theory and all the variables are real. The equations of motion resulting from the variation with respect to the triad and the connection are

$$\frac{\delta}{\delta A_a^i} \quad : \quad {}^3D_{[a}e_{b]}^i = 0, \tag{9.67}$$

$$\frac{\delta}{\delta e_a^i} \quad : \quad F_{ab}^i = 0. \tag{9.68}$$

The first equation tells us that A_a^i is the torsion free covariant derivative that annihilates the metric constructed with the triad. The second equation tells us that the curvature of the connection is zero and therefore spacetime is flat.

One can immediately perform a canonical decomposition of the action. The details can be seen in reference [153]. The resulting phase space consists of the pull-back of the connection A_a^i to the two-dimensional surface and the canonically conjugate momentum is the pull-back of the cotriad $\tilde{E}_i^a = \tilde{\epsilon}^{ab}e_{bi}$, where $\tilde{\epsilon}^{ab}$ is the Levi–Civita density on the two-dimensional spatial surface. The other variables are Lagrange multipliers whose variation enforces the constraints,

$$D_a \tilde{E}_i^a = 0, \tag{9.69}$$
$$F_{ab}^i = 0, \tag{9.70}$$

which can be obtained by pulling back to the spatial slice the equations of motion.

The first equation is the usual Gauss law that tells us that the theory is invariant under $SO(2,1)$ triad rotations in the tangent space. The second equation contains in a joint form the diffeomorphism and Hamiltonian constraint of the $2+1$ theory. This form of the constraints was first introduced (in a slightly different way) by Witten [46]. The relationship with the usual form of the (3+1)-dimensional Ashtekar constraints can be made explicit [158] by contracting the second equation with $\tilde{E}_i^a$ for the diffeomorphism and with $f_{ijk}\tilde{E}_i^a\tilde{E}_j^b$, where f_{ijk} are the structure constants of $SO(2,1)$. With these projections the form of the constraints becomes exactly the same as those in the $3+1$ theory (with the exception that the spatial indices run from 1 to 2 and the internal indices are $SO(2,1)$). One could choose either form of the constraints to study the theory, but it should be made clear that the two forms are inequivalent. The projections introduced to obtain the Ashtekar constraints from the Witten ones can become degenerate and therefore the former admit many more solutions. This inequivalence is far from academic. It can be shown that the Ashtekar form of the constraints classically allows configurations with an infinite number of degrees of freedom whereas the Witten one does not [159].

Counting the degrees of freedom, we have twelve variables in phase space and six first class constraints. The system has no local degrees of freedom. The constraints are either linear in momenta or independent of momenta. The situation is considerably simpler than in the $3 + 1$ theory. In particular we can find observables for the theory, i.e., quantities that commute with all the constraints, something we were unable to do in the $3 + 1$ theory. It is easy to see that the T^0 and T^1 quantities constructed with the canonical variables have vanishing Poisson brackets with the constraints (in the $2+1$ literature it is customary to integrate the T^1 along a loop using $T^1 = \int dy^a \epsilon_{ab} \text{Tr}(\tilde{E}^b(y) U(\gamma_y^y))$, we adopt this nomenclature for the rest of this section). These observables are closely related to those introduced by Martin [160] in the Witten language. Another difference with the $3+1$ case was that there the higher order T variables were needed to express in a convenient fashion physical quantities of interest, such as the Hamiltonian constraint. Here we can write the constraints exclusively in terms of the small T algebra. In order to gain an intuitive feeling for the meaning of the observables introduced, it is worthwhile mentioning that in three-dimensional gravity in the asymptotically flat case the metric of a point particle corresponds to a cone. The deficit angle of the cone is proportional to the mass of the particle. The T^0 measures that quantity. If the particle is spinning, the T^1 measures the rate of spin.

It may appear surprising that in a diffeomorphism invariant theory the T variables are observables. After all, they are not in the $3 + 1$ theory. The key to this difference lies in the flatness of the connection. For a flat connection, a loop and the same loop shifted by a diffeomorphism yield the same holonomy. Therefore quantities based on holonomies, in spite of their dependence on an external structure — the loop —, can be diffeomorphism invariant. Because the connection is flat, two homotopic loops lead to the same holonomy. If the loops are homotopic to the identity, the holonomy is the identity. That implies that the only loops that yield non-trivial holonomies are those that wrap around topologically non-contractible paths.

We can now proceed to the quantization of the theory following the same steps as for the $3+1$ theory. Let us start in the connection representation. We pick wavefunctionals of the connection $\Psi[A]$. The constraints are easily promoted in an unambiguous fashion to quantum operators. The Gauss law simply requires that we consider wavefunctions that are gauge invariant. The constraint $\hat{F}^i_{ab}\Psi[A] = 0$ simply requires that the wavefunctions have support on flat connections. Together they demand that the wavefunctions considered be wavefunctions on the moduli space of flat connections. This space can be endowed with a simplectic structure, and an inner product that makes the T operators self-adjoint [2].

Consider now the loop representation. We can find a representation of the T operators in terms of loops in exactly the same fashion as we did for the $3 + 1$ theory (the formulae differ slightly since the gauge group is $SO(2,1)$ instead of $SO(3)$). If one wants to make more progress it is convenient to pick a particular spatial topology. Let us consider the simplest non-trivial example: that in which the spatial manifold has the topology of a two-torus.

There are two possible avenues one could take to construct the loop representation. One is to promote the constraint equations to operators in loop space and study the space of solutions. The other one is to take advantage of the knowledge that the physical states in the connection representation have support on the moduli space of flat connections and directly build a loop representation for the flat connection.

If we take the latter approach let us start by characterizing the moduli space of flat connections on a torus. In this case any flat connection can be characterized by the value a_1, a_2 of its holonomy along two preferred families of loops η_1, η_2, the loops that encircle the two generatrices of the torus. As we discussed above, homotopically equivalent loops yield the same holonomy when the connection is flat, so the holonomies are really functions of the homotopy classes of loops. Since the homotopy group of the torus is Abelian, the holonomies along the two different families of loops commute. That is, they correspond to $SO(2,1)$ rotations around the same axis. Depending on the null, time-like or spatial character of this axis in the $SO(2,1)$ manifold one has three distinct sectors of the theory.

Let us discuss in detail the time-like sector. For this case, the two non-trivial holonomies can be written,

$$U(a_b) = \exp(2a_b t^i \tau_i), \qquad b = 1, 2, \tag{9.71}$$

where t^i is the time-like rotation axis in the internal space, a_b are the rotation parameters and τ_i are the $SO(2,1)$ matrices. From here it is immediate to compute the value of the T quantities,

$$T^0(\vec{n}) = 2\cos(\vec{n} \circ \vec{a}), \tag{9.72}$$
$$T^1(\vec{n}) = 2\sin(\vec{n} \circ \vec{a})\vec{n} \times \vec{p}, \tag{9.73}$$

where $\vec{n} = (n_1, n_2)$ and $\vec{n} \circ \vec{a} = n_1 a_1 + n_2 a_2$, $\vec{n} \times \vec{p} = n_1 a_2 + n_2 a_1$, and $\vec{p}$ are the variables canonically conjugate to the $\vec{a}$.

We now consider a quantum representation of this algebra on a space of functions $\Psi(a_1, a_2)$. This space can be endowed with an inner product such that $\vec{a}$ and $\vec{p}$ are self-adjoint operators. The measure is simply given by $da_1 da_2$. The T operators are [157]

$$\hat{T}^0(\vec{n})\Psi(a_1, a_2) = 2\cos(\vec{n} \circ \vec{a})\Psi(a_1, a_2), \tag{9.74}$$

$$\hat{T}^1(\vec{n})\Psi(a_1, a_2) = 2i\sin(\vec{n} \circ \vec{a})\vec{n} \times \frac{\partial}{\partial \vec{a}}\Psi(a_1, a_2), \qquad (9.75)$$

and with the inner product introduced above the Ts become self-adjoint operators.

Let us now build a loop representation from this "connection representation". Recalling the expression for the T^0 variable, the loop transform then reads

$$\Psi(n_1, n_2) = \int_{-1}^{1}\int_{-1}^{1} da_1 da_2 \cos((n_1 a_1 + n_2 a_2))\Psi(a_1, a_2), \qquad (9.76)$$

and we see that it reduces to a familiar Fourier transform.

Wavefunctions in the loop representation are given by functions of two integers n_1 and n_2. The only Mandelstam identity left (because of the Abelian nature of the homotopy group of the torus) is given by $\Psi(n_1, n_2) = \Psi(-n_1, -n_2)$.

In this space of wavefunctions one can particularize the usual expressions for the action of the T operators,

$$\hat{T}^0(\eta)\Psi(\gamma) \equiv \Psi(\gamma \circ \eta) + \Psi(\gamma \circ \eta^{-1}), \qquad (9.77)$$

$$\hat{T}^a(\eta_x^x)\Psi(\gamma) \equiv i\sum_{\epsilon=-1}^{1} \epsilon \oint dy^a \delta(x - y)\Psi(\gamma \circ \eta^\epsilon), \qquad (9.78)$$

to the following expressions for the basic operators associated with η_1,

$$\hat{T}^0(\eta_1)\Psi(n_1, n_2) = \Psi(n_1 + 1, n_2) + \Psi(n_1 - 1, n_2), \qquad (9.79)$$

$$\hat{T}^1(\eta_1)\Psi(n_1, n_2) = in_2(\Psi(n_1 + 1, n_2) - \Psi(n_1 - 1, n_2)), \qquad (9.80)$$

which are Hermitian in terms of the inner product

$$< \Psi|\Phi > = \sum_{n_{1,2}=-\infty}^{\infty} \bar{\Psi}(n_1, n_2)\Phi(n_1, n_2). \qquad (9.81)$$

The loop transform (9.76) is symmetric in $a_{1,2}$. Therefore antisymmetric functions are mapped to zero. The transform defines an isomorphism between the space of symmetric functions of $a_{1,2}$ and the loop representation[‡].

Up to now we have considered the case of holonomies that are $SO(2, 1)$ rotations around a time-like internal vector. Let us now briefly consider the case of space-like rotations. In that case, the holonomy is again given by (9.71), which particularized to the space-like sector of $SO(2, 1)$ (where

---

[‡] As we mentioned in chapter 3, we are disregarding symmetries that are not generated by constraints. Peldán [161] discusses the role that large diffeomorphisms play in this sector of 2+1 gravity.

the τ matrices are anti-Hermitic) gives

$$T^0(\vec{n}) = 2\cosh(\vec{n} \circ \vec{a}), \tag{9.82}$$

$$T^1(\vec{n}) = 2\sinh(\vec{n} \circ \vec{a})\vec{n} \times \vec{p}. \tag{9.83}$$

Notice that the topology of the phase space is different from that in the time-like case: $|\vec{a}|$ is now unbounded.

From here on one could construct a loop representation in exactly the same fashion as in the time-like case. The quantum T algebras are exactly the same and the inner product is the same.

It should be emphasized that the T algebras can be completely realized in terms of either the space-like or the time sectors of the phase space (the null sector has also been considered in detail in references [19, 162]). Therefore these sectors correspond to separate — in principle, inequivalent — quantum theories. In fact, the theories are quite inequivalent. As pointed out before, for the time-like case the T^0 operator was bounded as a classical quantity, and being a multiplicative quantum operator, its eigenvalues are bounded. This is not the case for the space-like holonomies. Moreover, it is the space-like sector that is equivalent to the ADM quantization [162].

This leads to a disquieting picture: both the space-like and the time-like cases in the connection representation seem to give rise to the same loop representation; however, at the level of connections they are quite distinct. How could it be that two inequivalent representations in terms of connections give rise to the same loop representation?

The answer lies in the precise relationship between the connection and loop representations. For the time-like case we saw that the loop representation was isomorphic to the symmetric connection representation. We will see that for the space-like case it is not. This solves the contradiction. Let us discuss the situation in some detail. The loop transform for the spatial case reads

$$\Psi(n_1, n_2) = \int_{-\infty}^{\infty} \int_{-\infty}^{\infty} da_1 da_2 \cosh((n_1 a_1 + n_2 a_2))\Psi(a_1, a_2), \tag{9.84}$$

and using the symmetries of the connections in the spatial sector which imply $\Psi(\vec{a}) = \Psi(-\vec{a})$, we can reduce it to a two-sided Laplace transform,

$$\Psi(n_1, n_2) = \int_{-\infty}^{\infty} \int_{-\infty}^{\infty} da_1 da_2 \exp((n_1 a_1 + n_2 a_2))\Psi(a_1, a_2). \tag{9.85}$$

The problem is that this would be a usual Laplace transform if the parameters n_1 and n_2 were real numbers. Being integers, one immediately finds that the transform is not an isomorphism. It turns out that it has a large non-trivial kernel. To give an intuitive idea of the problem it is instructive to construct one of the elements of the kernel of the transform. Simply consider a function $f(s_1, s_2)$ with $s_{1,2} \in \mathbf{R}$ such that it vanishes for $s_{1,2}$ integer. Such a function is mapped by the inverse Laplace transform to a non-trivial element in the connection representation that has a vanishing image in the loop representation.

One may ask if this problem is just a technicality or if it is a serious drawback. It turns out that the kernel we found is dense in the space of connections [19]. That means that *any* wavefunction in terms of connections can be obtained as a limit of a sequence of elements that are all in the kernel. This is a serious problem: the sector of the theory which corresponds to usual geometrodynamic quantization is not properly described by the loop representation we introduced.

Does this example imply that loop representations are inadequate to describe theories based on non-compact gauge groups? At least it should be viewed as a warning. It turns out that the problem is not fatal and one can deal with it if proper care is exercised. Up to now, two solutions have been proposed: the use of a non-trivial measure in the transform and the use of extended loops. In a sense, this difficulty can be seen as added motivation for the consideration of such ideas in the case of gravity.

Ashtekar and Loll [163] pointed out that if one introduces a non-trivial measure in the space of connections, the resulting loop transform does not have a non-trivial kernel. The explicit construction of the measure has been found for surfaces of lower genus and it implies a non-trivial change in the quantum realization of the T operators. This just corresponds to one of the many different choices one has when quantizing a field theory. For the case of the torus the non-trivial measure is

$$d\mu[A] = \exp(-c[T^0(\eta_1) + T^0(\eta_2)]) \qquad (9.86)$$

where c is a positive constant and η_1 and η_2 are two fixed loops belonging to independent homotopy classes. The measure depends quite non-trivially on the genus and apparently cannot be derived from any general principle in the loop representation. One has to know that the difficulty arises and then carefully examine the correspondence between the loop and the connection representation to construct a measure that solves it. This is somewhat discouraging since one does not hope to have all that information available in more complicated cases.

Another solution that has been proposed by Marolf [19] is the use of extended loops. From the point of view that is presented in this book, this would be a very satisfactory solution since it uses the same general principles that are advocated for use in the non-trivial theories. The idea has not been analyzed in great detail, but the basic concept is very simple: if one considers extended loops, the holonomies of a connection in the torus are labeled by two real numbers instead of integers. If one writes the extended loop transform in the spatial case it corresponds to

(9.85) but with $n_{1,2}$ real numbers, which defines an isomorphism between the spaces.

Many other issues have been explored in the (2+1)-dimensional theory. Our intention here was not to be exhaustive but to show that the program in general lines can be carried along in a theory in which one has control over the mathematical issues. We encountered expected difficulties that can be resolved with ideas that are being applied in more complicated cases. For more details see the reviews by Carlip [164] and Ashtekar [165] and the papers by Marolf and Louko [19, 162].

9.7 Conclusions

We have discussed several applications of the loop representation of general relativity coupled to matter fields and the definition of physically meaningful quantities in terms of loops. We have shown how to define regularized operators that could be used through matter couplings to give an idealized picture of the measurement process in quantum gravity. Evidently there is a long way to go before we can make actual physical predictions in quantum gravity. In particular it is expected that currently unknown approximation techniques will be crucially needed due to the complexity of the Einstein equations. The hope is that the ideas presented in this chapter may act as building blocks of such a measurement theory. We ended this chapter by discussing a rather disjoint application: general relativity in $2+1$ dimensions, which plays the important role of showing that the ideas being advocated in this book can actually be applied to a theory of quantum gravity in a simplified setting. The ideas in this chapter were mainly built either at a kinematical level or heuristically implemented in the space of smooth non-intersecting loops. In the following chapter we will be concerned with the dynamics of quantum gravity and therefore we will make progress towards finding physical states of the theory. Many, though not all, of the ideas of this present chapter go through for the intersecting solutions that we will present in the next two chapters. Some others will need a revision. We expect that in the next few years progress will be made in this direction.

10

Knot theory and physical states of quantum gravity

10.1 Introduction

In the previous two chapters we developed several aspects of the loop representation of quantum gravity. One of the main consequences of these developments is a radically new description of one of the symmetries of the theory: because of diffeomorphism invariance wavefunctions in the loop representation must be invariant under deformations of the loops, they have to be knot invariants. This statement is much more than a semantical note. Knot invariants have been studied by mathematicians for a considerable time and recently there has been a surge in interest in knot theory. Behind this surge of interest is the discovery of connections between knot theory and various areas of physics, among them topological field theories. We will see in this chapter that such connections seem to play a crucial role in the structure of the space of states of quantum gravity in the loop representation. As a consequence we will discover a link between quantum gravity and particle physics that was completely unexpected and that involves in an explicit way the non-trivial dynamics of the Einstein equation. Such a link could be an accident or could be the first hint of a complete new sets of relationships between quantum gravity, topological field theories and knot theory.

We will start this chapter with a general introduction to the ideas of knot theory. We will then develop the notions of knot polynomials and the braid group. In the third section we will discuss the connection between knot theory and topological field theories, through the Chern–Simons theory. In section 10.4 we will show how to use the previous notions to construct states of quantum gravity in the loop representation related to the Kauffman and Jones polynomials. In the last section we will present a simple explanation for the existence of the Jones polynomial state and a discussion on the possibility of generating new solutions.

10.2 Knot theory

The study of invariants of closed curves under smooth deformations is quite old. One of the first examples was the introduction by Gauss (1820) of the linking number. The linking number is an invariant of two closed curves that measures the number of times one of the curves winds around the other. This is obviously an invariant since the only way to change that number is to cut one of the closed curves and therefore it is not a smooth deformation. Although such an invariant may appear quite trivial, we will see it plays an important role in topological field theories and quantum gravity. In particular, it admits an integral expression, as we discussed in chapter 3,

$$L(\gamma_1, \gamma_2) = \frac{1}{4\pi} \oint_{\gamma_1} dx^a \oint_{\gamma_2} dy^b \epsilon_{abc} \frac{(x-y)^c}{|x-y|^3}. \tag{10.1}$$

Such an expression was considered by Maxwell [174] in connection with electromagnetic theory. If one builds a thin solenoid with the shape of each loop the above integral measures the magnetic flux produced by each solenoid across the other [175] in appropriate units. In particular Maxwell gives a good explanation for why that expression gives one or zero as result. It measures the solid angle that one of the loops subtends from the point of view of the other as one traverses along the latter. Therefore the result is either an integer multiple of 4π or 0 depending on how the loops are linked. Notice that the expression we give for the linking number depends explicitly on a background metric and yet the result is diffeomorphism invariant.

It is evident that there is much more to knot theory than the linking number as can be illustrated by the Borromean rings which we show in figure 10.1.

This example illustrates a usual difficulty with trying to distinguish knots through the values of a particular knot invariant. Every time one introduces an invariant it is able to detect up to a certain degree of knotting. For each invariant one can construct complicated links or knots such that the invariant does not detect the linking.

The fundamental problem of knot theory is the classification of knots and links*. The main question is how to tell apart two knots that are not smoothly deformable to each other.

Historically, there was a surge of interest in knot theory towards the end

---

* Usually "knot" refers to a single curve and "link" to many curves. We will loosely use them indistinguishably whenever the context allows. We will also use the word "loop" in the precise sense introduced in chapter 1 whenever applicable. For instance the Gauss linking number is a genuine function of loops.

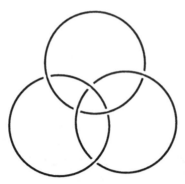

Fig. 10.1. The Borromean rings. An example of loops that have a non-trivial linking but zero linking number if taken in pairs.

of last century due to a failed theory of atoms of James Clerk Maxwell, Lord Kelvin and Peter Guthrie Tait [176]. After the discovery of the complete theory of electromagnetic phenomena, the outstanding unsolved problem in physics was the explanation of atomic spectra. In the proposed theory atoms were depicted as knotted lines of aether (this predated special relativity and quantum mechanics). The theory had several attractive features, among which it associated the stability of atoms with the topological nature of knots. The main lasting impact of it, however, was that through its development many of the central issues of modern knot theory were brought to the forefront. Among these was the classification of knots and their representations. It is remarkable that 100 years later, although the physical motivations are quite different, the interest in knot theory remains basically the same.

The typical depiction of a knot is through its projection on a plane, as we did when depicting the Borromean rings. This adds an additional complication in the sense that a single knot admits a number of different projections. Smooth deformations of knots in three-dimensional space translate themselves in a series of motions in terms of the projections. Such motions are known as Reidemeister moves. There are three types of Reidemeister moves, which are depicted in figure 10.2. If two knot projections can be mapped into each other through a finite number of Reidemeister moves, they are projections of the same knot.

In the knot theory literature two knots that are connected through a finite number of Reidemeister moves are called ambient isotopic. Strictly from our point of view, it is this kind of equivalence that we are interested

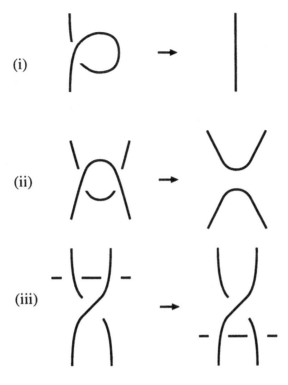

Fig. 10.2. Reidemeister moves.

in, since it corresponds to the usual diffeomorphism invariance. For several reasons that we will discuss shortly it will be useful to consider quantities invariant under a slightly different set of transformations leading to a notion called regular isotopy. Two knots are called regular isotopic to each other if they can be connected through a finite set of Reidemeister moves of types (ii) and (iii). Such an idea is important in the following context. Suppose instead of dealing with knots made of strings of zero width we were considering knots made of ribbons. It is clear that the first Reidemeister move does not correspond to a smooth deformation of a ribbon, since the elimination of a "curl" can only be attained through the introduction of a twist, as shown in figure 10.3. The justification for the consideration of regular isotopy in quantum gravity will be related to regularization issues. As we have done in previous chapters, in many contexts one needs to point-split expressions and in such splitting the resulting objects resemble ribbons rather than loops. We will give details in chapter 11.

At this point the reader may be wondering what is the connection with quantum gravity. To put it in a different way: one knows that the wavefunctions of quantum gravity are knot invariants. Which of all the possible

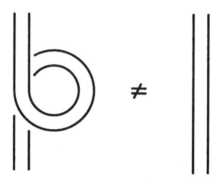

Fig. 10.3.   Reidemeister moves of type (i) do not leave invariant functions of ribbons.

knot invariants are of interest for quantum gravity? At the moment the answer is quite open. We will later introduce some invariants that solve the Hamiltonian constraint. Previous to that there are three main points to be remembered: (a) any knot invariant of interest to quantum gravity has to be a function of the group of loops rather than a function of curves; (b) it should satisfy the Mandelstam identities; (c) it has to be a well defined function of intersecting loops.

Very little needs to be said about points (a) and (b); since one is interested in a loop representation that is obtained via a loop transform from the connection representation only functions of the group of loops that satisfy the Mandelstam identities should be allowed. Point (c) stems from the discussion in chapter 8. As we saw there, the Mandelstam identities related the value of the function on loops with intersections with the value on loops without. Therefore for consistency one has to consider loops with intersections. Furthermore we saw that non-intersecting loops solved the constraints for all values of the cosmological constant: they corresponded to degenerate geometries.

Intersecting knot theory is a quite novel subject. A surge of interest has arisen as a consequence of the theory of Vassiliev invariants [188, 189]. Most of the studies of knot invariants, however, were done for non-intersecting, smooth curves. It turns out several ideas can be easily generalized. We will do so in section 10.3.4.

10.3   Knot polynomials

As we mentioned above, the main problem in knot theory is to classify knots. The obvious solution to this problem is to try to generate a large number of knot invariants. The hope is that through the computation of their values one could distinguish knots, since knots with different values of their invariants are necessarily different. This program is at present incomplete. Though we know a large number of knot invariants they are not enough to classify knots. An important step towards the generation of knot invariants was the construction of certain polynomials associated with knots. In this section we will sketch some ideas of the theory of knot polynomials. We start with a discussion of the braid group. We then construct polynomials and their skein relations. We end with a discussion of the extension of these ideas to intersecting loops. There are many good references on the subject of knot polynomials and braids and we many passages of this chapter are modeled after these ideas. As an example we can cite the books by Kauffman [177, 178] and the review article by Guadagnini [179] and his book [180]. A more elementary but very readable treatment is given in the books by Adams [182] and Baez and Muniain [181].

10.3.1   *The Artin braid group*

A useful way to represent knots and links is through the braid group, B_n. Consider a set of n vertical strings starting and ending in two rows of n horizontally aligned points. The lines can cross each other an arbitrary number of times, forming a braid. Now arrange the lines in such a way that at each horizontal level there is only one crossing at the ith strand, which we denote g_i. One can describe such a braid by a sequence $g_i g_j g_k \ldots$. Such an ordered sequence states that if one follows the braid from the top to bottom (or vice-versa) one encounters a twist of the strings at the ith and $(i+1)$th positions followed by a twist of the strings at the jth and $(j+1)$th positions and so on, as shown in figure 10.4. Each twist has two possible orientations, denoted by g_i and g_i^{-1}.

The twists g_i form a group structure, called the Artin braid group. For n strings B_n has $n-1$ generators $g_1, \ldots, g_{n-1}$ that satisfy the relations

$$g_i g_j = g_j g_i, \qquad |i - j| > 1, \tag{10.2}$$

$$g_i g_{i+1} g_i = g_{i+1} g_i g_{i+1}, \qquad i < n - 1, \tag{10.3}$$

which can be easily checked by drawing n strings and applying the twists.

The strings involved in the braid group can be thought of as the space-time trajectory of particles in $2 + 1$ dimensions as they orbit around each other. This suggests an immediate connection between the braid group and $(2+1)$-dimensional physics. This connection has been explored in several contexts, including particle [190] and solid state physics. In particular it is the root of unusual statistics in $2+1$ dimensions connected with the idea of anyons [191, 192, 193].

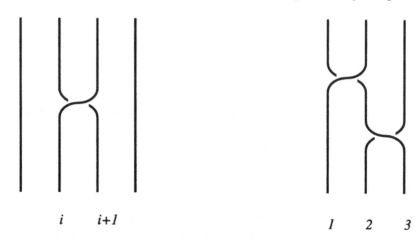

i $i+1$ 1 2 3

Fig. 10.4. Graphical representation of g_i and $g_1 g_2^{-1}$.

What is the relation with knots? One simply obtains a knot or a link by gluing together the ends of a braid in an order preserving manner (this is called a closure of the braid). Conversely, one can associate to each knot a braid. Therefore several properties of knots can be coded in the language of braids. The first question that arises is: given two braids, what are the conditions for their closure to yield the same knot or link up to ambient isotopy? For knot diagrams the answer is given by the Reidemeister moves. In terms of braids they translate themselves into a set of moves called the Markov moves. Reidemeister moves of type (iii) are already included in the braid group relation (10.3). Reidemeister moves of type (ii) are almost included in the relation (10.2), except for the fact that $g_1 \neq g_2 g_1 g_2^{-1}$, whereas both g_1 and $g_2 g_1 g_2^{-1}$ yield the same link under closure. The message is that to implement fully the second Reidemeister move in terms of braids, one has to identify elements that are conjugate under the adjoint action of the braid group. Two elements of the braid group that are conjugate are said to be related by a Markov move of type 1. Reidemeister moves of type (i) imply that a link diagram associated with the closure of a certain braid $b \in B_n$ and the closure of the braid $b g_n^{\pm 1} \in B_{n+1}$ are equivalent. These two elements are said to be related by a Markov move of type 2.

The advantage of the description of links in terms of braids is that one can present several properties of link diagrams in terms of algebraic notions. One can define link invariants as functionals of the elements of the braid group that are invariant under Markov moves. This implies the introduction of representations of the braid group. The closure of braids is represented by taking traces of expressions in terms of the group. We will explicitly use this kind of construction to derive expressions for some of the knot polynomials of relevance for quantum gravity. Before going into the details, we will discuss some general notions related to knot polynomials.

10.3.2 Skein relations, ambient and regular isotopies

A knot (or link) polynomial is an assignment of a finite set of numbers to a knot (or link) that is invariant under ambient or regular isotopies. Given a knot γ one gets a polynomial[†] $P(\gamma)_q$ in an arbitrary variable q such that all the coefficients $p_i(\gamma)$ of the polynomial are knot invariants. An important point is that for each knot the polynomial is of a finite order, but the order depends on the particular knot. An intuitive picture is that the lower coefficients of the polynomial represent "more naive" knot invariants that sense the simpler kinds of knottings whereas the higher coefficients are sensitive to more sophisticated kinds of knottings. Therefore for a simple kind of knot the lower coefficients of the polynomial are non-zero and the higher ones vanish. For more complicated knottings the lower coefficients fail to "see" the knottiness and the higher coefficients are the ones that sense it up to a certain order where again the knottiness is perceived as "trivial" by the more sophisticated higher coefficients. Therefore the order of a knot polynomial is finite and depends on the particular knot considered.

Why are these objects interesting? The reason is they are an ordered way of assigning an unlimited number of invariants to knots according to their complexity. There is therefore the expectation that they could constitute a systematic procedure for classifying knots. Moreover, some of the polynomials are defined by quite succinct recursion relations called the skein relations. The price for all this is high: there are only a handful of polynomials explicitly known at present.

The first polynomial was introduced in the 1920s by Alexander [197]. We present here a modification of that polynomial due to Conway [198] known as the Alexander–Conway polynomial $C(\gamma)_q$. It is defined by the skein relations,

$$C(U)_q = 1, \tag{10.4}$$

$$C(L_+)_q - C(L_-)_q = q\,C(L_0)_q, \tag{10.5}$$

where U is the unknot (a knot isotopic to a circle) and $L_\pm, L_0$ refer to the crossings shown in figure 10.5.

[†] In general they are Laurent polynomials. Sometimes it is convenient to write them as functions of a certain fractionary power of a variable, as we will see. Some polynomials may depend on several variables.

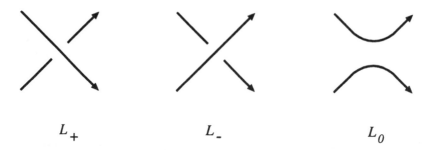

Fig. 10.5. The crossings $L_\pm$ and L_0.

The way in which the skein relations are to be interpreted is the following. The first one is simply a normalization condition that states that the polynomial evaluated for the unknot is 1. To read the second relation consider a specific knot and focus on a point where there is a line crossing. Excise a ball around the crossing so as to leave four incoming strands. The relation (10.5) states that if one evaluates the polynomial for the knot where the crossing we excised is replaced by the crossing L_+ and subtracts from it the polynomial evaluated for the same crossing replaced by L_- one gets as a result q (the polynomial variable) times the polynomial evaluated for the crossing replaced by L_0. The resulting equation is a relationship between the polynomials associated with three different knots. The strategy is to apply the relationship recursively combined with the Reidemeister moves until one gets a system of equations for the coefficients with a unique solution.

For a particular set of relations it is very difficult to prove that they determine the value of the polynomial for all knots unless one generates the skein relation in such a way as to guarantee it. The same consideration is true with respect to the diffeomorphism invariance of the objects constructed. The skein relations are relations between projections of the knots and it is quite non-trivial that the polynomial they define is independent of the projection.

Another important polynomial is the one due to Jones [199], $J(\gamma)_q$. The skein relations that define it are

$$J(U)_q = 1, \tag{10.6}$$

$$qJ(L_+)_q - q^{-1}J(L_-)_q = (q^{1/2} - q^{-1/2})J(L_0)_q. \tag{10.7}$$

The Jones polynomial is more "selective" than the Alexander–Conway one. However, there exist non-isotopic knots that have associated the same Jones polynomial, i.e., it fails to provide a classification for knots. There are other known polynomials, such as the HOMFLY [200] polynomial, which are slightly more general and contain Jones and Alexander–

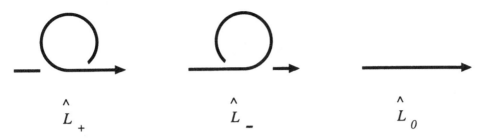

Fig. 10.6. Crossings for the skein relations of regular isotopic invariants.

Conway polynomials as particular cases. However, no polynomial known at present is sufficient to distinguish all knots.

Let us now concentrate on regular isotopy invariants. As we mentioned before, these are invariants that are sensitive to the first Reidemeister move, i.e., they "see" the additions of curls in the knots. Another way to put it is that they are invariants of (oriented) ribbons rather than of curves. Knot polynomials that are regular invariants can be defined. Their definition requires the introduction of a new set of crossings in their skein relations, $\hat{L}_{\pm,0}$, as shown in figure 10.6.

As an example of a regular invariant polynomial let us consider the Kauffman bracket, which can be viewed as a regular generalization of the Jones polynomial. The skein relations that define it are

$$K(U)_q = 1, \tag{10.8}$$

$$q^{1/4}K(L_+)_q - q^{-1/4}K(L_-)_q = (q^{1/2} - q^{-1/2})K(L_0)_q, \tag{10.9}$$

$$K(\hat{L}_+)_q = q^{3/4}K(\hat{L}_0)_q, \tag{10.10}$$

$$K(\hat{L}_-)_q = q^{-3/4}K(\hat{L}_0)_q. \tag{10.11}$$

Regular isotopic invariants of curves can be associated with ambient isotopic invariants of oriented ribbons if one gives a prescription to associate a ribbon to each curve. Such prescriptions are called "framings". Technically they correspond to an assignment of a vector to each point of the curve, such that one obtains a second curve by infinitesimally shifting the original one along the vector.

We now introduce some concepts that are useful in the discussion of regular isotopic invariants. The first of them is the writhe of a knot diagram, $w(\gamma)$, defined by

$$w(\gamma) = \sum_{\text{crossings}} \epsilon(\text{crossing}), \tag{10.12}$$

where $\epsilon(L_\pm) \equiv \pm 1$. This quantity measures the number of "curls" in the diagram. It is clearly not invariant under Reidemeister moves of type (i)

Fig. 10.7. A twist can be exchanged by a curl through a Reidemeister (i) move

but it is a regular isotopic invariant.

Another regular invariant, in this case of bands, is the twist. Assume one paints the two sides of the band with different colors. The twist measures how many times the color changes as seen from the planar projection. It can also be defined in terms of an analytic expression, but we will not discuss this here [196]. It is evident that if one performs a Reidemeister move of type (i) one can exchange a twist in a band by a curl, as shown in figure 10.7.

Using the fact that Reidemeister moves of type (i) exchange curls and twists in bands, one can combine the previous two quantities into an ambient isotopic invariant of the two curves that form the band. The resulting invariant is given by their linking number, which now can be viewed as a quantity associated with the knot diagram through a framing procedure. To reflect that association it is usually called the "self-linking" number of a knot diagram. One can summarize this result in a formula called White's theorem [194],

$$SL(\gamma) = T(\gamma) + w(\gamma), \qquad (10.13)$$

where $SL(\gamma)$ stands for self-linking number of the knot diagram. Explicit expressions for all the terms in White's theorem can be given. For the Gauss linking number, apart from the integral formula we have already discussed, a definition can be introduced terms of the plane projection of two curves. This is given by

$$L(\gamma_1, \gamma_2) = \tfrac{1}{2} \sum_{\text{crossings}(\gamma_1, \gamma_2)} \epsilon(\text{crossing}), \qquad (10.14)$$

where the summation is only over the crossings of one curve with the other. The reader can check that this expression gives the usual result for the linking of two curves. White's theorem has found important applications in biology, where one has to count the twists of DNA structures through the plane projections one gets when viewing it through a microscope [195] and also in Polyakov's description of the Fermi–Bose transmutation in the context of anyons [196].

There are many prescriptions for framing. One of them is the "vertical framing", in which the twist of the ribbon is set to zero, i.e., all twists are converted to curls. Due to White's theorem, in this framing the linking number coincides with the writhe. This is also called "blackboard framing" [178] since it corresponds to considering the projection of the knot and drawing a parallel knot along it. The resulting ribbon has no twist. Another common framing is the "standard" or "canonical" framing. This is defined by setting the self-linking number to zero. This is a "natural" framing in the sense that it does not depend on particular projections.

The value of the self-linking number in different framings differs by an integer corresponding to the number of twists introduced in the band associated with the loop by the framing procedure. The existence of the natural framing may appear as reassuring since it would seem to restore diffeomorphism invariance to the discussion. Unfortunately, the natural framing only exists in certain manifolds, e.g., S^3 manifolds, since in other cases the linking number may be ill defined or be a non-integer number [45].

The explicit relation between the Kauffman and Jones polynomials is given by

$$K(\gamma)_q = q^{\frac{3}{4}w(\gamma)} J(\gamma)_q, \tag{10.15}$$

and we will offer a proof of this in the next section. It is remarkable that all the framing dependence of the Kauffman bracket is concentrated in the prefactor involving the writhe.

10.3.3   *Knot polynomials from representations of the braid group*

At present a complete classification of the irreducible representations of the braid group is not known. Finding representations for the braid group is a non-trivial matter. We will present here a construction that yields the representation that gives rise to the Jones polynomial. This representation is the simplest one of the family that can be constructed with a method called the R matrix approach [201, 202].

Assume that a two-dimensional linear space V_i is associated with the ith string so that the total linear space associated with the n strings is given by the tensor product $V(n) = V_1 \otimes V_2 \otimes \cdots V_n$. In each space V_i introduce a basis e_i^A, $A = 1, 2$. Each generator is represented by a $2^n \times 2^n$ matrix of the form

$$G_i = q^{1/4}(I \otimes \dots \otimes R \otimes \dots \otimes I), \tag{10.16}$$

where q is an arbitrary complex number, I is the 2×2 identity matrix and the matrix R, which acts on $V_i \otimes V_{i+1}$, is given by

$$R = \begin{pmatrix} 1 & 0 & 0 & 0 \\ 0 & 1-q^{-1} & q^{-1/2} & 0 \\ 0 & q^{-1/2} & 0 & 0 \\ 0 & 0 & 0 & 1 \end{pmatrix} \tag{10.17}$$

in the basis of $V_i \otimes V_{i+1}$ given by $\{e_i^1 e_{i+1}^1, e_i^1 e_{i+1}^2, e_i^2 e_{i+1}^1, e_i^2 e_{i+1}^2\}$.

It is a straightforward calculation to show that the relations defining the braid group (10.2),(10.3) are satisfied by the matrices G_i and their corresponding inverses. Therefore they define a representation of the braid group on the vector space $V(n)$.

In order to construct knot polynomials starting from a representation of the braid group we need to construct quantities that are invariant under Markov moves. As we discussed in section 10.3.1 by taking traces of the representation one constructs invariants under Reidemeister moves of types (ii) and (iii), i.e., regular isotopic invariants. In order to implement invariance under type (i) moves we will introduce a matrix in $V(n)$ called "enhancement matrix". This is defined by

$$\mu^n = \mu_1 \otimes \ldots \otimes \mu_n, \tag{10.18}$$

where

$$\mu_i = \begin{pmatrix} q^{-1/2} & 0 \\ 0 & q^{1/2} \end{pmatrix}. \tag{10.19}$$

The enhancement matrix has two main properties. First, it commutes with all the generators of the braid group G_i. To introduce the second property we recall that in a tensor product of spaces one can introduce a partial trace operation on one of the factor spaces. For instance, if one considers the trace in V_{i+1} of a tensor product $V_1 \otimes \ldots \otimes V_{i+1}$ one gets as a result an element of $V_1 \otimes \ldots \otimes V_i$. Taking this into account one can check that for the enhancement matrix

$$\mathrm{Tr}|_{V_{i+1}}(R\,\mu_{i+1}) = q^{1/2} \mathbf{1}|_{V_i}, \tag{10.20}$$

where the product $R\mu_{i+1}$ is defined in the space $V_i \otimes V_{i+1}$ as R times $\mathbf{1}|_{V_i} \otimes \mu_{i+1}$. A similar result holds for the inverse,

$$\mathrm{Tr}|_{V_{i+1}}(R^{-1}\mu_{i+1}) = q^{-1/2} \mathbf{1}|_{V_i}. \tag{10.21}$$

We can use this property to construct quantities that are invariant under all Reidemeister moves. Consider a matrix B representing an arbitrary element b of the braid group B_n and define the quantity

$$F(B) = q^{-\frac{3}{4}w(\hat{b})}\mathrm{Tr}|_{V(n)}(B\mu^n), \tag{10.22}$$

where $\hat{b}$ is the link obtained as a closure of the element b and $w(\hat{b})$ is its writhe. We will now prove that this quantity is a link invariant under ambient isotopy associated with the closure $\hat{b}$ of the braid b. We prove this by showing that it is invariant under Markov moves. Since μ^n commutes with all the generators of the braid group, it is immediate to show that $F(B_2^{-1}B_1B_2) = F(B_1)$. Moreover, if B_1 and B_2 are the matrices representing the elements $b_1 \in B_n$ and $b_2 = b_1 g_n^{\pm 1} \in B_{n+1}$ in the spaces $V(n)$ and $V(n+1)$ we have

$$
\begin{aligned}
F(B_2) &= q^{-\frac{3}{4}w(\hat{b}_2)}\mathrm{Tr}|_{V(n+1)}(B_2\mu^{n+1}) \\
&= q^{-\frac{3}{4}w(\hat{b}_1)}q^{\mp\frac{3}{4}}\mathrm{Tr}|_{V(n)\otimes V_{n+1}}(B_1 G_n^{\pm 1}\mu^n \otimes \mu_{n+1}) \\
&= q^{-\frac{3}{4}w(\hat{b}_1)}\mathrm{Tr}|_{V(n)}(B_1\mu^n) \\
&= F(B_1)
\end{aligned}
\tag{10.23}
$$

which can be straightforwardly checked relating the trace operation in $V(n+1)$ with that in $V(n)$. The writhes of b_1 and b_2 differ by a factor ± 1 since $G_n^{\pm 1}$ introduces an additional curl in the loop. The extra power of q this introduces exactly cancels a factor that arises when relating the traces in $V(n+1)$ and $V(n)$.

Therefore F is associated with an ambient isotopic invariant. To see which invariant it is we compute its skein relations. One can check that the matrix G_i satisfies the relation

$$
q^{1/4}G_i - q^{-1/4}G_i^{-1} - (q^{1/2} - q^{-1/2})I_i = 0,
\tag{10.24}
$$

which combined with the definition of the invariant F gives

$$
qF(BG_i) - q^{-1}F(BG_i^{-1}) = (q^{1/2} - q^{-1/2})F(B),
\tag{10.25}
$$

which is the skein relation for the Jones polynomial.

Equation (10.24) yields, multiplying by μ^n and taking traces, the skein relation for the Kauffman bracket polynomial. As a consequence we immediately have that the Kauffman bracket polynomial is a regular isotopy invariant and is related to the Jones polynomial by expression (10.15) which we introduced in the previous section (they only differ by a factor depending on the writhe). This will have important consequences in quantum gravity.

10.3.4 Intersecting knots

Up to now we have studied the construction of knot polynomials based on smooth loops without intersections. As we have argued before, in the case of gravity we need to consider knots with intersections, because the Mandelstam identities naturally introduce them and because they are associated with non-degenerate metrics. There is no fundamental

Fig. 10.8. The additional element needed in the braid group to generate invariants of links with double intersections.

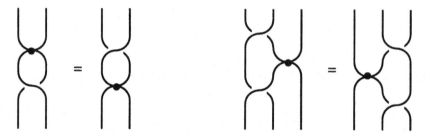

Fig. 10.9. The relations satisfied by the intersecting element of the braid group.

difficulty in adding intersections to the constructions of the braid group and the Jones polynomial we introduced.

The main idea is to extend the braid group with the introduction of an additional element that represents the crossing of two strands. The resulting structure is not a group but an algebra. If one wants to consider intersections of more than two lines additional elements need to be added. Though technically more complicated, the generalization is straightforward [189, 183]. The additional element needed to include double intersections is denoted a_i and we depict it in figure 10.8. It satisfies the relations

$$a_i g_i = g_i a_i, \tag{10.26}$$
$$g_i^{-1} a_{i+1} g_i = g_{i+1} a_i g_{i+1}^{-1}, \tag{10.27}$$

and

$$[g_i, a_j] = 0 \qquad [a_i, a_j] = 0 \qquad |i - j| > 1, \tag{10.28}$$

and the graphical representation of equations (10.26),(10.27) is given in figure 10.9.

The element a_i has no inverse (one cannot remove intersections) and that is the reason why the resulting structure of extending the braid group to intersections is not a group but an algebra.

A matrix representation including the intersecting elements is given by the $2^n \times 2^n$ matrices [183],

$$A_i = I \otimes \dots \otimes A \otimes \dots \otimes I, \qquad (10.29)$$

where A is given by the matrix acting on $V_i \otimes V_{i+1}$:

$$A = \begin{pmatrix} 1 & 0 & 0 & 0 \\ 0 & a & (1-a)q^{1/2} & 0 \\ 0 & (1-a)q^{1/2} & 1-(1-a)q & 0 \\ 0 & 0 & 0 & 1 \end{pmatrix}, \qquad (10.30)$$

where a is another complex parameter. We see that the generalization of a polynomial to (double) intersections requires the introduction of a new variable in the polynomial. The skein relations for the new matrix are

$$A_i = q^{1/4}(1-a)G_i^{-1} + aI_i, \qquad (10.31)$$

and we see that the extension of a polynomial to intersecting loops preserves the usual skein relations for the polynomial but requires additional skein relations that involve intersections. For the Kauffman bracket polynomial the additional skein relation derived from the above expression is

$$K(L_I)_{q,a} = q^{1/4}(1-a)K(L_-)_q + aK(L_0)_q. \qquad (10.32)$$

For triple intersections a generalization of the braid group can also be given in terms of an algebra. There are three new added elements corresponding to triple intersections since different (unrelated by diffeomorphisms) spatial orientations of the incoming strands are possible. A generalization of the HOMFLY polynomial to this case was given by Armand-Ugon, Gambini and Mora [183] and it coincides with the construction we gave for the doubly intersecting case. The generalized polynomials depend on a number of extra variables due to the presence of intersections.

It should be emphasized that there exist *many* non-equivalent extensions of a given polynomial to intersecting knots. General expressions taking into account this fact are present in reference [183]. The extension that we presented above is a particular one, corresponding to the use of R matrix techniques. It is remarkable that this particular extension turns out to be connected with the knot polynomials that appear in topological field theories, as we will discuss in the next section.

10.4 Topological field theories and knots

The previous derivations concerning the Artin braid group and the knot polynomials, as attractive as they may appear in their own right, seem to have little connection with the rest of this book. Throughout this book we have always considered functions of curves defined through explicit analytic expressions. In this chapter dependences on loops have up to this point been implicit and the resulting formulation seems ill suited to be mixed with the loop calculus developed in chapter 1. The missing link is provided in this section.

Topological field theories are field theories that do not require the introduction of a background structure (in particular, no background metric) for their definition. They are therefore naturally diffeomorphism invariant. If one formulates these theories in terms of loops, the resulting quantities should be knot invariants. This may not appear to be a great surprise or advantage: after all the wavefunctions of quantum gravity are diffeomorphism invariant as well. There is, however, an important difference in the case of topological field theories. Most of these theories have only a finite number of topological degrees of freedom and as a consequence are exactly solvable. As a result they provide concrete computable expressions that are invariants of knots.

This was precisely the insight of Witten [45] who noticed that computing expectation values of loop dependent quantities in Chern–Simons and other topological theories one could come up with explicit, analytic, expressions for knot invariants. In the following section we will exploit these results to construct explicit quantum states of the gravitational field. Here we discuss the connection between Chern–Simons theory and the Jones polynomial in some detail.

10.4.1 Chern–Simons theory and the skein relations of the Jones polynomial

A Chern–Simons theory is a gauge theory in $2 + 1$ dimensions where the action is given by the Chern–Simons form of a connection,

$$S_{CS} = \frac{k}{4\pi} \int d^3x \, \tilde{\epsilon}^{abc} \text{Tr}(\mathbf{A}_a \partial_b \mathbf{A}_c + \tfrac{2i}{3} \mathbf{A}_a \mathbf{A}_b \mathbf{A}_c), \qquad (10.33)$$

where k is the coupling constant of the theory.

In contrast to the usual Yang–Mills action, the Chern–Simons action does not require the introduction of a metric or any other background structure for its definition. The Chern–Simons action is invariant under diffeomorphisms and (small) gauge transformations [59]. It is not invariant under large gauge transformations (not connected with the identity). Moreover, the integral is crucial in providing the gauge invariance: the integrand itself is not invariant. The classical equations of motion of this action require that the connection be flat and the theory be gauge invariant. The Chern–Simons action can be written for an arbitrary compact simple gauge group; however, we will restrict our attention to the $SU(2)$

case.

We can now proceed and perform a computation similar to the one we did for Yang–Mills theory in chapter 5, where we computed the value of the Wilson loop average. We recall that the Wilson loop average was identified as the generating functional of the Green functions of the theory. The main difference is that in the present case we will be able to perform the computation explicitly. Because the system does not depend on any external structure the result for the Wilson loop average will be a topological invariant. Notice that we are talking about a Euclidean formulation and the loops will exist in three dimensions. The expectation value of a Wilson loop is given by

$$< W(\gamma) >= \int DA \, \exp{(iS_{CS})} \, W_A(\gamma). \qquad (10.34)$$

This quantity is a knot invariant since S_{CS} is invariant under diffeomorphisms and we assume the measure DA has been chosen to be invariant as well. Which knot invariant is it? We will show it satisfies the skein relations of the Kauffman bracket polynomial. The proof goes along the same lines as in the Makeenko–Migdal formulation of gauge theories that we introduced in chapter 5.

In order to check the skein relations satisfied by the expectation value of the Wilson loop we will consider its change under the addition of an infinitesimal loop. If one considers a straight strand $\hat{L}_0$ in the notation of the previous section and one adds a loop one obtains a crossing $\hat{L}_\pm$, the plus or minus sign being determined by the orientation of the loop added. Similar considerations apply to the other types of crossings; upper and under crossings are related through the addition of a loop to an intersection. We are well equipped to study the change of expressions that are functions of loops under the addition of an infinitesimal loop, so we will do the calculation in this limit. If one wants to consider the addition of a finite loop, a resummation of all orders of perturbation can be formally done, as is discussed in reference [184].

The change of the expectation value of a loop under the addition of a small loop can be computed simply by evaluating the loop derivative. A derivation along these lines was first introduced (for the non-intersecting case) by Cotta-Ramusino, Guadagnini, Martellini and Mintchev [185]. For the intersecting case it was generalized in reference [137]. Smolin [186] introduced a slightly different perturbative derivation. The first proof of the skein relation was introduced by Witten [45] using rational conformal field theory techniques.

We now consider the variation of the expectation value of a Wilson loop when a small loop of area σ^{ab} is appended to the loop γ. Let us first consider the case without intersections. We get

$$\sigma^{ab}\Delta_{ab}(x) < W(\gamma) >= i \int dA \quad \sigma^{ab} F^k_{ab}(x) \text{Tr}(\tau^k U(\gamma^x_x)) \exp(iS_{CS}),$$

(10.35)

where Δ_{ab} is the loop derivative and we have used

$$\Delta_{ab}(x)\text{Tr}(U(\gamma)) = iF^k_{ab}(x)\text{Tr}(\tau^k U(\gamma^x_x)),$$

(10.36)

in which γ^x_x is a loop with origin at the point x.

The exponential of the Chern–Simons action has the property that the quantum electric field acting on it is equal to the magnetic field,

$$\hat{\tilde{E}}^a_k \exp\left(iS_{CS}\right) = -i\frac{\delta}{\delta A^k_a} \exp\left(iS_{CS}\right) = \frac{k}{4\pi} \hat{B}^a_k \exp\left(iS_{CS}\right).$$

(10.37)

Using this relation and integrating by parts, one obtains

$$-\frac{4i\pi}{k} \int dA \, \sigma^{ab} \epsilon_{abc} \int dy^c \delta(x-y) \text{Tr}(\tau^k U(\gamma^y_x)\tau^k U(\gamma^x_y)) \exp(iS_{CS}).$$

(10.38)

The integral is proportional to the volume factor

$$\sigma^{ab}\epsilon_{abc}dy^c\delta(x-y),$$

(10.39)

which, depending on the relative orientation of the two-surface Σ^{ab} and the differential dy^c (which is tangent to γ), can lead to ± 1 or zero. (This expression is only formal, a regularization is needed. We have absorbed appropriate divergent factors in the definition of the coupling constant in order to normalize the volume to ± 1, see reference [184] for details.) Consequently, depending on the value of the volume, there are three possibilities

$$\delta < W(\gamma) >= 0,$$

(10.40)

$$\delta < W(\gamma) >= \mp\frac{3\pi i}{k} < W(\gamma) >.$$

(10.41)

These equations can be interpreted diagrammatically in the following way,

$$< W(\hat{L}_\pm) > - < W(\hat{L}_0) >= \mp\frac{3\pi i}{k} < W(\hat{L}_0) >,$$

(10.42)

and when the volume element vanishes it corresponds to a variation that does not change the topology of the crossing.

We therefore see that, to first order in the area of the added loop, the expectation value of a loop in Chern–Simons theory satisfies one of the skein relations of the Kauffman bracket polynomial. This is a quite non-trivial result that is the root of the renewed interest in knot theory in the past decade.

What about intersections? We introduced in the previous section skein relations for knot polynomials with intersections. Is Chern–Simons theory

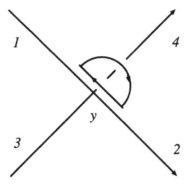

Fig. 10.10. The addition of a small loop at an intersection in the derivation of the skein relation

associated with knot invariants for intersections as well? The answer is yes. It is quite remarkable that of the many possible extensions of knot invariants to intersecting loops, the one that is most naturally picked by Chern–Simons theory coincides with the one we introduced in the previous section.

In order to derive the skein relation for intersections we consider as before an infinitesimal deformation of the loop consisting of the addition of a small closed loop, in this case at the point of intersection (see figure 10.10),

$$\sigma^{ab}\Delta_{ab}(y) < W(\gamma) >= \frac{4\pi}{k}\int dA\sigma^{ab}\epsilon_{dab}\mathrm{Tr}(\tau^k U_{23}(\gamma_y^y)U_{41}(\gamma_y^y))$$
$$\times \frac{\delta}{\delta A_d^k(y)}\exp(iS_{CS}). \qquad (10.43)$$

Again, integrating by parts and choosing the element of area σ^{ab} parallel to the segment 1–2 so that the contribution of the functional derivative corresponding to the action on the segment 1–2 vanishes (since the volume element is zero) we get

$$\sigma^{ab}\Delta_{ab} < W(\gamma) >= -\frac{4i\pi}{k}\int dA\sigma^{ab}\epsilon_{abc}\int dv^c\delta(y-v)$$
$$\times \mathrm{Tr}(\tau^k U_{23}(\gamma_y^y)\tau^k U_{41}(\gamma_y^y))\exp(iS_{CS}). \qquad (10.44)$$

Making use of the Fierz identity for the usual $SU(2)$ matrices (the convention for τ differs by a factor $i/\sqrt{2}$ from the ones considered in chapter 8),

$$\tau^{kA}{}_{B}\tau^{kC}{}_{D} = \frac{1}{2}\delta^{A}_{D}\delta^{C}_{B} - \frac{1}{4}\delta^{A}_{B}\delta^{C}_{D}, \qquad (10.45)$$

one finally gets

$$\sigma^{ab}\Delta_{ab} < W(\gamma) >=$$

$$-\frac{2i\pi}{k}\int dA \,\sigma^{ab}\epsilon_{abc}\int dv^{c}\delta(y-v)\mathrm{Tr}(U_{23}(\gamma^{y}_{y}))\mathrm{Tr}(U_{41}(\gamma^{y}_{y}))\exp(iS_{CS})$$

$$+\frac{i\pi}{k}\int dA \,\sigma^{ab}\epsilon_{abc}\int dv^{c}\delta(y-v)\mathrm{Tr}(U_{23}(\gamma^{y}_{y})U_{41}(\gamma^{y}_{y}))\exp(iS_{CS}), \quad (10.46)$$

where we have called $U_{ij}(\gamma^{x_2}_{x_1})$ the holonomy from point x_1 to x_2 traversing through lines i and j.

These relations can be interpreted as the following skein relation for the intersection:

$$< W(L_{\pm}) > = \left(1 \pm \frac{i\pi}{k}\right) < W(L_I) > \mp \frac{2i\pi}{k} < W(L_0) >, \quad (10.47)$$

$$< W(\hat{L}_{\pm}) > = \left(1 \mp \frac{3i\pi}{k}\right) < W(\hat{L}_0) > . \qquad (10.48)$$

In order to make a comparison with the link polynomials we must first notice that the results we have obtained correspond to a linear approximation, since we have only considered an infinitesimal deformation of the link. In order to consider a finite deformation we would have to consider higher order derivatives of the wavefunction.

It is convenient to rewrite the relations obtained in such a way that the correspondence with those of the Kauffman bracket polynomial in the intersecting case is manifest. To do this we notice that the factor $(1 - 3\pi i/k)$ plays the role of $q^{3/4}$ in the usual skein relation and therefore in the linearized case if we define q as $q = \exp(-4\pi i/k)$. Inverting relation (10.47) we get

$$< W(L_I) >= \left(1 \mp \frac{i\pi}{k}\right) < W(L_{\pm}) > \pm \frac{2i\pi}{k} < W(L_0) >, \qquad (10.49)$$

which allows us to recognize that the value of the variable a of the generalized Kauffman bracket polynomial is up to first order $a = -2\pi i/k$.

The expression relating $< W(L_+) >$ and $< W(L_-) >$ can be obtained in this case by combining equations (10.49). Again we emphasize that the above proofs are only to first order in the area of the loop; in order to prove the skein relations for the addition of a finite loop one can formally sum the perturbative series and confirm for the finite case the result we found infinitesimally. A detailed discussion of this is presented in the paper by Brügmann [184].

So we see that the generalized Kauffman bracket, introduced in the last section for loops with double self-intersections from the R matrix representation of the braid group, is actually the loop transforms of a physical non-degenerate quantum state of the gravitational field defined by values of q and a that to first order in perturbation theory coincide with the ones presented above.

It should be noticed that in order to recover exactly the expression for the polynomials introduced in the previous section we should normalize our results in such a way as to ensure that the value of the polynomials for the unknot is equal to one. This can easily be accomplished by dividing the above expressions by $< W(\text{unknot}) >$. This does not affect the skein relations and ensures the normalization condition.

At this point the reader may be confused. Our promise was to produce via Chern–Simons theory explicit expressions for knot invariants. As a result of our construction we almost obtained this objective, except for the fact that the resulting polynomial is not a genuine knot invariant, but rather a regular knot invariant. Why is the resulting expression not invariant under Reidemeister moves of type (i)?

The difficulty already arises if one considers the expectation value of a Wilson loop in the case of a $U(1)$ Chern–Simons theory. In that case the integral is a Gaussian and the result is the exponential of the self-linking number. The self-linking number is a quantity that involves a $0/0$ indeterminacy, which can be removed by considering a limit. The problem is that the limit is metric dependent. A way to view this is that the limit is a (metric dependent) regularization procedure and the result of it is not metric independent. Another way of viewing it is to consider a point-splitting regularization of the loop. In that case the final result is metric independent (it is the linking number of the split components of the loop) but depends on the particular way the loop is split.

Another difficulty is added in the non-Abelian case. Since the Chern–Simons form is not invariant under large gauge transformations and the Wilson loop is, the resulting integral is not expected to be invariant under large gauge transformations. Therefore, strictly speaking it cannot be a function only of a loop. How this problem relates to the framing ambiguity is not clear. However, it should be stressed that this problem does not arise in the Abelian case (in which all the transformations are small) but the framing ambiguity still persists. The fact that the non-Abelian Chern–Simons form is not invariant under large gauge transformations poses difficulties to doing computation in the non-Abelian case using the rigorous integration techniques of Ashtekar and collaborators [203].

The framing ambiguity issue completely disappears in the extended loop representation, since the extended holonomy is not invariant under large gauge transformations. This issue lies at the crux of the problem of how much is it needed to extend the group of loops to account for these kinds of issues. Is the extension to framed loops enough, as the Chern–Simons integral seems to suggest or does one really need to consider the full extended group of loops? These issues are at present not settled.

10.4.2 Perturbative calculation and explicit expressions for the coefficients

The original intention in connecting knot theory and topological field theories was that in this way one would obtain explicit expressions for knot invariants. Through the calculations of the last section we now know that there is an explicit connection between the expectation value of the Wilson loop in a Chern–Simons theory and the Kauffman bracket. Because Chern–Simons theories are perturbatively renormalizable, one can compute an explicit expression for the expectation value of the Wilson loop in terms of Feynman diagrams. Such an expression we know is equal to the Kauffman bracket. This equality will allow us to give explicit expressions for each of the coefficients of the Kauffman bracket.

We therefore consider the expression of the expectation value of the Wilson loop in a Chern–Simons theory,

$$< W(\gamma) >= \int DA \exp(iS_{CS}) W_A(\gamma), \qquad (10.50)$$

and expand it in powers of the coupling constant k. In order to do this, we write the Wilson loop explicitly,

$$W_A(\gamma) = \sum_{i=0}^{\infty} X^{a_1 \, x_1 \, ... a_i \, x_i}(\gamma) \mathrm{Tr}(A_{a_1 \, x_1} \cdots A_{a_i \, x_i}), \qquad (10.51)$$

and get as the result,

$$< W(\gamma) >= \sum_{i=0}^{\infty} X^{a_1 \, x_1 \, ... a_i \, x_i}(\gamma) < \mathrm{Tr}(A_{a_1 \, x_1} \cdots A_{a_i \, x_i}) > . \qquad (10.52)$$

Therefore by evaluating the n-point functions $< \mathrm{Tr}(A_{a_1 \, x_1} \cdots A_{a_i \, x_i}) >$ perturbatively we can get the expression we were seeking. In order to perform the perturbative expansion one needs to introduce a background metric in order to fix the gauge[‡].

The expression for the propagator is finally given by [187]

$$< A_a^i(x) A_b^j(y) >= \frac{i}{k} \delta^{ij} \epsilon_{abc} \frac{(x-y)^c}{|x-y|^3} + O(1/k^4), \qquad (10.53)$$

[‡] It can be seen that the background metric enters into the gauge fixed action as a commutator of an arbitrary gauge fixing function with the BRST charge and therefore drops out from expressions involving physical states since the BRST charge annihilates such states.

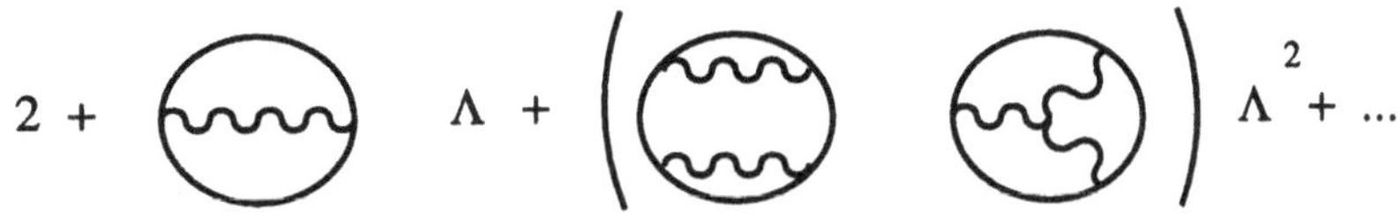

Fig. 10.11. The diagrammatic expansion of the expectation value of the Wilson loop. The circles with insertions correspond to the multitangents of order equal to the number of insertions. The wavy lines are the Chern–Simons propagators, which may be joined in triple vertices. The constant $\Lambda = 3\pi i/k$ is related in the gravitational case to the cosmological constant

where $O(1/k^4)$ may be vanishing but has not been carefully studied. We will not need explicit expressions at that order for our calculations. From it we define the quantity

$$g_{ax\,by} \equiv \frac{ik}{12\pi} < A_a^i(x)A_b^i(y) >, \qquad (10.54)$$

which we have already encountered in chapter 2 as the coordinate expression of the naturally defined metric in the space of transverse vector densities.

The vertex for the theory is given by

$$\frac{ik}{4\pi}\epsilon^{abc}\epsilon_{ijk}, \qquad (10.55)$$

which contracted with three propagators gives rise to the quantity,

$$\left(\frac{4\pi}{ik}\right)^2 h_{ax\,by\,cz} = \int d^3w\, g_{ax\,dw}g_{by\,ew}g_{cz\,fw}\epsilon^{def} + O(1/k^2). \qquad (10.56)$$

We can now proceed to write perturbatively an expansion for the polynomial (shown diagrammatically in figure 10.11),

$$\Psi_k(\gamma) = a_0(\gamma) + a_1(\gamma)\frac{3\pi i}{k} - a_2(\gamma)\frac{9\pi^2}{k^2} - a_3(\gamma)\frac{27\pi^3 i}{k^3} + O(1/k^4), \qquad (10.57)$$

where

$$a_0(\gamma) = 2, \qquad (10.58)$$
$$a_1(\gamma) = X^{ax\,by}g_{ax\,by}, \qquad (10.59)$$
$$a_2(\gamma) = \tfrac{1}{2}a_1(\gamma)^2 - \tfrac{2}{3}\mathcal{A}_2(\gamma), \qquad (10.60)$$
$$a_3(\gamma) = \tfrac{1}{6}a_1(\gamma)^3 + \tfrac{2}{3}a_1(\gamma)\mathcal{A}_2(\gamma) + \tfrac{8}{9}\mathcal{A}_3(\gamma), \qquad (10.61)$$

and

$$\mathcal{A}_2(\gamma) = h_{ax\,by\,cz}X^{ax\,by\,cz} + g_{ax\,cz}g_{by\,dw}X^{ax\,by\,cz\,dw}, \qquad (10.62)$$

$$\mathcal{A}_3(\gamma) = -2 \left[(h_{\mu_1\mu_2\alpha} g^{\alpha\beta} h_{\mu_3\mu_4\beta} - h_{\mu_1\mu_4\alpha} g^{\alpha\beta} h_{\mu_2\mu_3\beta}) X^{\mu_1\mu_2\mu_3\mu_4} \right.$$
$$+ g_{(\mu_1\mu_3} h_{\mu_2\mu_4\mu_5)_c} X^{\mu_1\mu_2\mu_3\mu_4\mu_5}$$
$$\left. + (2 g_{\mu_1\mu_4} g_{\mu_2\mu_5} g_{\mu_3\mu_6} + \tfrac{1}{2} g_{(\mu_1\mu_3} g_{\mu_2\mu_5} g_{\mu_4\mu_6)_c}) X^{\mu_1\mu_2\mu_3\mu_4\mu_5\mu_6} \right],$$

$$(10.63)$$

where as usual greek indices correspond to a pair of spatial index and a point in the manifold. Actually, if γ were a multiloop, $a_0(\gamma)$ would be two raised to the number of connected components of the loop. $a_1(\gamma)$ is the self-linking number of the loop that we have already discussed. $\mathcal{A}_2(\gamma)$ is an ambient isotopic invariant associated with the second coefficient of the Alexander–Conway knot polynomial (the precise expression is given by $\frac{1}{2}(\mathcal{A}_2 + \frac{1}{12})$) and is also related to the classical Arf and Casson knot invariants. This explicit expression was first obtained by Guadagnini, Martellini and Mintchev [205]. The third contribution has been obtained by Di Bartolo and Griego [47]. Central to finding the explicit form of the third order contribution has been the clear identification of the relations satisfied by the loop multitangents (algebraic constraints) which we discussed in chapter 2.

One could continue giving explicit expressions for higher order coefficients. However, one would need refined expressions for the propagators which consider the higher order contributions of ghosts in the diagrammatic expansion.

To summarize, we see that the use of the diagrammatic expansions allows us to construct explicit analytic expressions for the coefficients of the knot polynomials. These expressions provide the completion of the ideas we introduced in chapter 2 in which we suggested that the use of the loop coordinates was good for discussing knot invariants. At that point we were not able to construct the invariants explicitly due to the lack of a natural metric in the space of multitangents (the only natural structure was the kernel used to construct the linking number). We see that through the use of Chern–Simons theory we can construct quantities that contracted with the multitangents yield the knot invariants that we were intending to construct. We will see in the next section how to make use of these invariants to construct physical states of quantum gravity.

Let us end this section with a discussion of framing in the context of the perturbative expansions. In the previous section we showed that the expectation value of the Wilson loop gave rise to the Kauffman bracket. We also saw that the Kauffman bracket was related to the Jones polynomial through a framing dependent prefactor that condensed all the framing dependence of the Kauffman bracket. The prefactor was equal to the exponential of the writhe. Recall that in the vertical framing the writhe coincides with the self-linking number. In the perturbative context, we see that the self-linking number arises in all the coefficients of the expansion of the Kauffman bracket. From the few coefficients we have computed we can get a glimpse of how the different contributions precisely combine to give the prefactor we found in the previous subsection. Explicitly, if one writes

$$K(\gamma)_q = q^{3/4a_1(\gamma)} J(\gamma)_q$$

$$= (1 + \frac{3\pi i}{k}a_1(\gamma) - \frac{1}{2}\frac{9\pi^2}{k^2}a_1(\gamma)^2 - \frac{1}{6}\frac{27\pi^3}{k^3}a_1(\gamma)^3 + \ldots)$$

$$\times (1 + J_2(\gamma)(\frac{4\pi i}{k})^2 + J_3(\gamma)(\frac{4\pi i}{k})^3 + \ldots), \qquad (10.64)$$

where we have expanded the exponential of the self-linking number in powers of k and we have introduced an infinite expansion of the Jones polynomial (this corresponds to considering $q = \exp(4\pi i/k)$ as the variable in the polynomial and writing it as a Laurent expansion in powers of k). We have used the fact that the first coefficient of the Jones polynomial vanishes [177]. From this expression, and comparing with the explicit expansions we introduced before, we see that $J_2(\gamma)$, the second coefficient of the infinite expansion of the Jones polynomial, is proportional to the $\mathcal{A}_2(\gamma)$ invariant we introduced before. We also see that the presence of the terms involving the self-linking number in all the coefficients of the expansion just corresponds to the expansion of the prefactor introduced in the last subsection.

Notice that we get an expression for the coefficients of the polynomial in a particular framing (vertical). This is quite reasonable, the polynomials are defined in a framing independent manner by the skein relations but if one wants a concrete analytic expression for their coefficients one has to give it in a definite framing. The particular framing that appears is determined by the details of the regularization procedure (recall that when we computed the skein relations for the expectation value of the Wilson loop we absorbed divergent factors; the correspondence between that regularization and the one chosen for the perturbative expansion determines the particular framing).

It is not obvious to see explicitly from the expressions we introduced for $\mathcal{A}_2(\gamma)$ that it is an ambient isotopic quantity, as it should be if it is to represent the second coefficient of the Jones polynomial. The issue has been discussed (for the non-intersecting case) by Guadagnini, Martellini and Mintchev [205] and they reach the conclusion that the second coefficient is framing independent. Similar reasonings apply to the third coefficient, though the issue has not been studied in detail.

Do these analytic expressions apply for intersecting loops? Almost all of the expressions are ill defined if the loop has intersections. In order for them to be valid one has to add a prescription (for instance, a point-splitting regularization) at the intersections. The analytic expressions coincide with the coefficients of the extension of the polynomials to the intersecting case which we introduced through the extension of the braid group for some particular prescription for regularization at the intersections. This has only been analyzed for some simple cases and the issue deserves further study.

10.5   States of quantum gravity in terms of knot polynomials

We are now prepared to apply the notions of knot theory derived in the previous sections to the construction of quantum states of the gravitational field.

10.5.1   *The Kauffman bracket as a solution of the constraints with cosmological constant*

As we noticed in chapter 8, in the factor ordering in which triads appear to the left there exists a solution to all the constraints of quantum gravity with a cosmological constant given by the exponential of the Chern–Simons form of the Ashtekar connection,

$$\Psi_{CS}[A] = \exp\left(-\frac{6}{\Lambda}\int d^3x\tilde{\epsilon}^{abc}\mathrm{Tr}(A_a\partial_b A_c + \tfrac{2}{3}A_a A_b A_c)\right). \tag{10.65}$$

If one considers the loop transform of such a state one gets,

$$\Psi_{CS}(\gamma) = \int DA W_A(\gamma)\Psi_{CS}[A] = \int DA W_A(\gamma)\exp\left(-\frac{6}{\Lambda}S_{CS}[A]\right), \tag{10.66}$$

where with the conventions for the gravitational case

$$S_{CS} = \int d^3x\tilde{\epsilon}^{abc}\mathrm{Tr}(A_a\partial_b A_c + \tfrac{2}{3}A_a A_b A_c). \tag{10.67}$$

But this expression is precisely the same as the one we encountered when computing $< W(\gamma) >$ in the context of a Chern–Simons theory. The cosmological constant plays the role of the coupling constant k of the theory. We therefore know what the result is, it is given by the Kauffman bracket knot polynomial in the variable Λ. Therefore the implication is that the Kauffman bracket solves in the loop representation all the constraints of quantum gravity with a cosmological constant.

This suggestion appears as very striking and beautiful, since it allows us instantly to apply in quantum gravity elaborate results from Chern–Simons theory. Before becoming too enthusiastic about this result, we should point out several things that make the proof of the above statement far from solid. First of all, recall that in the Ashtekar formulation of quantum gravity the variables involved are complex. In Chern–Simons theory the connection is real. Therefore the analogy of the expressions presented is only formal. For instance, the expression of the propagator of the theory diverges if the connection is complex. There is no result supporting the existence of the path integral defining the expectation value of the Wilson loop if the connection is complex. The only expectation one can have is that whenever there is a well defined understanding of the complex loop transform, the final calculation will reduce to an analytic continuation of the real result of Chern–Simons theory. If that were the case, we would be justified in using the analogy. Another problem is that the state produced in the loop representation is not a genuine diffeomorphism invariant state, since a framing is required for its definition. At present, due to these difficulties, the results we present can only be taken as purely heuristic in terms of loops. The present attitude towards these problems is that loops may be insufficient to characterize all possible states in the quantum theory. The presence of a framing suggests that a formulation in terms of ribbons or thickened loops could be better suited to the treatment of these issues. At present, however, the only explored context in which they can be given some level of consistency is in terms of extended loops, where all quantities are regularized and the framing ambiguities disappear. We will devote the next chapter to the study of the extended representation and we will find that all the heuristic results that we introduced in this chapter will be mirrored — in a regularized context — in terms of extended loops.

Why should one pursue this avenue at all? Why not simply admit that the transform of the Chern–Simons state is ill defined and forget it as a means of constructing states in quantum gravity? The answer will be given by the next sections. We will see that in spite of the difficulties of putting these results in a rigorous setting a quite non-trivial number of consistent results can be achieved. In particular we will see that the action of the constraints we found in the loop representation on the transform of the Chern–Simons state yield a series of remarkable results that confirm that there is a certain amount of truth behind the formal manipulations we perform.

10.5.2 The Jones polynomial and a state with $\Lambda = 0$

One may have an unsatisfactory feeling about the result introduced in the last section. After all it depended on an arguably vague analogy of the loop transform of the Chern–Simons state and the expectation value of the Wilson loop in a Chern–Simons theory. However, given the developments of chapter 8 we are in a good position to check that the Kauffman bracket is a state of quantum gravity directly in the loop representation.

We have explicit expressions of the polynomial and of the constraints in terms of loops and it is a matter of applying the constraints and checking that the result holds. This will not be a rigorous proof either since the expression for the constraints in the loop representation was obtained through formal manipulations of either the loop transform or the elements of the T algebra. It is, however, quite reassuring that all these formal manipulations yield the same results. Moreover, we will find a remarkable surprise while doing this computational check: we will discover that some of the coefficients of the Jones polynomial must be annihilated by the Hamiltonian constraint of general relativity with $\Lambda = 0$.

The calculation will proceed order by order in the cosmological constant,

$$\hat{H}_\Lambda K(\gamma)_\Lambda = (\hat{H}_0 + \Lambda \hat{\text{det}} q) K(\gamma)_\Lambda. \tag{10.68}$$

The above expression is a polynomial in Λ. If it is to vanish, it has to do so order by order in Λ. To compute the different orders we substitute the expansion for the Kauffman bracket of the previous section. The result is

Order Λ^0:

$$\hat{H}_0 \, 2(\gamma) = 0, \tag{10.69}$$

Order Λ^1:

$$\tfrac{1}{4}\hat{H}_0 \, a_1(\gamma) + \hat{\text{det}} q \, 2(\gamma) = 0, \tag{10.70}$$

Order Λ^2:

$$\hat{H}_0 \, (\tfrac{1}{8}a_1(\gamma)^2 - \tfrac{1}{6}\mathcal{A}_2(\gamma)) + \hat{\text{det}} q \, a_1(\gamma) = 0 \tag{10.71}$$

and so on for higher orders. To obtain these formulae in the conventions we are using for gravity one should replace $ik/4\pi$ by $-6/\Lambda$ in the expressions derived in section 10.4.2.

Notice that we have written $2(\gamma)$ for the number 2 that appears as leading order of the perturbative expansion of the Wilson loop. This is to emphasize that this constant is to be viewed as a constant function in loop space. What we mean by this is that operators like the determinant of the metric, which is a multiplicative operator in loop space will have a non-trivial action on it.

Let us summarize the results we will find. We will mainly prove two things:

(a) One can check by straightforward calculation that the contributions to the three orders in Λ that we listed all vanish.

(b) We will see that in the contribution to order Λ^2, the quantity

$$\hat{H}_0 \, \mathcal{A}_2(\gamma) \tag{10.72}$$

vanishes independently and therefore the second coefficient of the expansion of the Jones polynomial is annihilated by the Wheeler–DeWitt equation for vacuum general relativity with cosmological constant equal to zero.

This last fact is one of the most remarkable results that arise from the loop representation. We find a new non-trivial, non-degenerate state of quantum gravity which we only know in terms of loops. We do not at present know its expression in terms of connections. We will see that its annihilation is the product of a very elaborate cancellation of terms. It may therefore be the manifestation of a very deep relationship between knot theory and the dynamics of quantum gravity of which we are unaware. There was no *a priori* reason to expect this coefficient to be a state and there is no simple explanation of why it is so. We will attempt an explanation in the next section.

Let us now proceed to show these results explicitly. We start with the order Λ^0. In that case we have the action of the Hamiltonian constraint with vanishing cosmological constant on the constant function in loop space $2(\gamma)$. The Hamiltonian constraint trivially annihilates this function since the loop derivative involved in its definition does, due to the fact that it is a constant function. Notice that the determinant of the metric does *not* annihilate this function. We have found the first solution ever of all the constraints of quantum gravity that is only a solution for $\Lambda = 0$ and therefore can be interpreted as associated with a non-degenerate metric. The function is just a constant in loop space. We do not know its form in the connection representation, though we can intuitively picture it as a "delta function" in connection space, requiring the connection to be flat. This would automatically be annihilated by the constraints in the connection representation if one ignores regularization issues.

In order to check that the other orders cancel we need to digress and consider in some detail the action of the constraints introduced in chapter 8. Let us start with the expression of the Hamiltonian constraint of the vacuum theory. As we saw, such an expression acts non-trivially only on the intersections of loops. We have no problem considering intersections in the expressions for the coefficients introduced in the previous section, since we have generalized the polynomials appropriately to the case of intersecting knots. In order to simplify the treatment we will consider the explicit action of the constraints for the case of a triple self-intersecting knot. We saw in chapter 8 that this is the minimal number of intersections one needs in order to produce states of quantum gravity that are not annihilated by the constraints for an arbitrary value of the cosmological constant. This is due to the fact that the definition of the determinant of the metric requires a loop with a triple tangent vector at (at least) one point in order to be non-vanishing.

The expressions for the constraints we introduced in chapter 8 are completely general, we only need to particularize them to the case of interest. As we have argued before it is, in general, incorrect to introduce limitations in the space of loops to consider states with loops with a certain number of intersections. This is not what we are doing here. We are just exhibiting the triple self-intersecting calculation for the sake of clarity but the calculation for an intersection of arbitrary order is done in exactly the same way, only additional terms arise. In particular, we will consider the calculation in the next chapter in terms of the extended representation (which includes all kinds of intersections, even non-isolated ones) and the result is the same.

Let us now consider the expression for the Hamiltonian constraint introduced in chapter 8,

$$H(x)\Psi(\gamma) = 2\oint_\gamma dy^{[b}\oint_\gamma dz^{a]}\delta(x-y)\delta(x-z)\Delta_{ab}(\gamma_o^x)\Psi(\gamma_y^z \circ \gamma_{y\,o}^z). \quad (10.73)$$

We consider a state that is a function of a loop with a triple self-intersection $\gamma = \gamma_1 \circ \gamma_2 \circ \gamma_3$, where γ_i are the petals forming the loop joined at the intersection point. The above expression particularizes to

$$\begin{aligned}
\hat{H}_0(x)\ \Psi(\gamma_1 \circ \gamma_2 \circ \gamma_3) = 2\{&2X^{bx}(\gamma_1)X^{ax}(\gamma_2)\Delta_{ab}(\gamma_{3\,o}^x)\Psi(\gamma_1 \circ \bar\gamma_3 \circ \bar\gamma_2)\\
+&2X^{bx}(\gamma_1)X^{ax}(\gamma_3)\Delta_{ab}(\gamma_{3\,o}^x)\Psi(\gamma_1 \circ \gamma_2 \circ \bar\gamma_3)\\
+&2X^{bx}(\gamma_2)X^{ax}(\gamma_3)\Delta_{ab}((\gamma_3 \circ \gamma_1)_o^x)\Psi(\gamma_2 \circ \bar\gamma_1 \circ \bar\gamma_3)\}, \quad (10.74)
\end{aligned}$$

where $\bar\gamma_i = \gamma_i^{-1}$.

The above particularization is obtained as follows. First notice that the action of the constraint is only non-trivial at the intersection point, which we label x. The point x arises several times when one traverses the loop from beginning to end and there are three different tangent vectors at it (we assume the loop has no kinks at the intersection, i.e., all lines go "straight through", as we discussed in chapter 8). The three non-trivial contributions arise when the loop derivative is contracted with the tangent vectors $1, 2$, $1, 3$ and $2, 3$. Each of these possibilities arises twice but it is easy to see that their contributions are the same as the ones we list here so we account for them by an overall factor of 2. We therefore start traversing the loop with the two integrals that appear in the constraint and compute the non-trivial contributions. The origin of the loop can be taken at an arbitrary point, which we fix at some point of the loop γ_3. The first contribution appears when the integral in y has traversed from the origin to the point x along γ_3 and therefore is at the origin of the loop γ_1, and the integral in z has traversed the first petal of the loop, γ_1, completely and is at the beginning of the loop γ_2. The contribution then has a multitangent corresponding to the origin of γ_1, one corresponding to the origin of γ_2 and the argument of the loop derivative is the portion of the loop γ_3 going from the origin to the intersection point. The second contribution is analogous to the first one but the integral in z has moved to the beginning of the third loop, γ_3. The last contribution has the integral in y moved to the beginning of the loop γ_2. The integral in z can only give a non-trivial contribution when reaching the beginning of γ_3 (we have already counted the possibility that it could be in γ_1). Since the variable in y is now at the beginning of γ_2 (or the end of γ_1) we denote so in the path dependence of the loop derivative. Since we are taking care explicitly of the ordering along the loop of the integrals, we denote the tangent vectors of the loops (and the associated distributions) simply through the first order multitangents evaluated at the corresponding loops.

We did not present in chapter 8 an explicit expression for the determinant of the metric, but it can be computed straightforwardly using the same techniques used for the Hamiltonian. The result is [206]

$$\hat{\det}q\Psi(\gamma) = -4\epsilon_{abc}X^{ax}(\gamma_1)X^{bx}(\gamma_2)X^{cx}(\gamma_3)$$
$$\times \left(\Psi(\gamma_1\gamma_3\bar{\gamma}_2) + \Psi(\gamma_2\gamma_1\bar{\gamma}_3) + \Psi(\gamma_2\gamma_3\bar{\gamma}_1)\right). \quad (10.75)$$

Both the expression for the Hamiltonian and the determinant of the metric are cyclic expressions in terms of the three petals of the loop, in spite of the fact that their immediate appearance is that they are not.

Let us now consider the expression to order Λ^1. First let us concentrate on the action of the determinant of the metric on $1(\gamma)$. As we argued, it is non-vanishing and immediately we can see it is equal to $\epsilon_{abc}\dot{\gamma}_1^a\dot{\gamma}_2^b\dot{\gamma}_3^c 1(\gamma)$.

To compute the action of the Hamiltonian constraint on $a_1(\gamma)$ we consider the explicit form of the wavefunction, the linking number, for a triple self-intersection. This is given by

$$a_1(\gamma_1 \circ \gamma_2 \circ \gamma_3) = g_{\mu\nu}X^\mu(\gamma_1 \circ \gamma_2 \circ \gamma_3)X^\nu(\gamma_1 \circ \gamma_2 \circ \gamma_3)$$
$$= g_{\mu\nu}(X^\mu(\gamma_1) + X^\mu(\gamma_2) + X^\mu(\gamma_3))(X^\nu(\gamma_1) + X^\nu(\gamma_2) + X^\nu(\gamma_3)),$$
$$(10.76)$$

and as usual greek indices refer to a pair of spatial index and spatial point $\mu_1 = a_1\,x_1$.

We now recall the techniques that we used in the calculation in chapter 4 of the action of the Hamiltonian of Maxwell theory on the vacuum state. The loop derivative acts on each first order multitangent producing the derivative of a delta function. Explicitly,

$$\Delta_{ab}(\gamma_o^x)X^{cy}(\gamma) = \delta_{[b}^c\partial_{a]}\delta(x-y). \quad (10.77)$$

Care should be exercised when one considers the particularization of this expression for the petals of the loop. For instance, $\Delta_{ab}(\gamma_{3o}^x)\Psi(\gamma_1)$ is non-vanishing for the loop considered since the deformation introduced by the loop derivative acts at the beginning of the petal γ_1. As a consequence

$\Delta_{ab}(\gamma_{3o}^x)\Psi(\gamma_{2,3}) = 0$, and similarly for the other petals.

We can therefore write the action of the loop derivative in the first term of the Hamiltonian,

$$
\begin{aligned}
\Delta_{ab}(\gamma_{3o}^x)\ a_1(\gamma_1 \circ \bar\gamma_3 \circ \bar\gamma_2) = & \\
\Delta_{ab}(\gamma_{3o}^x)[g_{\mu_1\mu_2}(X^{\mu_1}(\gamma_1) & - X^{\mu_1}(\gamma_2) - X^{\mu_1}(\gamma_3)) \\
\times (X^{\mu_2}(\gamma_1) - X^{\mu_2}(\gamma_2) & - X^{\mu_2}(\gamma_3))] = \\
2\delta_{[b}^{a_1}\partial_{a]}^x\delta(x - x_1)g_{a_1\,x_1\,\mu_2}X^{\mu_2} & (\gamma_1 \circ \bar\gamma_2 \circ \bar\gamma_3),
\end{aligned}
\tag{10.78}
$$

where we have used $X^\mu(\gamma) = -X^\mu(\bar\gamma)$ (as discussed in chapter 2).

We can now integrate by parts the derivative of the delta function. In order to do this, it is useful to introduce the following relation, which can be directly obtained from the definition of the propagator g:

$$
\partial_{[a}^x g_{b]x\,cy} = \delta(x - y)\epsilon_{abc} - g_{ax\,by}\partial_c^x
\tag{10.79}
$$

which together with the transverse character of the first order multitangents implies

$$
\begin{aligned}
2\delta_{[b}^{a_1}\partial_{a]}^x\delta(x - x_1)g_{a_1x_1\,a_2x_2}X^{a_2x_2}(\gamma_1 \circ \bar\gamma_2 \circ \bar\gamma_3) = & \\
-2\epsilon_{a_2ba}\delta(x - x_2)X^{a_2x_2}(\gamma_1 \circ \bar\gamma_2 \circ \bar\gamma_3). &
\end{aligned}
\tag{10.80}
$$

Similar contributions are obtained from the other terms in the Hamiltonian, which combined with the multitangents that multiply the loop derivative yield

$$
\hat{H}_0(x)a_1(\gamma_1 \circ \gamma_2 \circ \gamma_3) = 24\epsilon_{abc}X^{ax}(\gamma_1)X^{bx}(\gamma_2)X^{cx}(\gamma_3).
\tag{10.81}
$$

This expression exactly cancels out the contribution from the determinant of the metric on $2(\gamma)$, which implies that the contribution to order Λ^1 vanishes.

We now consider the Λ^2 contribution. The determinant of the metric on the linking number produces a contribution of five first order multitangents contracted with an ϵ_{abc} and a propagator of Chern–Simons theory. If one considers the action of the Hamiltonian on the linking number squared the loop derivative acts on the linking number and produces ϵ_{abc} contracted with three multitangents, as in the contribution of order Λ^1, times a linking number. The two contributions cancel each other and the Λ^2 contribution vanishes if and only if

$$
\hat{H}_0\mathcal{A}_2(\gamma) = 0.
\tag{10.82}
$$

This calculation can be checked explicitly in exactly the same way as the others. The whole calculation is just more tedious since the different reroutings affect $\mathcal{A}_2(\gamma)$ in a less trivial fashion and the loop derivative acts in various points. There also appear loop derivatives of higher order multitangents, which we presented in chapter 2. Many terms are generated by the action of the Hamiltonian, involving multitangents of order three, four and five. In the end they all cancel [209]. We will present an explicit proof of this in the next chapter since in terms of the extended loop coordinates the resulting expressions are more concise.

The remarkable fact is that in order for the expression of order Λ^2 to vanish we see that $\mathcal{A}_2(\gamma)$, which was the second coefficient of the infinite expansion of the Jones polynomial, has to be annihilated by the Hamiltonian constraint with vanishing cosmological constant. It can easily be seen that it is not annihilated by the determinant of the metric and therefore is the second solution we find to all the constraints of quantum gravity that is non-degenerate in the sense that we discussed in chapter 8. It is the first non-trivial one, in the sense that the previous one we found was just a constant. It is quite remarkable that this highly non-trivial expression is annihilated by the Wheeler–DeWitt equation in loop space.

If one continues this analysis to higher order one checks that at third order the contribution also vanishes, but the "miracle" that happens at the second order is not repeated: the different contributions cancel among themselves but one cannot identify any portion that is annihilated alone by the vacuum Hamiltonian constraint. The reason why something "special" happens at order two will be discussed in the next section. It is possible that it repeats at higher orders, but this has not yet been checked. An important point to be stressed is that any candidate to solution of the Hamiltonian constraint should also be compatible with the Mandelstam identities. This happens to second order, it does not happen at third and is yet to be investigated at higher orders.

We will see in the section 10.5.3 why the second order coefficient seems to play a special role and we will see that it is related to the role that the Gauss linking number plays in the theory.

It is interesting to notice that the above calculations have been performed for a loop with a triple self-intersection but they actually work for any loop. In particular for loops with double self-intersections, one can check the calculations very rapidly: any expression involving ϵ_{abc} contracted with three tangents automatically vanishes, and therefore all the terms that canceled among themselves in the above proof vanish independently.

We have therefore checked perturbatively that the Kauffman bracket is a solution of the constraints of quantum gravity with cosmological constant, as the conjunction of the loop transform and the Witten argument had suggested. The verification has been order by order for only the first four orders, but we see that even at that level several non-trivial cancellations had to occur. Remarkably, we found as a by-product a completely new solution to the vacuum constraints that we did not know *a priori* and which at present we cannot connect with any known expression in terms of connections. We can therefore see the power of working in the loop representations from the point of view of generating solutions of the constraints.

The new solution generated is given by the second coefficient of an infinite expansion of the Jones polynomial. Since the first coefficient ($2(\gamma)$) is also a solution, this led to the conjecture [52] that maybe the whole polynomial was a solution of the constraints with $\Lambda = 0$. It seems at present that this is not the case. Detailed calculations [210] for the third order show that the third coefficient of the expansion is not a solution and a generic argument shows that if Kauffman being a solution with Λ had to imply that Jones was a solution with $\Lambda = 0$, Jones should satisfy several relations it is known not to satisfy. It seems therefore that the construction singles out the second coefficient as a very special quantity. We will show in section 10.5.3 an argument as to why the second coefficient vplays such a singular role.

10.5.3   The Gauss linking number as the key to the new solution

As we have seen, there is evidence that the Kauffman bracket is a solution of the Hamiltonian constraint of quantum gravity with cosmological constant. The Kauffman bracket is given by the loop transform of the exponential of the Chern–Simons form,

$$K(\gamma)_\Lambda = \int DA \exp\left(-\frac{6}{\Lambda}S_{CS}\right) W_\gamma[A]. \qquad (10.83)$$

As we argued, due to the results of Witten and others we know how to compute this quantity explicitly for any gauge group. It is interesting to notice that if the group is $U(1)$ [196, 45],

$$\exp\left(-\frac{\Lambda}{24}a_1(\gamma)\right) = \int DA \exp\left(-\frac{6}{\Lambda}S_{CS}\right) W_\gamma[A], \qquad (10.84)$$

and $S_{CS} = \int d^3x \,\tilde{\epsilon}^{abc} A_a \partial_b A_c$ and the convention for the Abelian Wilson loop is $W_\gamma[A] = \exp(i \oint_\gamma dy^a A_a)$.

So we see that the prefactor that relates the Kauffman and Jones polynomials arises like the "Abelian limit" of the Kauffman bracket. (There is a difference in the numerical factor 24 due to the fact that conventions are slightly different and the Abelian limit of an $SU(2)$ theory yields three $U(1)$ contributions). In particular, it is easy to see that in the perturbative expansion if the group is Abelian all the vertex terms drop out and one gets a resummation of the exponential of the linking number.

Now, the Kauffman bracket solves the Wheeler–DeWitt equation with a cosmological constant. Is there any sense in which one could take the

Abelian limit of this fact and argue that the exponential of the linking number does too? The quick answer to this question is no. There is no systematic way of considering "Abelian limits" in terms of the loop representation, since the non-Abelian nature of the group is embodied from the beginning in the kinematic structure of the theory. Moreover, the expressions for the Hamiltonian constraint and the determinant of the metric collapse in the Abelian limit in terms of connections. However, this idea of exploring the Abelian limit of the Kauffman bracket will lead us to a new solution of the constraints of quantum gravity.

Consider the action of the Hamiltonian constraint on the exponential of the self-linking number. The calculation can be immediately done based on the experience of section 10.5.2. Due to the Abelian nature of the self-linking number, the reroutings have a trivial action and the loop derivative has the effect we discussed when acting on the self-linking number. It is not difficult to see that the total action of the vacuum Hamiltonian constraint on the exponential of the self-linking number is equal to the action of the determinant of the metric [206]. We therefore have the remarkable fact

$$(\hat{H}_0 + \Lambda \hat{\det q}) \exp\left(-\frac{\Lambda}{4} a_1(\gamma)\right) = 0. \tag{10.85}$$

We have therefore found another non-trivial solution of all the constraints of quantum gravity in the loop representation. This solution is completely novel: we do not know its counterpart in the connection representation. It can be loosely understood in terms of the Abelian limit ideas that we introduced, which have no apparent counterpart in the connection representation. It is unfortunate that these ideas cannot be given a more concrete implementation, since they could possibly serve as a basis to construct other solutions to the constraints by considering "expansions in terms of Abelianness".

The remarkable fact is that this solution can be viewed as the root of the results we introduced in section 10.5.2. Since the exponential of the Gauss linking number is a solution with cosmological constant and so is the Kauffman bracket, we could consider their difference, divided by Λ^2,

$$D(\gamma)_\Lambda = \frac{K(\gamma)_\Lambda - \exp(\Lambda a_1(\gamma))}{\Lambda^2}, \tag{10.86}$$

and this quantity solves the Hamiltonian constraint with cosmological constant.

Each polynomial solution with a cosmological constant corresponds, in the limit $\Lambda \to 0$, to a solution of the constraint $\hat{H}_0$. For instance, the Kauffman bracket produces in that limit $2(\gamma)$, which we showed was a solution of $\hat{H}_0$. In the case of D we have

$$\mathcal{A}_2(\gamma) = \lim_{\Lambda \to 0} D(\gamma). \tag{10.87}$$

So we see that the fact that the exponential of the self-linking number is a solution of the Hamiltonian constraint with a cosmological constant has the direct consequence that $\mathcal{A}_2(\gamma)$ has to be a solution of $\hat{H}_0$.

Unfortunately, there is no simple way of constructing a similar argument for the higher coefficients. The root of this difficulty is that the motivation for finding this solution, based on notions of Abelian limit, was quite vague and cannot be embodied in an approximation scheme. Our lack of understanding of the Abelian limit in the loop representation also prevents us from making a clear connection with expansions of the theory in terms of Newton's constant ("weak" [207] and "strong" [208] limits) and should be studied more carefully.

10.6 Conclusions

We have seen that the developments in knot theory, in particular the ideas of knot polynomials, can be successfully extended to the case of intersecting loops and be used in practice to construct quantum states of gravity. We have succeeded in constructing two different states with cosmological constant and two states of the vacuum Hamiltonian constraint. They all solve the constraints in very non-trivial fashion and several of them have no simple counterpart in terms of the connection representation that we know of at present. In a sense this chapter has unleashed the full power of the loop representation in that it allows us to make effective use of the notions of knot theory to solve the constraints. All the solutions that we have discussed here were presented in a formal fashion and only exhibited explicitly for the case of a triply self-intersecting loop. One could try to regularize them using point-splitting or loop-thickening techniques such as the ones we introduced in chapter 8 for the non-intersecting solutions and also generalize the results to loops with more intersections. It is intriguing that all solutions with cosmological constant are regular isotopic invariants whereas the solutions with $\Lambda = 0$ are ambient isotopic. We will postpone the discussion of all these issues to the next chapter where we will discuss these solutions in terms of the extended loop representation in which all regularization issues can be analyzed in a clear fashion. We will see that the solutions survive the scrutiny of a careful regularization.

11

The extended loop representation of quantum gravity

11.1 Introduction

In chapter 2 we saw that the extended loops arise as natural extensions of the group of loops into a Lie structure. We also saw in chapter 4 that the use of extended loops provided a natural framework for the regularization for Maxwell theory. The intention in this chapter is to explore to what extent they can be useful for addressing regularization issues in quantum gravity. As an important by-product we will find that they are also an efficient computational tool for discussing several issues related to the solution space of quantum gravity and the action of the constraints.

Regularization issues in quantum gravity are considerably more involved than those of Maxwell theory. It is therefore remarkable that there is a formal similarity with the case of Maxwell theory. In that case one of the regularization difficulties that we confronted in the loop representation was that the vacuum of the theory,

$$\Psi_0(\gamma) = \exp(-\tfrac{1}{2} \oint_\gamma dx^a \oint_\gamma dy^b K_{ab}(x-y)), \qquad (11.1)$$

where $K_{ab}(x-y)$ was the (distributional) Feynman propagator, was an ill defined quantity. Apart from this difficulty in the definition of the wavefunctions one also had the expected regularization problems of the Hamiltonian, which was quadratic in momenta.

The ill definition of the vacuum in Maxwell theory appears remarkably similar to the problem of framing that we confronted in the loop representation of quantum gravity in the previous chapter. As we saw there, the exponential of the self-linking number,

$$\Psi_0(\gamma) = \exp\left(-\frac{\Lambda}{4} \oint_\gamma dx^a \oint_\gamma dy^b g_{ax\,by}\right) \qquad (11.2)$$

where $g_{ax\,by}$ is the (distributional) propagator of Chern–Simons theory, was a solution to all the constraints and embodied all the framing ambiguities that are present in the Kauffman bracket. The similarity of the two expressions, the one corresponding to Maxwell theory and the one corresponding to gravity is quite striking.

A word of caution should be said about jumping to the conclusion that the similarity of these two problems necessarily implies their solutions should be the same. It is true that going to extended loops fixes the regularization problems of Maxwell theory and allows us to recover the Fock structure of the theory. However, one expects that in quantum gravity, due to the diffeomorphism invariance, the structure of the theory will be quite different from a Fock structure. Intuitively, one expects diffeomorphism invariance will yield some sort of discrete structure, possibly better suited for a description in terms of loops, which are essentially discrete, than extended loops, which are inherently continuous. At the moment, however, the picture is far from clear and the attitude should be to explore all possible avenues to regularize the theory in order to be able to decide which is the better strategy. Because of its natural formulation in terms of objects to which we can apply the usual rules of functional calculus, the extended loop representation presents an attractive formulation in which we can set many of the unsolved questions about regularization raised in the previous chapter.

Another issue related to the use of extended loops is that part of the geometric flavor that representations in terms of loops have is lost. For instance, we saw in chapter 8 how the diffeomorphism invariance of general relativity was naturally coded in the ideas of knot theory. In the extended representation this connection is lost and the diffeomorphism constraint has to be treated as a functional equation. Not everything is lost, since as we will see, several of the ideas of knot theory can be generalized to the extended representation. These issues, connected with the problem that extended holonomies may have convergence problems, have led to a general feeling that some intermediate avenue between ordinary loops and extended loops could be the genuine framework for quantizing gravity. At present, however, such a framework has not been developed.

The proposal to use extended loops to build a representation for quantum gravity was first advanced in references [224, 225].

The structure of this chapter is as follows. We start with a discussion of wavefunctions and their identities in terms of extended loops. We then write the constraints in terms of the extended representation via the loop transform. We then proceed to find the extended version of the solutions to the constraints that we discussed in chapter 10. The usual loop representation is then obtained as a limit of the extended representation. We end with a discussion of the regularization of constraints and solutions in terms of this representation.

11.2 Wavefunctions

We start by discussing general properties that wavefunctions in the extended representations must satisfy. Wavefunctions are related to those in the connection representation by the extended transform,

$$\Psi(\mathbf{X}) = \int DA\,\Psi[A]\,W_{\mathbf{X}}[A],\tag{11.3}$$

with $W_{\mathbf{X}}[A]$ the extended Wilson loop,

$$W_{\mathbf{X}}[A] = \mathrm{Tr}(H_A[\mathbf{X}]) = \mathrm{Tr}[A_{\underset{\sim}{\mu}}]\,X^{\underset{\sim}{\mu}},\tag{11.4}$$

where the notation is as usual, indices with tildes represent sets of pairs of vector indices and space points $\underset{\sim}{\mu} = (a_1\,x_1 \ldots a_n\,x_n)$ and repeated indices with tildes imply integrations over the x_is, Einstein convention summations on the a_i's and a summation on n from zero to infinity. The notation $A_{\underset{\sim}{\mu}}$ denotes the product $A_{a_1}(x_1)\cdots A_{a_n}(x_n)$.

In order to have a gauge invariant Wilson loop, the multitensors $\mathbf{X}$ must satisfy the differential constraint,

$$\frac{\partial}{\partial x_i^{a_i}}\,X^{a_1 x_1 \ldots a_i x_i \ldots a_n x_n} =$$
$$\left(\delta(x_i - x_{i-1}) - \delta(x_i - x_{i+1})\right) X^{a_1 x_1 \ldots a_{i-1} x_{i-1}\,a_{i+1} x_{i+1} \ldots a_n x_n},\tag{11.5}$$

and we call the space of such multitensors $\mathcal{D}_o$. Notice that we do not require the algebraic constraints that we introduced in chapter 2. At this point one has a choice of which precise kind of extended representation one wants to consider. The choice to ignore the algebraic constraint has the payoff that the resulting representation is simpler, because one avoids dealing with non-linear constraints. The price is that the degree of redundancy in the description is higher.

As in the case of loops, the structure of the particular gauge group imprints on the wavefunctions in the extended representation a series of relations, the Mandelstam identities. When we introduced the Mandelstam identities in chapter 3 for usual loops we did it by considering the properties of the traces of products of group elements, which in that case were the holonomies. In the extended case, this is not possible, since the holonomies no longer belong to the gauge group, as we discussed in chapter 2. It turns out that the Mandelstam identities in the extended case arise as a consequence of the properties of the traces of products of the connections $\mathrm{Tr}(A_{a_1}(x_1)\cdots A_{a_n}(x_n))$ combined with the linearity of the extended holonomies in terms of the multitensors. Their explicit form is

$$W_{\mathbf{X}_1 \times \mathbf{X}_2}[A] = W_{\mathbf{X}_2 \times \mathbf{X}_1}[A], \tag{11.6}$$

$$W_{\mathbf{X}}[A] = W_{\bar{\mathbf{X}}}[A], \tag{11.7}$$

$$W_{\mathbf{X}_1}[A]W_{\mathbf{X}_2}[A] = W_{\mathbf{X}_1 \times \mathbf{X}_2}[A] + W_{\mathbf{X}_1 \times \bar{\mathbf{X}}_2}[A]. \tag{11.8}$$

The first identity corresponds to the usual cyclic property of traces. The second one corresponds to the inversion of loops $W_\gamma[A] = W_{\gamma^{-1}}[A]$ which in terms of extended loops corresponds to inversion of the indices,

$$\overline{X}^{\mu_1 \cdots \mu_n} \equiv (-1)^n X^{\mu_n \cdots \mu_1}. \tag{11.9}$$

This equation corresponds (when particularized to loops and making use in that case of the algebraic constraint) to the expression for the inverse of a multitensor that we introduced in chapter 2. Notice that in general it is *not* the inverse multitensor.

In terms of wavefunctions the identities translate into

$$\Psi(\mathbf{X}_1 \times \mathbf{X}_2) = \Psi(\mathbf{X}_2 \times \mathbf{X}_1), \tag{11.10}$$

$$\Psi(\mathbf{X}) = \Psi(\overline{\mathbf{X}}), \tag{11.11}$$

$$\Psi(\mathbf{X}_1 \times \mathbf{X}_2 \times \mathbf{X}_3) + \Psi(\mathbf{X}_1 \times \mathbf{X}_2 \times \overline{\mathbf{X}}_3) =$$
$$\Psi(\mathbf{X}_2 \times \mathbf{X}_1 \times \mathbf{X}_3) + \Psi(\mathbf{X}_2 \times \mathbf{X}_1 \times \overline{\mathbf{X}}_3). \tag{11.12}$$

The identity corresponding to loop inversions (11.11) implies in the extended representations that wavefunctions must depend on the extended coordinates through the combination

$$R^{\mu_1 \cdots \mu_n} = \frac{1}{2}[X^{\mu_1 \cdots \mu_n} + (-1)^n X^{\mu_n \cdots \mu_1}], \tag{11.13}$$

where the **R**s satisfy the following symmetry property under the inversion of the indices

$$R^{\mu_1 \cdots \mu_n} = (-1)^n R^{\mu_n \cdots \mu_1}. \tag{11.14}$$

An important property of the wavefunctions in the extended representation is that they are *linear* functions of the extended coordinates. This is due to the fact that the extended Wilson loop is also a linear function of the extended coordinates. The general form of a wavefunction in the extended representation is therefore given by

$$\Psi(\mathbf{X}) = D_\mu X^\mu, \tag{11.15}$$

and all the information of the particular wavefunction is coded in the coefficients D. In turn, the properties that the wavefunctions have as a consequence of the Mandelstam identities are translated into properties of the coefficients D,

$$D_{\mu_1 \cdots \mu_n} = D_{(\mu_1 \cdots \mu_n)_c}, \tag{11.16}$$

$$D_{\mu_1\ldots\mu_n} = (-1)^n D_{\mu_n\ldots\mu_1}, \tag{11.17}$$

$$D_{\mu_1\ldots\mu_k\mu_{k+1}\ldots\mu_n} + (-1)^k D_{\mu_k\ldots\mu_1\mu_{k+1}\ldots\mu_n} =$$
$$\tfrac{1}{k}D_{(\mu_1\ldots\mu_k)_c\,\mu_{k+1}\ldots\mu_n} + (-1)^k\tfrac{1}{k}D_{(\mu_k\ldots\mu_1)_c\,\mu_{k+1}\ldots\mu_n}\ \ \forall k, \tag{11.18}$$

where c indicates the cyclic combination of indices,

$$D_{(\mu_1\ldots\mu_k)_c} = D_{(\mu_1\,\mu_2\ldots\mu_k)} + D_{(\mu_2\,\mu_3\ldots\mu_k\,\mu_1)} + \cdots D_{(\mu_k\,\mu_1\ldots\mu_{k-1})}. \tag{11.19}$$

The linearity is a remarkable property of the wavefunctions in the extended representation. Notice that all the wavefunctions explicitly known in the loop representation for quantum gravity have this property when they are written in terms of the multitangent fields. Moreover, this property will also be inherited by the operators that we can construct in the extended representation. In general, the linearity of the wavefunctions could be imposed by means of the "linearity constraint" $\mathcal{L}$

$$\mathcal{L}(\mathbf{X}')\,\Psi(\mathbf{X}) \equiv X'^{\,\mu\nu}_{\,\,\mathclap{\underset{\sim}{}}} \frac{\delta^2}{\delta X^{\mu}_{\,\mathclap{\underset{\sim}{}}}\delta X^{\nu}_{\,\mathclap{\underset{\sim}{}}}}\,\Psi(\mathbf{X}) = 0, \tag{11.20}$$

where $\mathbf{X}'$ is any object that satisfies the differential constraints. The functional derivatives produce elements of the extended group of loops and therefore the second functional derivative is the group product of the resulting elements. The addition of the element $\mathbf{X}'$ is to ensure that the result is a function of multitensors that satisfy the differential constraint (i.e., it makes the linearity constraint a well defined operator on the space of wavefunctions with support on $\mathcal{D}_o$).

Any observable of the theory has to commute with the linearity constraint. This means that the action of any quantum observable on a wavefunction reduces to a shift in the argument of the wavefunction. The linearity in the wavefunctions is in correspondence with the proliferation of arguments. One trades the non-linearity of the wavefunctions in terms of a connection for an increased number of arguments in the extended representation. This is a technique that is applied in constructive quantum field theories for non-linear theories, where non-linearities are traded for an increase in the number of variables.

An example that clarifies these issues of linearity and proliferation of variables is given by the usual Fourier representation of the quantum mechanics of a free particle in one dimension. The usual theory has wavefunctions in the position representation $\Psi(x)$ and momentum representation $\Psi(k)$ related by the usual Fourier transform. The idea of extended representation is to substitute the basis of the Fourier transform by an infinite parameter basis,

$$\exp(ikx) \to k_0 + k_1 x + k_2 x^2 + k_3 x^3 + \cdots \tag{11.21}$$

and the resulting wavefunctions in the "extended" representation are given by linear functions of an infinite tower of ks, $\Psi(\vec{k})$. The linearity is imposed by a linearity constraint $\partial^2/\partial k_i \partial k_j \Psi(\vec{k}) = 0$

One can write the physical operators of the theory in terms of such a representation, and they all become linear operators,

$$\hat{x} = \sum_{n=0}^{\infty} k_{n-1} \frac{\partial}{\partial k_n}, \tag{11.22}$$

$$\hat{p} = \sum_{n=0}^{\infty} (n+1) k_{n+1} \frac{\partial}{\partial k_n}, \tag{11.23}$$

$$\hat{\mathcal{H}} = \frac{\hat{p}^2}{2m} = \frac{1}{2m} \sum_{n=0}^{\infty} (n+1)(n+2) k_{n+2} \frac{\partial}{\partial k_n}, \tag{11.24}$$

which commute with the linearity constraint.

How is the usual theory recovered? Since one has first class constraints (the linearity constraints), one can fix the gauge generated by them. In particular one can choose $k_n = k_1^n/n!$ and one recovers the usual theory free of constraints. If one decides to quantize the theory before fixing the gauge, the usual theory is recovered by considering analytic functions of the tower of ks and introducing an inner product that implements a gauge fixing similar to the one discussed.

At the moment this seems like a futile exercise: we have converted the simplest quantum mechanical problem into a field theory with an infinite number of variables and constraints. It is true that for the example of a free particle nothing is gained in solving the theory in this way. In the case of gauge theories, however, one knows that fixing the gauge is not necessarily the easiest way of solving a theory. The attractiveness of having a theory cast in terms of linear functions and first order differential operators may well compensate for the proliferation of variables (a less obvious problem in a theory that from the outset has an infinite number of degrees of freedom).

An intriguing point is that the resulting quantum theory with linear wavefunctions and first order operators could, in principle, be obtained as the canonical quantization of a classical theory with constraints and operators linear in momenta. The classical theory involved has an infinite number of degrees of freedom and the linearity implies the use of Grassmann variables in its formulation. These classical theories have not been studied in detail at present.

11.3   The constraints

We now proceed to write the constraints of quantum gravity in terms of the extended representation. We will proceed formally via the loop transform exactly as we did in chapter 8. We could also proceed via the construction of a non-canonical algebra which is the natural generalization of the T algebra to the extended case. As we argued at length, the results one gets are equivalent to those of a loop transform and involve a similar number of formal manipulations. We will therefore concentrate on the loop transform approach.

11.3.1 The diffeomorphism constraint

We start with the diffeomorphism constraint. The action of this constraint on the wavefunctions $\Psi(\mathbf{R})$ is defined by

$$\hat{C}_{ax}\Psi(\mathbf{R}) = \int DA\, W_{\mathbf{R}}[A]\, [\hat{C}_{ax}\Psi(A)]. \tag{11.25}$$

The constraint acting on $\Psi(A)$ can be applied on the generalized Wilson functional integrating (formally) by parts. As a result we get

$$\hat{C}_{ax}\Psi(\mathbf{R}) = \int DA\, \Psi[A]\, [F_{ab}^i(x)\frac{\delta}{\delta A_{bx}^i} W_{\mathbf{R}}[A]]. \tag{11.26}$$

At this point it is useful to introduce some notation that will prove beneficial in the calculations. Let $\delta_{\underset{\sim}{\beta}}^{\alpha}$ be defined as

$$\delta_{\underset{\sim}{\beta}}^{\alpha} = \begin{cases} \delta_{\beta_1}^{\alpha_1}\cdots\delta_{\beta_n}^{\alpha_n}, & \text{if } n(\underset{\sim}{\alpha}) = n(\underset{\sim}{\beta}) = n \geq 1 \\[2mm] 1, & \text{if } n(\underset{\sim}{\alpha}) = n(\underset{\sim}{\beta}) = 0 \\[2mm] 0, & \text{otherwise,} \end{cases} \tag{11.27}$$

where $n(\underset{\sim}{\alpha})$ is the number of indices of the set $\underset{\sim}{\alpha}$. The δ matrix allows us to write the group product defined in chapter 2 as

$$(\mathbf{E}_1 \times \mathbf{E}_2)^{\underset{\sim}{\ell}} = \delta_{\underset{\sim}{\nu\beta}}^{\underset{\sim}{\ell}}\, E_1^{\underset{\sim}{\nu}} E_2^{\underset{\sim}{\beta}}. \tag{11.28}$$

Notice that in particular

$$(\delta_{\underset{\sim}{\nu}} \times \delta_{\underset{\sim}{\beta}})^{\underset{\sim}{\ell}} = \delta_{\underset{\sim}{\nu\beta}}^{\underset{\sim}{\ell}}, \tag{11.29}$$

where $\delta_{\underset{\sim}{\alpha}}$ are the "vectors" with components $(\delta_{\underset{\sim}{\alpha}})^{\underset{\sim}{\ell}} = \delta_{\underset{\sim}{\alpha}}^{\underset{\sim}{\mu}}$.

The functional derivative of any product of As can be written with the help of the δ matrix as

$$\frac{\delta}{\delta A_{bx}^i}(A_{\underset{\sim}{\alpha}}) = A_{\underset{\sim}{\mu}}\tau^i A_{\underset{\sim}{\nu}}\, \delta_{\underset{\sim}{\alpha}}^{\underset{\sim}{\mu}\,bx\,\underset{\sim}{\nu}}, \tag{11.30}$$

where the τs are the generators of the SU(2) algebra with the conventions of chapter 8. Taking the trace of the above expression we get

$$\frac{\delta}{\delta A_{bx}^i}\mathrm{Tr}(A_{\underset{\sim}{\alpha}}) = \mathrm{Tr}(\tau^i A_{\underset{\sim}{\beta}})\,\delta_{\underset{\sim}{\nu\mu}}^{\beta}\,\delta_{\underset{\sim}{\alpha}}^{\mu\,bx\,\nu} = \mathrm{Tr}(\tau^i A_{\underset{\sim}{\beta}})\,\delta_{\underset{\sim}{\alpha}}^{(bx\,\beta)c}. \qquad (11.31)$$

The curvature tensor involved in the definition of the constraint can be written as

$$F_{ab}(x) = A_{\underset{\sim}{\nu}}\mathcal{F}_{ab}^{\nu}(x), \qquad (11.32)$$

where $\mathcal{F}_{ab}$ represents the element of the algebra of the extended loop group with non-vanishing components,

$$\mathcal{F}_{ab}{}^{a_1 x_1}(x) = \delta_{ab}^{a_1\,d}\,\partial_d\,\delta(x_1 - x), \qquad (11.33)$$

$$\mathcal{F}_{ab}{}^{a_1 x_1,\,a_2 x_2}(x) = \delta_{ab}^{a_1\,a_2}\,\delta(x_1 - x)\,\delta(x_2 - x). \qquad (11.34)$$

Using (11.31) and (11.32) we obtain the following expression for the action of the diffeomorphism constraint on the generalized Wilson functional:

$$F_{ab}^i(x)\frac{\delta}{\delta A_{bx}^i}\mathrm{Tr}(A_{\underset{\sim}{\alpha}})\,R^{\underset{\sim}{\alpha}} = \mathrm{Tr}(F_{ab}(x)\,A_{\underset{\sim}{\beta}})\,\delta_{\underset{\sim}{\alpha}}^{(ax\,\beta)c}\,R^{\underset{\sim}{\alpha}}$$

$$= \mathrm{Tr}(A_{\underset{\sim}{\rho}})\,\delta_{\underset{\sim}{\nu\beta}}^{\rho}\,\mathcal{F}_{ab}(x)^{\underset{\sim}{\nu}}R^{(bx\,\underset{\sim}{\beta})c}. \qquad (11.35)$$

Putting expression (11.35) in the expression of the differential constraint and using (11.28) we obtain

$$\mathcal{C}_{ax}\Psi(\mathbf{R}) = \int DA\,\Psi[A]\,\mathrm{Tr}(A_{\underset{\sim}{\rho}})\left[\mathcal{F}_{ab}(x) \times \mathbf{R}^{(bx)}\right]^{\underset{\sim}{\rho}}$$

$$= \Psi(\mathcal{F}_{ab}(x) \times \mathbf{R}^{(bx)}), \qquad (11.36)$$

where we have introduced the element of the group $\mathbf{R}^{(bx)}$ which has components defined by

$$[\mathbf{R}^{(bx)}]^{\underset{\sim}{\mu}} = R^{(bx)\underset{\sim}{\mu}} \equiv R^{(bx\,\underset{\sim}{\mu})c} \qquad (11.37)$$

and satisfies the differential constraint (on the $\underset{\sim}{\mu}$ indices) basepointed at x.

We therefore see that the action of the diffeomorphism constraint reduces to a shift in the argument of the wavefunction, as we suggested, due to the linearity of the operator. The operator can, of course, be written as a first order differential operator,

$$\mathcal{C}_{ax}\Psi(\mathbf{R}) = [\mathcal{F}_{ab}(x) \times \mathbf{R}^{(bx)}]^{\underset{\sim}{\mu}}\frac{\delta}{\delta\mathbf{R}^{\underset{\sim}{\mu}}}\Psi(\mathbf{R}). \qquad (11.38)$$

11.3.2 The Hamiltonian constraint

Let us now consider the construction of the Hamiltonian constraint in the extended representation.

In this case we have to use the properties of the $SU(2)$ algebra in order to take into account the two derivatives that appear in $\hat{\mathcal{H}}(x)$. We have now

$$\hat{\mathcal{H}}(x)\,\Psi(\mathbf{R}) = \int DA\,\Psi[A]\,\epsilon^{ijk}\left[F^i_{ba}(x)\frac{\delta}{\delta A^j_{bx}}\frac{\delta}{\delta A^k_{ax}}W_A(\mathbf{R})\right]. \quad (11.39)$$

From (11.31) we get the following expression for the second functional derivative

$$\frac{\delta}{\delta A^j_{bx}}\frac{\delta}{\delta A^k_{ax}}\mathrm{Tr}(A_{\underset{\sim}{\alpha}}) = \mathrm{Tr}(\tau^k\frac{\delta}{\delta A^j_{bx}}A_{\underset{\sim}{\beta}})\,\delta^{(ax\,\beta)c}_{\underset{\sim}{\alpha}} =$$

$$\mathrm{Tr}(\tau^k A_\mu\tau^j A_{\underset{\sim}{\nu}})\,\delta^{\mu\,bx\,\nu}_{\underset{\sim}{\beta}}\,\delta^{(ax\,\beta)c}_{\underset{\sim}{\alpha}} = \mathrm{Tr}(\tau^k A_\mu\tau^j A_{\underset{\sim}{\nu}})\,\delta^{(ax\,\mu\,bx\,\nu)c}_{\underset{\sim}{\alpha}}. \quad (11.40)$$

To put this result in a useful form we need the following well known property of the $SU(2)$ matrices

$$\epsilon^{ijk}\mathrm{Tr}(\tau^k A_\mu\tau^j A_\nu) = \mathrm{Tr}(\tau^i A_{\underset{\sim}{\nu}})\,\mathrm{Tr}(A_{\underset{\sim}{\mu}}) - \mathrm{Tr}(A_{\underset{\sim}{\nu}})\,\mathrm{Tr}(\tau^i A_{\underset{\sim}{\mu}}), \quad (11.41)$$

which allows us to write the product between traces of $SU(2)$ matrices as a combination of traces in the following way:

$$\mathrm{Tr}(A_{\underset{\sim}{\mu}})\,\mathrm{Tr}(A_{\underset{\sim}{\nu}}) = \mathrm{Tr}(A_{\underset{\sim}{\mu}}A_{\underset{\sim}{\nu}}) + (-1)^{n(\underset{\sim}{\nu})}\mathrm{Tr}(A_{\underset{\sim}{\mu}}A_{\underset{\sim}{\nu^{-1}}}), \quad (11.42)$$

where if $\underset{\sim}{\nu} = (\nu_1,\ldots,\nu_n)$, then $\underset{\sim}{\nu^{-1}} = (\nu_n,\ldots,\nu_1)$. This allows us to rearrange the expression of interest as

$$\epsilon^{ijk}\mathrm{Tr}(\tau^k A_\mu\tau^j A_{\underset{\sim}{\nu}}) = (-1)^{n(\underset{\sim}{\mu})}\mathrm{Tr}(\tau^i A_{\underset{\sim}{\nu}} A_{\underset{\sim}{\mu^{-1}}}) - (-1)^{n(\underset{\sim}{\nu})}\mathrm{Tr}(\tau^i A_{\underset{\sim}{\nu^{-1}}} A_{\underset{\sim}{\mu}}). \quad (11.43)$$

We then have for the action of the constraint on the product of connections,

$$\epsilon^{ijk}F^i_{ba}(x)\frac{\delta}{\delta A^j_{bx}}\frac{\delta}{\delta A^k_{ax}}\mathrm{Tr}(A_{\underset{\sim}{\alpha}}) =$$

$$(-1)^{n(\underset{\sim}{\mu})}\,\mathrm{Tr}(F_{ba}(x)A_{\underset{\sim}{\nu\mu}})\left\{\delta^{(ax\,\mu^{-1}\,bx\,\nu)c}_{\underset{\sim}{\alpha}} - (-1)^{n(\underset{\sim}{\mu}+\underset{\sim}{\nu})}\,\delta^{(ax\,\mu\,bx\,\nu^{-1})c}_{\underset{\sim}{\alpha}}\right\} =$$

$$(-1)^{n(\underset{\sim}{\mu})}\,\mathrm{Tr}(F_{ba}(x)A_{\underset{\sim}{\nu\mu}})\left\{\delta^{(bx\,\nu\,ax\,\mu^{-1})c}_{\underset{\sim}{\alpha}} + (-1)^{n(\underset{\sim}{\mu}+\underset{\sim}{\nu})}\,\delta^{(\mu\,ax\,\nu^{-1}\,bx)c}_{\underset{\sim}{\alpha}}\right\} =$$

$$(-1)^{n(\underset{\sim}{\mu})}\,\mathrm{Tr}(A_{\underset{\sim}{\beta\nu\mu}})\,\mathcal{F}_{ab}^{\underset{\sim}{\gamma}}(x)\,\delta^{(ax\,\nu\,bx\,\mu^{-1})c}_{\underset{\sim}{\gamma}}\left\{\delta^{\gamma}_{\underset{\sim}{\alpha}} + (-1)^{n(\underset{\sim}{\gamma})}\,\delta^{\gamma^{-1}}_{\underset{\sim}{\alpha}}\right\}, \quad (11.44)$$

where the combination that arises in curly braces gives rise exactly to the element $\mathbf{R}$ that we introduced before when contracted with $\mathbf{X}$. This contraction is exactly what we need to do to get the expression of the action of the constraint on an extended holonomy,

$$\epsilon^{ijk} F_{ba}^i(x) \frac{\delta}{\delta A_{bx}^j} \frac{\delta}{\delta A_{ax}^k} W_A(\mathbf{R}) =$$

$$2(-1)^{n(\mu)} \operatorname{Tr}(A_{\beta\nu\mu}) \mathcal{F}_{ab}^{\beta}(x) \delta_{\chi}^{(ax\,\nu\,bx\,\mu^{-1})c} R^{\chi} =$$

$$2(-1)^{n(\mu)} \operatorname{Tr}(A_{\alpha}) \delta_{\beta\varrho}^{\alpha} \mathcal{F}_{ab}^{\beta}(x) \left[\delta_{\nu\mu}^{\varrho} R^{(ax\,\nu\,bx\,\mu^{-1})c} \right], \quad (11.45)$$

where in the first step we have used the symmetry property (11.14) of the Rs under the inversion of the indexes. The expression in square brackets defines a specific combination of Rs, that we denote

$$[\mathbf{R}^{(ax,\,bx)}]^{\varrho} = R^{(ax,\,bx)\varrho} \equiv (\delta_{\nu} \times \delta_{\mu})^{\varrho} (-1)^{n(\mu)} R^{(ax\,\nu\,bx\,\mu^{-1})c}. \quad (11.46)$$

Explicitly,

$$R^{(ax,\,bx)\rho_1\ldots\rho_n} = \sum_{k=0}^{n} (-1)^{n-k} R^{(ax\,\rho_1\ldots\rho_k\,bx\,\rho_n\ldots\rho_{k+1})c}. \quad (11.47)$$

An important fact is that this combination satisfies the differential constraint with respect to the ϱ indices basepointed at x. It also satisfies the following property

$$R^{(ax,\,bx)\varrho^{-1}} = (-1)^{n(\varrho)} R^{(bx,\,ax)\varrho}. \quad (11.48)$$

Equation (11.45) can then be written

$$\epsilon^{ijk} F_{ba}^i(x) \frac{\delta}{\delta A_{bx}^j} \frac{\delta}{\delta A_{ax}^k} W_A(\mathbf{R}) = 2 \operatorname{Tr}(A_{\alpha}) (\delta_{\beta} \times \delta_{\varrho})^{\alpha} \mathcal{F}_{ab}^{\beta}(x) R^{(ax,\,bx)\varrho}$$

$$= 2 \operatorname{Tr}(A_{\alpha}) (\mathcal{F}_{ab} \times \mathbf{R}^{(ax,\,bx)})^{\alpha}, \quad (11.49)$$

and from this we conclude that

$$\hat{\mathcal{H}}(x) \Psi(\mathbf{R}) = 2 \Psi(\mathcal{F}_{ab}(x) \times \mathbf{R}^{(ax,\,bx)}). \quad (11.50)$$

Also in this case the action of the Hamiltonian constraint reduces to evaluating the wavefunction on a new argument. As was already mentioned, this is a general property of the operators in the extended representation due to the linearity of the wavefunctions. In fact, the last expression can be written as the action of a single functional derivative with respect to the $\mathbf{R}$ variables

$$\hat{\mathcal{H}}(x) \Psi(\mathbf{R}) = 2 \left[\mathcal{F}_{ab}(x) \times \mathbf{R}^{(ax,\,bx)} \right]^{\mu} \frac{\delta}{\delta R^{\mu}} \Psi(\mathbf{R}). \quad (11.51)$$

Notice that in order for this expression to be well defined on the space of wavefunctions considered it is necessary that the term contracted with the functional derivative satisfies the differential constraint, as is the case here.

The new element of the extended group of loops on which the wave-function is evaluated involves a combination of multitensor fields with two indices fixed at the point where the Hamiltonian is acting and the other indices having a specific alternating order. We will show in the next section that this alternating order of the indexes is related to the reroutings of a loop when the above expression is particularized to loops. The appearance of a rerouting is typical of the loop representation and plays a crucial role in the quantum gravity case.

The presence of a multitensor with two indices evaluated at the same point in the Hamiltonian constraint implies that the resulting expression for the operator is divergent. This is due to the distributional character of the multitensors. A multitensor satisfying the differential constraint (2.11) diverges when two successive indices are evaluated at the same spatial point. This divergence of the formal expression of the constraint will have to be regularized, and we will return to it in detail in section 11.6.2.

11.4   Loops as a particular case

As we discussed in chapter 2, the extended group of loops includes the group of loops as a particular case. We should therefore be able to partic-ularize the extended representation to the loop representation by substi-tuting $\mathbf{R} \rightarrow \mathbf{R}(\gamma)$. We analyze here in detail the case of the Hamiltonian constraint.

In order to evaluate $\mathbf{R}^{(ax,\,bx)}(\gamma_o)$ we have to use the explicit expression of this object in terms of the multitangents fields. We have

$$
\begin{aligned}
R^{(ax,bx)\mu_1\ldots\mu_n} = {} & \tfrac{1}{2}\sum_{k=0}^{n}\sum_{l=0}^{k}(-1)^{n-k}[X^{\mu_{l+1}\ldots\mu_k\,bx\,\mu_n\ldots\mu_{k+1}\,ax\,\mu_1\ldots\mu_l} \\
& + (-1)^n X^{\mu_l\ldots\mu_1\,ax\,\mu_{k+1}\ldots\mu_n\,bx\,\mu_k\ldots\mu_{l+1}}] \\
& + \sum_{k=0}^{n}\sum_{l=k}^{n}(-1)^{n-k}[X^{\mu_l\ldots\mu_{k+1}\,ax\,\mu_1\ldots\mu_k\,bx\,\mu_n\ldots\mu_{l+1}} \\
& + (-1)^n X^{\mu_{l+1}\ldots\mu_n\,bx\,\mu_k\ldots\mu_1\,ax\,\mu_{k+1}\ldots\mu_l}].
\end{aligned}
\tag{11.52}
$$

One can write the above expression in a more compact and useful form introducing the following combinations of Xs,

$$
X^{(ax,\overrightarrow{bx})\mu} \equiv \sum_{k=0}^{n}(-1)^{n-k}X^{(ax\,\mu_1\ldots\mu_k\,bx\,\mu_n\ldots\mu_{k+1})c},
\tag{11.53}
$$

and

$$X^{(ax,\overleftarrow{bx})\mu} \equiv \sum_{k=0}^{n} (-1)^k X^{(ax\,\mu_k\cdots\mu_1\,bx\,\mu_{k+1}\cdots\mu_n)_c}. \qquad (11.54)$$

These objects have definite symmetry properties under the inversion of the indices, which we will use later. Basically, the inversion of the order of the indices flips the direction of the arrow and multiplies the object by $(-1)^{n(\mu)}$,

$$(-1)^{n(\mu)} X^{(ax,\overleftarrow{bx})\mu^{-1}}(\gamma_o) = X^{(ax,\overrightarrow{bx})\mu}(\gamma_o). \qquad (11.55)$$

In terms of these combinations, $\mathbf{R}^{(ax,bx)}$ simply reads

$$R^{(ax,bx)\mu} = \tfrac{1}{2}[X^{(ax,\overrightarrow{bx})\mu} + (-1)^{n(\mu)} X^{(ax,\overleftarrow{bx})\mu^{-1}}]. \qquad (11.56)$$

As we discussed extensively in chapter 8, the Hamiltonian constraint in the loop representation has only a non-trivial action on intersecting loops. We suppose then that at the point x the loop γ intersects itself p times; i.e., γ has "multiplicity" p at x. We start with some suitable notation to take this fact into account.

If the loop γ has multiplicity p at x one can write it in the following way

$$\gamma_{xx} = \gamma_{xx}^{(1)} \circ \gamma_{xx}^{(2)} \circ \cdots \circ \gamma_{xx}^{(p)}. \qquad (11.57)$$

We denote by $[\gamma_{xx}]_i^{i+j}$ the following composition of loops basepointed at x

$$[\gamma_{xx}]_i^{i+j} = \gamma_{xx}^{(i)} \circ \cdots \circ \gamma_{xx}^{(i+j)}. \qquad (11.58)$$

Let us suppose that the loop named $\gamma_{xx}^{(1)}$ contains the origin o of the loops. Then

$$\gamma_o = \gamma^{(1)x}_{\ o} \circ [\gamma_{xx}]_2^p \circ \gamma^{(1)o}_{\ x}. \qquad (11.59)$$

Here, $\gamma^{(1)x}_{\ o}$ represents the portion of $\gamma^{(1)}$ from the origin o to the point x. The loop γ_o is completely described by the multitangent fields $X^\mu(\gamma_o)$ of all ranks. As we know, these fields satisfy both algebraic (2.10) and differential (2.11) constraints. Moreover, these objects have another property derived from the fact that one can write a loop as a composition of open paths. This reads

$$X^{\mu_1\cdots\mu_n}(\gamma_o) = \int_{\gamma_o} dz^{a_i} \delta(x_i - z) X^{\mu_1\cdots\mu_{i-1}}(\gamma_o^z) X^{\mu_{i+1}\cdots\mu_n}(\gamma_z^o), \qquad (11.60)$$

which can be derived simply from the properties of the ordered integrals that appear in the definition of the multitangent.

Suppose now that the index μ_i is fixed at the point x. Then

$$X^{\mu_1\cdots\mu_i\,ax\,\mu_{i+1}\cdots\mu_n}(\gamma_o) =$$

$$\sum_{m=1}^{p} X^{\mu_1\cdots\mu_i}(\gamma^{(1)x}_{\ o}\circ[\gamma_{xx}]_2^m)\,X_m^{ax}(\gamma)\,X^{\mu_{i+1}\cdots\mu_n}([\gamma_{xx}]_{m+1}^p\circ\gamma^{(1)o}_{\ x}),$$

$$\tag{11.61}$$

where $X_m^{ax}(\gamma)$ is the tangent at x when the loop goes through that point on the mth occasion. The following convention is assumed: $[\gamma_{xx}]_{m+1}^m \approx \iota_{xx}$, with ι_{xx} the null path. The above expression can easily be generalized to the case of any number of indices fixed at x. The above two expressions are exactly the same, except that in the second one we have written explicitly the case in which the point x is at an intersection, partitioning the integral of the first formula in a summation on the different petals of the loop with the intersection at x.

We are now ready to compute $\mathbf{R}^{(ax,\,bx)}(\gamma_o)$. We have

$$X^{(ax,\overrightarrow{bx})\mu}(\gamma_o) =$$

$$\sum_{m=1}^{p-1}\sum_{q=m+1}^{p}[X_m^{bx}(\gamma)\,X_q^{ax}(\gamma)\,X^\mu([\gamma_{xx}]_1^m\circ\overline{[\gamma_{xx}]}_{m+1}^q\circ[\gamma_{xx}]_{q+1}^p)$$

$$+(-1)^{n(\mu)}X_m^{ax}(\gamma)\,X_q^{bx}(\gamma)\,X^{\mu^{-1}}([\gamma_{xx}]_1^m\circ\overline{[\gamma_{xx}]}_{m+1}^q\circ[\gamma_{xx}]_{q+1}^p)],$$

$$\tag{11.62}$$

where $\overline{[\gamma_{xx}]}_{m+1}^q = \overline{\gamma}_{xx}^{(q)}\circ\cdots\circ\overline{\gamma}_{xx}^{(m+1)}$ and $\overline{\gamma}$ denotes the loop γ with opposite orientation. The inversion of the orientation of the loop (rerouting) in (11.62) comes from the property (11.9) of the multitangent fields. We then use the properties of the arrowed objects under inversion of the indices (11.55) and obtain for the action of the Hamiltonian,

$$\hat{\mathcal{H}}(x)\Psi(\gamma_o) = 2\Psi[\mathcal{F}_{ab}(x)\times\mathbf{R}^{(ax,\,bx)}(\gamma_o)] =$$

$$2\int DA\,\Psi(A)\,\mathrm{Tr}(A_{\alpha\mu})\,\mathcal{F}_{ab}(x)^\alpha_{\underset{\sim}{\alpha}}X^{(ax,\overrightarrow{bx})\mu}(\gamma_o) =$$

$$4\sum_{m=1}^{p-1}\sum_{q=m+1}^{p}X_m^{[bx,}(\gamma)X_q^{ax]}(\gamma)\int DA\Psi(A)$$

$$\times\mathrm{Tr}\Big[F_{ab}(x)H_A\big\{\mathbf{R}([\gamma_{xx}]_1^m\circ\overline{[\gamma_{xx}]}_{m+1}^q\circ[\gamma_{xx}]_{q+1}^p)\big\}\Big],\quad(11.63)$$

where we have arranged the product of connections contracted with the multitangents as the holonomy, and its contraction with $\mathcal{F}_{ab}$ as the field tensor F_{ab} using formulae we introduced at the beginning of this chapter.

We can now recover the loop derivative through the usual expression,

$$\text{Tr}(F_{ab}(x)H_A\{\mathbf{R}(\gamma_{xx})\}) = \Delta_{ab}(x)\text{Tr}(H_A\{\mathbf{R}(\gamma_{xx})\}). \tag{11.64}$$

The final result is

$$\hat{\mathcal{H}}(x)\Psi(\gamma_o) = 4 \sum_{m=1}^{p-1} \sum_{q=m+1}^{p} X_m^{[bx,}(\gamma)X_q^{ax]}(\gamma)$$

$$\times \Delta_{ab}(x)\,\Psi\left([\gamma_{xx}]_1^m \circ \overline{[\gamma_{xx}]}_{m+1}^q \circ [\gamma_{xx}]_{q+1}^p\right). \tag{11.65}$$

This expression corresponds to the usual Hamiltonian constraint of quantum gravity in the loop representation introduced in chapter 8. For the diffeomorphism constraint we obtain a similar result. Equation (11.36) reduces to the usual expression of the diffeomorphism constraint in the loop representation when one particularizes this constraint to the case of loops.

It is important to stress the relationship between the solutions of the constraints in both representations. Since loops are a particular case of multitensors, any solution found in the extended representation can be particularized to loops and would yield in the limit a solution to the usual constraints of quantum gravity in the loop representation. The converse is not necessarily true. Given a solution in the loop representation, it may not generalize to a solution in the extended representation. An example is the solutions to the Hamiltonian based on smooth non-intersecting loops, which find no analogue in the extended representation.

The process by which one obtains a solution in the loop representation from a solution in the extended representation may be ill defined. In that sense, one can always obtain a solution in terms of loops from the extended representation only at a formal level. In particular we will see that the solutions we find in the next section are only well defined in the extended space if one excises from it certain multitensors, including those which correspond to loops. Therefore such solutions do not have a rigorous meaning in terms of loops, only a formal one, which corresponds to the level of discussion of the solutions that we have maintained up to now.

The fact that the solutions we will present do not include loops as a particular case does not preclude obtaining them through a suitable limiting process. These limiting processes may include additional structures —such as framings— and the end result may be a formulation in terms of some generalization of the idea of loops.

A simple example of the situation is given by the exponential of the self-linking number. Its extended form is $\exp(g_{ax\,by}X^{ax}X^{by})$. If the Xs are smooth, this is a well defined quantity in spite of the fact that $g_{ax\,by}$ is distributional. However, if one considers the Xs that correspond to a loop it is not, as we have discussed, and an ambiguity appears. Therefore if one wants to have the self-linking number as a well defined function in the extended loop space one has to restrict it to smooth first order multitensors, which exclude those of loops. If one defines a limiting process in which the multitensors of (framed) loops arise as a limit of smooth multitensors, the self-linking number is well defined.

11.5   Solutions of the constraints

As we have seen, the expressions for the constraints in the extended representation are very compact: they amount to the evaluation of the wavefunctions in a shifted argument. The compactness of these expressions allows us to compute in a very efficient way their action on specific states. In particular it allows us to compute very efficiently the action of the Hamiltonian constraint on the second coefficient of the Jones polynomial, which we claimed without presenting an explicit proof in chapter 10 was annihilated by the constraint. The discussion in this section serves both as proof of that fact and as an illustration of the computational economy attained by the use of the extended representation. Even if the extended representation does not in the end have intrinsic value for representing quantum gravity it is a powerful computational framework for doing calculations in the loop representation. The computation presented here will be unregularized, we will discuss the regularization of it in section 11.6.2.

The expression for the coefficient $\mathcal{A}_2(\gamma)$ in terms of the multitangent fields is

$$\mathcal{A}_2(\gamma) = h_{\mu_1\mu_2\mu_3} X^{\mu_1\mu_2\mu_3}(\gamma) + g_{\mu_1\mu_3} g_{\mu_2\mu_4} X^{\mu_1\mu_2\mu_3\mu_4}(\gamma), \qquad (11.66)$$

where

$$h_{\mu_1\mu_2\mu_3} = \epsilon^{\alpha_1\alpha_2\alpha_3} \, g_{\mu_1\alpha_1} \, g_{\mu_2\alpha_2} \, g_{\mu_3\alpha_3}, \qquad (11.67)$$

with

$$\epsilon^{\alpha_1\alpha_2\alpha_3} = \epsilon^{c_1c_2c_3} \int d^3t \, \delta(z_1 - t) \, \delta(z_2 - t) \, \delta(z_3 - t). \qquad (11.68)$$

The generalization of this knot invariant to extended loops is straightforward

$$\mathcal{A}_2(\gamma) = \mathcal{A}_2[\mathbf{X}(\gamma)] \to \mathcal{A}_2(\mathbf{X}) = \mathcal{A}_2(\mathbf{R}), \qquad (11.69)$$

where $\mathbf{X}$ is now an element of the extended group $\mathcal{D}_o$. We now analyze the application of the Hamiltonian constraint to this state in the extended representation. By (11.50) we have

$$\mathcal{H}(x)\,\mathcal{A}_2(\mathbf{R}) = 2\,h_{\mu_1\mu_2\mu_3} \left[\mathcal{F}_{ab}{}^{\mu_1}(x) \, R^{(ax,\,bx)\mu_2\mu_3} + \mathcal{F}_{ab}{}^{\mu_1\mu_2}(x) \, R^{(ax,\,bx)\mu_3} \right]$$

$$+2\,g_{\mu_1\mu_3}g_{\mu_2\mu_4}\left[\mathcal{F}_{ab}{}^{\mu_1}(x)\,R^{(ax,\,bx)\mu_2\mu_3\mu_4} + \mathcal{F}_{ab}{}^{\mu_1\mu_2}(x)\,R^{(ax,\,bx)\mu_3\mu_4}\right].\quad(11.70)$$

The contraction of the element of the extended algebra $\mathcal{F}_{ab}$ with the propagators leads to integrations by parts similar to those we encountered in chapter 10 while analyzing the action of the Hamiltonian constraint on the Gauss linking number. Explicitly, we have

$$\mathcal{F}_{ab}{}^{\mu_1}(x)\,g_{\mu_1\mu_3} = -\epsilon_{aba_3}\delta(x-x_3) - \partial_{a_3}g_{ax\,bx_3},\quad(11.71)$$

$$\mathcal{F}_{ab}{}^{\mu_1\mu_2}(x)\,g_{\mu_1\mu_3}g_{\mu_2\mu_4} = g_{\mu_3[ax}\,g_{bx]\,\mu_4},\quad(11.72)$$

$$\mathcal{F}_{ab}{}^{\mu_1}(x)\,h_{\mu_1\mu_2\mu_3} = -g_{\mu_2[ax}g_{bx]\,\mu_3} + (g_{ax\,bx_2} - g_{ax\,bx_3})g_{\mu_2\mu_3},$$
$$+\tfrac{1}{2}g_{ax\,bz}\epsilon^{def}[g_{\mu_3\,dz}\partial_{a_2}\,g_{ex_2\,fz} - g_{\mu_2\,dz}\partial_{a_3}\,g_{ex_3\,fz}],$$
$$(11.73)$$

$$\mathcal{F}_{ab}{}^{\mu_1\mu_2}(x)\,h_{\mu_1\mu_2\mu_3} = 2\,h_{ax\,bx\,\mu_3}.\quad(11.74)$$

In the last term of equation (11.73) an integral in z is assumed. The derivatives that appear in the above expressions can be integrated by parts and as a consequence act on the **R**s. Using the differential constraint we generate from them terms of lower multitensor rank. For example, from (11.71) we have

$$g_{\mu_2\mu_4}\partial_{a_3}g_{ax\,bx_3}R^{(ax,\,bx)\mu_2\mu_3\mu_4} = g_{\mu_2\mu_4}(g_{ax\,bx_2} - g_{ax\,bx_4})R^{(ax,bx)\mu_2\mu_4}.$$
$$(11.75)$$

Performing these calculations, the following partial results are obtained for the four expressions quoted above

$$-\epsilon_{abc}g_{\mu_1\mu_2}R^{(ax,bx)\mu_1\,cx\,\mu_2} - (g_{ax\,bx_1} - g_{ax\,bx_2})g_{\mu_1\mu_2}R^{(ax,\,bx)\mu_1\mu_2},\quad(11.76)$$

$$g_{\mu_1[ax}\,g_{bx]\,\mu_2}R^{(ax,bx)\mu_1\mu_2},\quad(11.77)$$

$$-g_{\mu_1[ax}g_{bx]\,\mu_2}R^{(ax,bx)\mu_1\mu_2} + (g_{ax\,bx_1} - g_{ax\,bx_2})g_{\mu_1\mu_2}R^{(ax,\,bx)\mu_1\mu_2}$$
$$-\epsilon^{def}g_{ax\,bz}g_{\mu_1\,dz}g_{ex\,fz}R^{(ax,\,bx)\mu_1},\quad(11.78)$$

$$2\,h_{ax\,bx\,\mu_1}R^{(ax,\,bx)\mu_1}.\quad(11.79)$$

After some cancellations we finally obtain

$$\mathcal{H}(x)\,\mathcal{A}_2(\mathbf{R}) = -2\,\epsilon_{abc}g_{\mu_1\mu_2}R^{(ax,bx)\mu_1\,cx\,\mu_2}$$
$$+\,2[2h_{ax\,bx\,\mu_1} - \epsilon^{def}g_{ax\,bz}g_{\mu_1\,dz}g_{ex\,fz}]R^{(ax,\,bx)\mu_1}.\quad(11.80)$$

One can check that the terms in the bracket are identical and of opposite signs, so the bracket vanishes. One can also see that the term of rank five vanishes. To see this, expand $R^{(ax,\,bx)\mu_1\,cx\,\mu_2}$ and as a result one gets,

$$R^{(ax,\,bx)\mu_1\,cx\,\mu_2} = -2\,R^{(ax\,bx\,\mu_1\,cx\,\mu_2)c} + R^{(cx\,ax\,\mu_1\,bx\,\mu_2)c} + R^{(bx\,cx\,\mu_1\,ax\,\mu_2)c}.$$
$$(11.81)$$

which implies the contribution vanishes due to symmetry considerations when contracted with ϵ_{abc}.

We therefore conclude that

$$\mathcal{H}(x)\,\mathcal{A}_2(\mathbf{R}) = 0. \tag{11.82}$$

We see that the explicit computation of this formal result in the extended representation involves only a few simple steps that basically amount to integrations by parts and application of the constraints satisfied by the multitensors. This should be compared with the lengthy computation in terms of loops outlined in reference [209].

An interesting point is that the computational efficiency that is obtained in the extended representation may be useful at the level of the diffeomorphism constraint. It is straightforward to show, for instance, that $\mathcal{A}_2$ is diffeomorphism invariant simply by checking that it is annihilated by the diffeomorphism constraint. This may find useful applications as a technique for searching for knot invariants.

11.6 Regularization

The extended representation provides a new scenario for analyzing the regularization problems in quantum gravity. In the loop representation regularization ambiguities appear at the level of both quantum operators *and* quantum states. Whereas the first problem is common to all the representations that one can construct for quantum gravity (and lies in the fact that the constraints involve the product of operators evaluated at the same point), the second is typical of the loop representation. In the case of quantum gravity the loop wavefunctions are knot invariants and their analytic expressions require the introduction of a regularization (framing). This difficulty does not only arise for the gravitational case. As we discussed in section 11.1 it is suggestive that even in the simple case of a free Maxwell field it is known that the quantum states in the loop representation are ill defined and a regularization is needed.

We will see that in the extended representation the problems in the definition of the wavefunctions can be solved. We are going to show that with an adequate restriction of the domain of dependence, the extended wavefunctions are well defined functionals. In the regularization of the constraints, we shall limit the analysis to the case of wavefunctions with a totally specified analytical dependence. More precisely, we shall study the action of the regularized Hamiltonian constraint over the wavefunctions that are formally annihilated by the constraint. The regularization of the constraint on the space of all wavefunctions has not yet been studied in detail.

11.6.1   *The smoothness of the extended wavefunctions*

Let us consider now the regularity properties of the extended wavefunctions. Generically the multitensors $X^{\vec{\mu}}$ are distributional, as is directly inferred from the differential constraint (their derivative is a delta function). As we saw in chapter 2 any multitensor that satisfies the differential constraint can be written in the form $\mathbf{X} = \sigma[\phi] \cdot \mathbf{Y}$, where the $\mathbf{Y}$ fields satisfy the homogeneous differential constraint. For example, for the rank two component we have

$$X^{ax\,by} = Y^{ax\,by} + \phi^{ax}_{\ \ y}\, Y^{by} - \phi^{by}_{\ \ x}\, Y^{ax} - \phi^{ax}_{\ \ z}\phi^{by}_{\ \ z,c}\, Y^{cz} + \phi^{[by}_{\ \ o}\, Y^{ax]}.$$
(11.83)

As we discussed in chapter 2 the function ϕ fixes a prescription for the decomposition of the multitensors in transverse and longitudinal parts, $\mathbf{Y} = \delta_T \cdot \mathbf{X}$ with

$$\delta_T^{\ \mu_1\cdots\mu_n}_{\ \ \nu_1\cdots\nu_m} = \delta_{n,m}\ \delta_T^{\ \mu_1}_{\ \ \nu_1}\cdots\delta_T^{\ \mu_n}_{\ \ \nu_n},$$
(11.84)

$$\delta_T^{\ ax}_{\ \ by} = \delta^{ax}_{\ \ by} - \phi^{ax}_{\ \ y,b}.$$
(11.85)

As the $\mathbf{Y}$s satisfy the homogeneous differential constraint, they can be *chosen* to be smooth functions. In that case, all the divergent behavior of the $\mathbf{X}$ is concentrated in the function ϕ. The σs control the divergent character of the group elements.

Let us define the following set of elements of the extended space: $\mathbf{X} \in \{\mathbf{X}\}_s$ if, and only if, there exists a prescription function ϕ such that $\delta_T[\phi] \cdot \mathbf{X} = \mathbf{Y}$ is a smooth function. We shall show that the wavefunctions defined on this domain are smooth in the extended variables and that this property is invariant under diffeomorphism transformations.

Given a diffeomorphism transformation $\mathbf{\Lambda}_D$ defined by $x'^a = D^a(x)$ it can be shown that $\delta_{DT} \equiv \mathbf{\Lambda}_{D^{-1}} \cdot \delta_T \cdot \mathbf{\Lambda}_D$ is a transverse projector in the prescription

$$\phi_D^{\ ax}_{\ \ y} = J(x)\, \frac{\partial x^a}{\partial D^b(x)}\phi^{bD(x)}_{\ \ \ \ \ D(y)},$$
(11.86)

where $J(x)$ is the Jacobian of the coordinate transformation and ϕ is the function that fixes the prescription of the projector δ_T. In this prescription $\mathbf{X} = \sigma \cdot \mathbf{Y} = \mathbf{\Lambda}_{D^{-1}} \cdot \sigma_{D^{-1}} \cdot \mathbf{\Lambda}_D \cdot \mathbf{Y}$. For any diffeomorphism transformation $\mathbf{\Lambda}_D$, the transverse part of $\mathbf{\Lambda}_D \cdot \mathbf{X}$ is a smooth function with the prescription $\phi_{D^{-1}}$. In effect

$$\delta_{D^{-1}T} \cdot (\mathbf{\Lambda}_D \cdot \mathbf{X}) = \delta_{D^{-1}T} \cdot \sigma_{D^{-1}} \cdot \mathbf{\Lambda}_D \cdot \mathbf{Y} = \mathbf{\Lambda}_D \cdot \mathbf{Y},$$
(11.87)

and we therefore see that there is a prescription $\phi_{D^{-1}}$ in which $\delta_T[\phi_{D^{-1}}]$ is a smooth function since $\mathbf{Y}$ is and its character is unchanged by the action of the diffeomorphism. The set $\{\mathbf{X}\}_s$ is then invariant under diffeomorphism transformations.

Let us now consider the specific wavefunctions we introduced in chapter 10. The extended loop transform of the exponential of the Chern–Simons form

$$\Psi_\Lambda(\mathbf{X}) = \int DA \, \exp(S_\Lambda[A]) \, \mathrm{Tr}(\mathbf{A} \cdot \mathbf{X}) = \sum_{n=0}^{\infty} (\mathbf{g}^{(n)} \cdot \mathbf{X}) \, \Lambda^n, \qquad (11.88)$$

where the dot indicates the contraction of indexes. We denote by $\mathbf{g}$ the products of propagators that arise in the perturbative expansion of the functional integral. As we have argued, they play the role of one of the diffeomorphism invariant metrics in the space of multitensors we were seeking in chapter 2. We recall that those metrics were, in general, objects that depended on the particular prescription one took for defining transverse and longitudinal parts.

Now, for any $\mathbf{X} \in \{\mathbf{X}\}_s$ we have

$$\Psi(\mathbf{X}) = \mathbf{g} \cdot \mathbf{X} = \mathbf{g} \cdot \sigma[\phi] \cdot \mathbf{Y} \equiv \mathbf{g}_\phi \cdot \mathbf{Y}, \qquad (11.89)$$

where one can see that $\mathbf{g}_\phi$ is a well defined distributional object that corresponds to the metric $\mathbf{g}$ in a particular prescription determined by ϕ.

This is a very important result. It implies that all the distributional character of the multitensors that is embodied in the σs is incorporated in the distributional nature of the $\mathbf{g}$s. Therefore if one chooses smooth $\mathbf{Y}$s, the wavefunctions are well defined. This fact is invariant under diffeomorphisms. One can always find a prescription in terms of which the wavefunction is written as $\mathbf{g} \cdot \mathbf{Y}$.

It is satisfying to check that by going to the extended representation and suitably restricting the domain of dependence of the wavefunctions one can remove the divergences in their definition. However, there is a price to be paid for this. As we argued before, ordinary loops are included in extended loops. The price we pay for limiting the domain of the extended wavefunctions in order to make them smooth is that we *exclude ordinary loops from the representation*. Ordinary loops *do not* correspond to smooth $\mathbf{Y}$s.

This is consistent with what we discussed before. Written purely in terms of ordinary loops the expressions for the knot invariants are divergent. Therefore they could never have arisen as a restriction of a smooth expression in terms of extended loops. The consistency goes beyond this fact. We saw that one could to a certain extent make sense of the knot invariants in terms of ordinary loops if one supplemented them with an additional structure: a framing. What this is suggesting is that in order to obtain the ordinary loop expressions from the expression of the knot invariants in terms of extended loops one has to go outside their domain of well behavedness. In order to obtain well behaved expressions, that limit should involve a choice of a prescription or regularization which translates itself in the notion of framed loops. The details of how to take this limit and derive a consistent framing from the extended representation have only been studied for particular cases and should be studied further.

11.6.2   The regularization of the constraints

As we discussed in section 11.3, the expressions for the constraints in the extended representation we have introduced are ill defined. They involve a multitensor with indices with a repeated spatial dependence. Due to the distributional character of multitensors imposed by the differential constraint (2.11) a repetition of a spatial dependence implies a divergence. Furthermore, the expression also involves an element of the algebra $\mathcal{F}_{ab}$ which may lead upon contraction to a distribution. Similar arguments apply to the diffeomorphism constraint.

To regularize the constraints we will proceed to point-split them. This is one of the simplest regularization methods one can consider. It may introduce difficulties due to its dependence on a background metric as we argued in chapter 8. It is straightforward to point-split the formal expressions for the constraints introduced in section 11.3. One takes expressions (11.36),(11.50) and point-splits the dependence on the variable x. The result is

$$\mathcal{C}^{\epsilon}_{ax}\,\Psi(\mathbf{R}) = \int d^3w \int d^3v\, f_{\epsilon}(w,x)\, f_{\epsilon}(v,x)\, \Psi(\mathcal{F}_{ab}(w) \times \mathbf{R}^{(bv)}), \quad (11.90)$$

$$\hat{\mathcal{H}}^{\epsilon}(x)\,\Psi(\mathbf{R}) =$$
$$2\int d^3w \int d^3u \int d^3v\, f_{\epsilon}(w,x)\, f_{\epsilon}(u,x)\, f_{\epsilon}(v,x)\, \Psi(\mathcal{F}_{ab}(w) \times \mathbf{R}^{(au,\,bv)}),$$

$$(11.91)$$

where f_{ϵ} is any appropriate symmetric smearing of the delta function. Notice that this point-splitting regularization is not uniquely determined by the formal factor ordered expression. Several sources of ambiguities arise, one of which is related to the background metric used in the smearing functions. It is also possible, but not mandatory, to preserve the gauge invariance in the regularization process. Gauge invariance is easily preserved in the extended representation by a procedure analogous to "closing the loops" in the usual representation. It has been checked that this procedure yields the same result as the non-invariant calculation we will perform here [210]. Finally additional factor ordering problems may arise due to the distributional character of the fundamental fields. We will see that distributional connections will appear naturally in the discussion.

We shall proceed as follows: we will introduce a naive point-splitting and study the action of the regularized and renormalized operators on the formal solutions. We will prove that there is a factor ordering that ensures consistency between the known results in the connection and the loop representation.

In section 11.5 we have shown that the invariance under diffeomorphisms of the coefficients of the expansion of the generalized transform (11.88) is ensured by construction. We also saw that with an appropriate definition of the domain of dependence, the wavefunctions can be endowed with convenient regularity properties (in particular, the smoothness dependence on the extended variables can be ensured in a diffeomorphism invariant way). All this can be explicitly confirmed by checking that the wavefunctions are annihilated by the regularized diffeomorphism constraints. Let us explicitly perform one of these calculations. This will also serve as a warm-up for the Hamiltonian case. Let us check the behavior of the regularized diffeomorphism constraint for the particular case of the extended Gauss linking number. From (11.90) we obtain

$$\mathcal{C}_{ax}^{\epsilon}\, a_1(\mathbf{R}) = \int d^3w \int d^3v\, f_\epsilon(w,x)\, f_\epsilon(v,x)\, g_{\mu_1\mu_2}\, \mathcal{F}_{ab}{}^{\mu_1}(w)\, R^{(bv)\mu_2}. \quad (11.92)$$

This result is valid for any prescription. Due to practical computational reasons we shall restrict the domain of the wavefunctions to those prescriptions connected by diffeomorphisms to the "transverse" prescription, given by

$$\phi_o{}^{ax}{}_y = \frac{1}{4\pi}\, \frac{\partial}{\partial x_a}\, \frac{1}{|x-y|}. \quad (11.93)$$

In the transverse prescription the free Chern–Simons propagator $g_{ax\,by}$ takes the form introduced in chapter 10. Then using (11.71) we get

$$\mathcal{C}_{ax}^{\epsilon}\, a_1(\mathbf{R}) = -\epsilon_{abc} \int d^3w \int d^3v\, f_\epsilon(w,x)\, f_\epsilon(v,x)\, R^{(bv)\,cw}, \quad (11.94)$$

where

$$R^{(bv)\,cw} = Y^{bv\,cw} + Y^{cw\,bv}, \quad (11.95)$$

is a smooth function symmetric under the interchange of the indices b and c (using the fact that the integration points are indistinguishable) contracted with an antisymmetric tensor. The last expression is well defined and we therefore have

$$\mathcal{C}_{ax}^{\epsilon}\, a_1(\mathbf{R}) = 0. \quad (11.96)$$

Notice that no divergences occur in (11.94) and we do not need to take the limit when ϵ goes to zero. The diffeomorphism constraint is perfectly well defined and no renormalization is needed. A similar result holds for $\mathcal{A}_2$ in the sense that no renormalization is needed, although the expression only vanishes when the regulator is removed. This situation is likely to be repeated for all other invariants constructed from Chern–Simons theory.

Let us analyze now the action of the regularized Hamiltonian constraint. This will allow us to put on a rigorous footing the formal results introduced in chapter 10 concerning the transform of the Chern–Simons state. We will not present a complete account here, but we will concentrate on the most elaborate calculation, the action of the Hamiltonian constraint on the second coefficient of the infinite expansion of the Jones polynomial, $\mathcal{A}_2(R)$. This result is of interest in itself since $\mathcal{A}_2(R)$ is the first non-trivial non-degenerate solution to the Wheeler–DeWitt equation with vanishing cosmological constant. We will end this section with some discussion of the rest of the calculation of the action of the Hamiltonian with cosmological constant on the extended Kauffman bracket.

The action of the regularized Hamiltonian constraint on the second coefficient $\mathcal{A}_2(R)$ is

$$
\begin{aligned}
\hat{\mathcal{H}}^\epsilon(x)\,\mathcal{A}_2(\mathbf{R}) = \int d^3w \int d^3u \int d^3v\, f_\epsilon(w,x)\, f_\epsilon(u,x)\, f_\epsilon(v,x) \\
\times \{ - \epsilon_{abc} g_{\mu_1\mu_2} R^{(au,\,bv)\mu_1\,cw\,\mu_2} \\
+ [2\,h_{aw\,bw\,\mu_1} - \epsilon^{def} g_{aw\,bz} g_{\mu_1\,dz} g_{eu\,fz}] R^{(au,\,bv)\mu_1} \\
+ (g_{aw\,bu} - g_{aw\,bv})\, g_{\mu_1\mu_2} R^{(au\,\mu_1\,bv\,\mu_2)_c} \}.
\end{aligned} \tag{11.97}
$$

If we now compare this with the unregulated result that we obtained in section 11.5, equation (11.80), we notice that there is an extra term, the last one in (11.97). We call this the "anomalous term". In the unregulated calculation, the variable $\mathbf{R}^{(au,\,bv)}$ appeared as $\mathbf{R}^{(ax,\,bx)}$ and satisfied the differential constraint based at the point x. In the regulated case, the variable $\mathbf{R}^{(au,\,bv)}$ satisfies a similar equation,

$$
\begin{aligned}
\partial_{\mu_i} R^{(au,\,bv)\mu_1\dots\mu_i\dots\mu_n} = [\delta(x_i - x_{i-1}) - \delta(x_i - x_{i+1})]\, R^{(au,\,bv)\mu_1\dots\hat{\mu}_i\dots\mu_n} \\
+ [\delta(x_i - u) - \delta(x_i - v)]\,(-1)^{n-i} R^{(au\,\mu_1\dots\mu_{i-1}\,bv\,\mu_n\dots\mu_{i+1})_c},
\end{aligned} \tag{11.98}
$$

instead of the usual differential constraint. In the above expression one should identify $x_0 = u$ and $x_{n+1} = v$.

To consider the limit of (11.97) when one removes the regulators, one needs to take into account the divergences that come from the group elements (through the matrix σ) and from the **gs**. The first observation is that both types of contributions are of the same order.

In order to see this we compare the first term in (11.97), which has divergences due to σ (the repeated indices in **R**) and the anomalous term which has divergences due to **g** which in the limit means both indices are evaluated at the same point.

We start with the rank five group elements $R^{(au,\,bv)\mu_1\,cw\,\mu_2}$. If one recalls the definition of σ from chapter 2 and expands $\mathbf{X} = \sigma \cdot \mathbf{Y}$ one finds a large number of terms. One can see that all these terms have a structure of divergences that is characterized by

$$\phi_o{}^{au}{}_v \, Y^{(bv\,\mu_1\,cw\,\mu_2)c}, \tag{11.99}$$

with $Y^{(bv\,\mu_1\,cw\,\mu_2)c}$ a regular function in the limit $\epsilon \to 0$. This expression gives the leading divergence of the rank five term in (11.97).

These leading divergences are exactly the same as those that arise from the anomalous term. In order to see this first notice that

$$\epsilon_{bca}\phi_o{}^{au}{}_v \, Y^{(bv\,\mu_1\,cw\,\mu_2)c} = g_{bu\,cv} \, Y^{(bv\,\mu_1\,cw\,\mu_2)c}, \tag{11.100}$$

whereas in the anomalous term one has a contribution $g_{bu\,cv} \, Y^{(bv\,\mu_1\,cw\,\mu_2)c}$. This last expression apparently has a different divergence structure since it involves an $\mathbf{R}$ instead of a $\mathbf{Y}$ but it turns out that the contraction with $g_{\mu_1\mu_2}$ "erases" the extra divergences introduced by the $\mathbf{R}$ and the order is the same. Therefore in the limit $u \to v$ both the anomalous term and the first term of (11.97) only have divergences due to the presence of $g_{bu\,cv}$.

The result (11.100) ensures, due to the same symmetry properties used in the formal calculation, that the contribution of the first term in (11.97) vanishes. Indeed, one gets from (11.81)

$$-2\,R^{(ax\,bx\,\mu_1\,cx\,\mu_2)c} + R^{(cx\,ax\,\mu_1\,bx\,\mu_2)c} + R^{(bx\,cx\,\mu_1\,ax\,\mu_2)c}, \tag{11.101}$$

contracted with ϵ_{abc} and integrated in u, v, w. One can relabel the dummy indices a, b, c and the integration variables u, v, w in such a way that the three terms in the above expression are equal. The contribution from the first term in (11.97) therefore cancels before removing the regulator.

One can see that the second term in (11.97) also vanishes when one removes the regulator for exactly the same reasons mentioned in the formal calculation since no singularities are involved in the canceling terms.

In order to consider the anomalous term we rearrange slightly the form for it that appears in (11.97). First of all we notice that the contributions to the anomalous term of the two **g**s in the parenthesis actually are the same and add up, giving a single **g** and a factor of 2. The way to see this is to write the **g**s explicitly. Each includes an ϵ_{abc}, which contracted with the **R** yields an expression antisymmetric in u, v and therefore the terms add up. Moreover, we notice that the contraction of $g_{\mu_1\mu_2}$ with **R** is equivalent to the contraction with **Y** as we argued in section 11.6.1. We then have

$$2\int d^3w \int d^3u \int d^3v \, f_\epsilon(w,x)\, f_\epsilon(u,x)\, f_\epsilon(v,x)\, g_{aw\,bv}\, g_{\mu_1\mu_2} R^{(au\,\mu_1\,bv\,\mu_2)c} =$$

$$\frac{2}{\sqrt{2\pi}\,\epsilon}\,\epsilon_{abc}\, g_{\mu_1\mu_2}\, \partial^{cy} R^{(ax\,\mu_1\,by\,\mu_2)c}\big|_{y=x} + O(\epsilon), \tag{11.102}$$

where we have used a Gaussian regulator $f_\epsilon(\vec{z}) = (\sqrt{\pi}\epsilon)^{-3} \exp{(-z^2\epsilon^{-2})}$. This result is obtained by writing $g_{aw\,bv}$ as $\epsilon_{abc}\partial^c(1/|w-v|)$, expanding $R^{(au\,\mu_1\,bv\,\mu_2)\,c}$ in the limit $v \to u$, and explicitly performing the Gaussian integrals.

As we have already discussed, the contraction of **g** with **R** in non-contiguous indices is a regular expression and therefore the result is well defined without singularities.

We therefore see that in order to have a finite expression for the Hamiltonian we need to renormalize the point-split version by a factor ϵ. The end result for the regularized and renormalized Hamiltonian is

$$\mathcal{H}_r(x)\,\mathcal{A}_2(\mathbf{R}) = \lim_{\epsilon \to 0}\, \epsilon\, \mathcal{H}^\epsilon(x)\,\mathcal{A}_2(\mathbf{R})$$

$$= \sqrt{\frac{2}{\pi}}\, \epsilon_{abc}\, g_{\mu_1\mu_2}\, \partial^{cy}\, R^{(ax\,\mu_1\,by\,\mu_2)c}\big|_{y=x}. \qquad (11.103)$$

We conclude that the renormalized Hamiltonian constraint does *not* annihilate the generalized diffeomorphism invariant corresponding to the second coefficient of the Alexander–Conway knot polynomial in the point-splitting regularization procedure we have followed.

This leads immediately to an apparent contradiction. We argued in section 11.5 that as a consequence of the Kauffman polynomial being a state with cosmological constant, the vacuum Hamiltonian with $\Lambda = 0$ had to annihilate $\mathcal{A}_2(\gamma)$. We now see that in a regularized calculation it does not. But the Kauffman bracket arose as the transform of an exact state in the connection representation, independent of regularization problems, the exponential of the Chern–Simons form. How can all these apparently contradicting facts be compatible?

The answer lies in the hypotheses made in order to claim that the exponential of the Chern–Simons form was a solution of the Hamiltonian constraint of quantum gravity in the connection representation. As we argued in chapter 7 this result is quite robust, depending only on choosing a factor ordering with functional derivatives to the right. Because the cancellation between the vacuum Hamiltonian constraint and the cosmological constant term arose with the computation of only one functional derivative one expected the result to be quite robust under changes in regularization procedures. This is true. However, implicit assumptions are made in the domain of dependence of the wavefunctions. For instance, one typically assumes the connections to be smooth. If the connections are not smooth the definition of even apparently trivial multiplicative operators like the field tensor F^i_{ab} becomes problematic and has to be regularized.

Why should one consider distributional connections at all? The problem arises in the functional integrals used to define the loop transform.

Functional integrals have contributions from non-smooth fields. This can be seen even in simple examples of finite-dimensional quantum mechanics. If one considers the path integral formulation of a free particle, the integral has contributions from discontinuous paths when performing the partition to compute it. It is therefore natural to consider distributional connections if one is to perform the transform with usual functional integrals, such as the ones we explicitly used when performing the perturbative expansion.

It turns out that the anomaly we find when regulating the calculation of the action of the Hamiltonian on the $\mathcal{A}_2(\gamma)$ coefficient can be corrected with the introduction of a counterterm. A counterterm is a regularized term which vanishes when acting on an extended Wilson functional constructed with non-distributional, smooth connections. Consider, for example, the following expression, symmetric under the interchange of the internal indices,

$$(A^i_{aw} A^j_{bu} - A^i_{aw} A^j_{bv}) \frac{\delta}{\delta A^{(j}_{bv}} \frac{\delta}{\delta A^{i)}_{au}} W_A(\mathbf{R}) =$$

$$\{ \operatorname{Tr}(A_{(aw \,|\, \mu\!\!\!\downarrow bu)\, \nu}) - \operatorname{Tr}(A_{(aw \,|\, \mu\!\!\!\downarrow bv)\, \nu}) \} \, R^{(au\, \mu\!\!\!\downarrow bv\, \nu)c}. \quad (11.104)$$

It is clear that this term vanishes in the limit $\epsilon \to 0$ if the connections are regular functions, but it may have a non-trivial contribution if the connections are distributions. The corresponding regularized expression in the extended space is

$$\mathcal{C}^\epsilon = R^{(au\, \mu\!\!\!\downarrow bv\, \nu)c} \left(\frac{\delta}{\delta R^{(aw \,|\, \mu\!\!\!\downarrow bu)\, \nu}} - \frac{\delta}{\delta R^{(aw \,|\, \mu\!\!\!\downarrow bv)\, \nu}} \right). \quad (11.105)$$

This expression generates anomalous type contributions. For example,

$$\mathcal{C}^\epsilon(g_{\mu_1 \mu_2} R^{\mu_1 \mu_2}) = 2(g_{aw\, bu} - g_{aw\, bv}) R^{(au\, bv)c}. \quad (11.106)$$

Could it be that by adding expressions like the above one to the Hamiltonian one can cancel the anomalous terms? The answer is in the affirmative. The precise counterterm is given by the difference of two terms, $\mathcal{C}_2 - \mathcal{C}_1$,

$$\mathcal{C}_1 = R^{(au\, \mu\!\!\!\downarrow bv\, \nu)c} \left(\frac{\delta}{\delta R^{aw\, \mu\!\!\!\downarrow bu\, \nu}} - \frac{\delta}{\delta R^{aw\, \mu\!\!\!\downarrow bv\, \nu}} \right), \quad (11.107)$$

$$\mathcal{C}_2 = (R^{(au\, bv)\, \alpha} + \tfrac{1}{2} R^{[au\, bv]\, \alpha}) \left(\frac{\delta}{\delta R^{(aw\, bu)c\, \alpha}} - \frac{\delta}{\delta R^{(aw\, bv)c\, \alpha}} \right), \quad (11.108)$$

where $R^{[au\, bv]\, \alpha}$ is given by expression (11.56) without the $(-1)^{n(\mu)}$ factor and without the "rerouting" action (the index μ^{-1} is replaced by μ.) Remarkably, these expressions also have a simple form in the connection representation,

$$C_1 := (A^i_{aw} A^k_{bu} - A^i_{aw} A^k_{bv}) \frac{\delta}{\delta A^i_{au}} \frac{\delta}{\delta A^k_{bv}}, \tag{11.109}$$

$$C_2 := (A^i_{aw} A^i_{bu} - A^i_{aw} A^i_{bv}) \frac{\delta}{\delta A^k_{au}} \frac{\delta}{\delta A^k_{bv}}. \tag{11.110}$$

With this *single* counterterm all the anomalous contributions to the action of the Hamiltonian constraint on the a_2, the Kauffman bracket and the exponential of the self-linking number cancel. The fact that a single counterterm is responsible for all the cancellations is remarkable and shows that the construction is not just a gimmick to fix the anomaly problem, but might well be a genuine counterterm arising from quantum gravity. The fact that the counterterm has a simple and precise expression in the connection representation raises the hope that a better intuitive explanation of it could be gained by viewing it in this context. At present this issue is not settled: could it be that $C_2 - C_1$ is what one needs to add to the Hamiltonian in the connection representation in order to annihilate the exponential of the Chern–Simons form when distributional connections are allowed? Could it reflect the fact that in that case a non-trivial contribution to the measure arises? These issues are currently being studied.

11.7   Conclusions

We constructed a representation for quantum gravity based on extended loops. We studied the space of wavefunctions and promoted the constraints to wave equations. The wavefunctions are linear functionals of the multitensors and the constraints are first order functional differential operators. This introduces computational simplifications that allow to operate very efficiently with the constraints. The price paid for this is that one loses the simple geometric characterization of the solutions of the diffeomorphism constraint in terms of knot classes. One has to deal with that constraint as another functional equation. In spite of this, the knot invariants derived from Chern–Simons theory that were formal solutions of the constraints in terms of loops admit a straightforward extension to the space of multitensors. We checked formally that they solved the constraints. We then studied a regularization and showed that the solutions found also solved the constraints in a rigorous regularized way through the introduction of appropriate counterterms. The situation regarding the regularization of the constraints is still unsatisfactory, since although we can recover in a regularized fashion all of the formal results, we do not have a physical argument for the introduction of the counterterms. The fact that they have a simple expression in the connection representation raises the hope that some physical insight might be gained into their origin. The results obtained are just a first step in the regularization process, the next step being the computation of the algebra of constraints.

12

Conclusions, present
status and outlook

In this book we have attempted to present in a structured fashion the various aspects of the use of loops in the quantization of gauge theories and gravitation. The discussion mixed historical and current developments and we rewrote many results in a more modern language. In this chapter we would like to concentrate on the outlook arising from the material presented and focus on current developments and on possible future avenues of work. We will divide the discussion into gauge theories and gravity, since the kinds of developments in these two fields follow naturally somewhat disjoint categories.

12.1 Gauge theories

Overall, the picture which emerges is satisfying in the sense that the bulk of the techniques developed can be applied systematically to the construction of loop representations for almost any theory based on a connection as the main canonical variable, either free or interacting with various forms of matter. In this respect we must emphasize the developments listed in chapters 1, 2 and 3 which are the main mathematical framework that we used to understand the physical applications. Many of these aspects, as we have mentioned, have been studied with mathematical rigor by various authors in spite of the fact that the presentation we have followed here is oriented towards physicists.

The main conclusion to be drawn from this book is that loop techniques are at present a practical tool for the analysis of the quantum mechanics of gauge theories. There are three main lines of attack that are worthwhile discussing separately:

• Quantization of gauge theories in the continuum. Even though the loop representation has very appealing features such as the gauge invariance and its geometrical content, there has not been great improvement

302

over other computational methods up to now. Although as we saw in chapter 5 there is a very compact description of the theory, no exact solutions of the non-perturbative Hamiltonian eigenvalue equation known. The loop equations need to be renormalized, but we do not know how to introduce a non-perturbative renormalization and therefore the strategy has been to try to solve the regularized equations and renormalize at the end.

At present it seems that a significant revision of these issues could come from the new developments in loop techniques we mentioned in chapters 2 and 3 The use of rigorous methods to define the loop transform and the measure could shed new light on many issues in non-perturbative Yang–Mills theories such as the issue of confinement. The expectation is that with the introduction of a rigorous measure one could compute the functional integral involved in the expectation value of the Wilson loop that we discussed in chapter 6, and therefore provide an exact expression for all the Green functions of the theory. A less ambitious hope would be to at least obtain a rigorous characterization of the asymptotic behaviors of these expressions and therefore elucidate the issue of confinement in a solid mathematical framework. On the other hand, the use of extended loops could prove a powerful tool for producing solutions to the quantum Hamiltonian eigenvalue problem and set a better stage to address the non-perturbative regularization problem.

Historically there was a fair amount of activity in this area in the early 1980s, prior to the introduction of extended loops and the rigorous transform but then the level of activity decayed. The expectation is that in the future the old results will be revised in the light of the new techniques and the hope is that many exciting new developments will take place. Some results are already starting to appear such as the ones mentioned for Maxwell theory [72, 65] and also some non-Abelian results [212, 96, 211, 102, 214] (see also reference [213] for a review), but a greater increase of activity is needed in this area in order to exploit in full the possibilities offered by the new techniques.

An appealing aspect of the use of loop variables in gauge theories is that if one considers interactions with matter fields, the loop variables are naturally adapted to the physical degrees of freedom of the theory. In particular the confined lines of force give rise naturally to the physical excitations in the confinement phase. For instance, in QCD the physical state space is defined in terms of loops and open paths with two or three quarks at the end points. These variables are respectively associated with the physical excitations, gluons, mesons, and barions.

• Gauge theories on the lattice. In this area there has been sustained activity over the last decade. The main obstruction has always been the overcompleteness of the basis of loops. The proliferation of loops when one considers larger lattices and higher dimensions completely washes out the advantages provided by the formalism. Although this difficulty can be remedied in part by the use of cluster expansions, the fact that the approximations involved are uncontrolled discourages in part a systematic use of them for tackling more realistic problems in 3 + 1 dimensions. It is still the case that the use of loops offers overall advantages over other Hamiltonian methods. Monte Carlo techniques are, however, at present more efficient overall. Again, the use of extended loops motivated the introduction of classical loop actions for the lattice (as discussed in references [72, 215, 216]). These actions allow the use of Monte Carlo techniques in terms of loops and have a very simple expression, being proportional to the quadratic area associated with the world-sheet defined by the evolution of the loop. This may even suggest some connections with string theory.

• Topological field theories. Although topological field theories are a particular case of gauge theories, many special techniques and approaches have been developed for their study and a significant activity has taken place in this field in recent years. The main driving force is that, as we saw in chapter 10, topological field theories can be a powerful practical tool for the study of problems in mathematics. As an example of the recent results in this area, apart from the well known results of Witten we have mentioned in chapter 10, one can cite many results on triangulations (see [220] for references). For collected sets of articles see the books by Baez [217], Yetter [218] and Baadhio and Kauffman [219].

In summary, there are great opportunities for future developments in the application of loops to gauge theories with the possibility of obtaining concrete practical results that cannot be obtained by other methods. The future years will tell us if these expectations are fulfilled.

12.2   Quantum gravity

The impact of the introduction of loop techniques in quantum gravity has been quite great since their inception in the late 1980s. As opposed to gauge theories, where loop techniques are a minor part of the overall effort, in quantum gravity they have become one of the main approaches to the problem. In fact, they have significantly changed the perspective on many of the central issues of the field. The late 1980s and early 1990s has been a period of great excitement for the loop approach to quantum gravity and many formal, indicative results have been established: the use of knot theory to solve the diffeomorphism constraint, the observation that one could find solutions to the Wheeler–DeWitt equation, the connection with Chern–Simons theory and the Jones polynomial, the realization that one can define rigorously loop transforms and the discovery of a set of operators that are well defined without a renormalization, which led to the idea of weaves. It seems that present activity is concentrating on putting many of these results on a more solid footing and laying out the basis for a solid rigorous theory of quantum gravity. Current efforts are also being directed towards tackling the main problems of the field: those of observables and time, through different kinds of approximations. The activity in the field at present is channeled into three main approaches, which we would like to comment upon separately. Dividing any field in a set of efforts is usually only a partial characterization and we do so here only to order ideas in some way.

• The definition of a physical Hamiltonian and the use of the spin network basis. This approach involves choosing a matter time clock and the introduction of a topological Hamiltonian in the space of knots. One of the open issues at present is to provide a well defined characterization of the action of the Hamiltonian in the space of knots. Also checking the consistency of the constraints is a difficult task that has yet to be completed. The main effort is currently being directed towards the computation of possible topological Feynman rules arising from the proposed Hamiltonian coupled to matter. The main challenge once these technical issues have been settled is to connect these approximation techniques effectively with the situations in which there is interest in exploring quantum gravity effects, for instance, Hawking evaporation. A subject we have barely had any chance to discuss in this book due to its recent nature is the use of spin network states to construct a basis of independent loop states (free of Mandelstam identities) that could also help to diagonalize many physical operators in the theory. It is clear that these states may have implications for physical predictions of the theory. They may find use in Yang–Mills theory as well.

• The use of a rigorous measure to compute the loop transform and an inner product. This direction of research could potentially lead to a revision of several sections of this book. It is expected that the material we developed here will provide the basis on which to deal with the potential new expressions for constraints and wavefunctions that the introduction of non-trivial rigorous measures in the loop transform could produce. At present there is great excitement due to the possibility of incorporating in the measure the reality conditions through the results of Hall which we discussed in chapter 3. Not only would this allow us to define the transform but it would also allow us to introduce a physical inner product into the theory, which would be a major achievement in the case of quantum gravity and would allow us among other things to decide which solutions to the constraints are normalizable. The main challenge of this approach is to ensure that the measures introduced produce physical theories with the expected properties. Introducing measures in infinite-dimensional spaces is, as we discussed, not a trivial task and therefore one has to ensure that what one constructs has a non-trivial content. This approach may allow a rigorous definition of the constraints of quantum gravity and also of the functional integrals of interest to knot theorists and particle physicists. It may also allow us to establish to what extent the connection and loop representations are "functional duals" of each other.

• The use of extended loops and the extended representation. A powerful machinery has been set up to represent quantum gravity through extended loops. It is not clear at present if this machinery will simply be a calculational tool to perform computations in terms of loops or if extended loops are genuinely needed to represent quantum gravity. As a calculational tool it has proved powerful to generate solutions to the Wheeler–DeWitt equation and their regularization. It has introduced a systematic way to operate with loops, which is a valuable achievement in itself. In spite of the fact that there are several indications, as we commented in chapter 10, that loops may not be enough to represent gauge theories there is also evidence that extended loops are "too big" to represent a quantum theory. It has been pointed out [222] that simple examples can be constructed in which the use of extended loops leads to loss of gauge invariance and other serious pathologies. The root of these difficulties always lies in convergence problems of the extended expressions, which we have largely ignored in this book. It was inevitable that this should happen, since one is trying to build a "functional dual" of the space of connections modulo gauge transformations and it is therefore unavoidable to delimit proper domains of convergence. Is the proper domain given by just the ordinary loops or do we inevitably need some of the extended elements? This is the main challenge of this approach at present: to find a subset of extended loops (usually referred to as "thickened out loops") that is large enough to capture all information needed from the gauge theory in question and yet is not so large as to run into convergence problems. There are several proposals currently under study for such objects [21].

• Other ideas. Great possibilities also lie in the integration of all these efforts. For instance, at present it is not clear how to introduce rigorous measures for the extended loop transform. The root of this problem lies in the notion of independent loops that was crucial to introduce the cylindrical measure, which it is not clear how to generalize to extended loops. It appears that this difficulty is in no way fatal and some suitable generalization is likely to be found. Mixing the rigorous transforms with the extended loops could provide a powerful tool for addressing the convergence problems of extended loops. Another avenue is to try to fuse together the ideas used to define the finite diffeomorphism invariant observables and the notion of weaves with the extended representation. There does not seem to be any real obstruction to doing this and it seems to be a problem ready to be tackled. In particular the ideas of spin networks that serve naturally to diagonalize some of the operators are very likely to have a counterpart in algebraic properties of the multitensors of the extended representation. Finally, exhausting the permutations of the above points, one could apply the rigorous measure to give a proper definition of the physical Hamiltonian constraint in the space of knots that is being proposed using the square-root techniques. This would be an important point since the ambiguities introduced by the addition of small loops in the space of knots pervade all the loop formalism, not just this last approach.

• Open problems. The main open problems in canonical quantum gravity have for many years been the problems of time and of the interpretation in quantum gravity. Suppose we succeed in finding a large space of solutions to the Wheeler–DeWitt equation. Suppose we find non-local quantities that commute with the constraints. Suppose an inner product implementing the reality conditions is found. What next? Some authors hold the view that this is not the way to make physical sense of quantum gravity but that the proper way lies in identifying an "internal time" in the theory in terms of which to write it as a Schrödinger equation. There have been several attempts to do this, and even for simplified midi-superspace models [223] it has proved an elusive issue. The usual answer to these objections is that in order to obtain a notion of time from quantum gravity one should have recourse to the "complexity" and "many degrees of freedom" of the theory and therefore simplified models are too "frozen" to be a good arena to test these issues. Loops provide a framework in which "complexity" can be tabulated in a particular way, by specifying the degree of knottiness. As a consequence, a notion of time can be associated with the increase of complexity of the knottings that appear in the wavefunctions. This has been explored in some detail [221] with encouraging results, leading to a notion of evolving Hilbert spaces. In general, the problem of identifying an internal clock in a system is the problem of choosing a suitable set of variables to describe it in terms of which such a clock is manifest. It may be that the new variables are better suited for this problem than the geometrodynamical ones, as suggested by the results of Ashtekar [56], or that loops may present a better picture, as suggested by the evolving Hilbert space construction.

All in all we can say that several new avenues have been opened by the use of loops to represent quantum gauge theories and quantum gravity. There are many issues to be tackled before a final word can be given on the usefulness or otherwise of this approach. The new perspectives introduced in this process and the insights to several results that have been provided are probably going to be a lasting contribution to physics and mathematics regardless of what the final theory is.

References

[1] P. Ramond, "Field theory, a modern primer", 2nd edition, Addison-Wesley, Redwood City, CA (1989).

[2] A. Ashtekar (Notes prepared in collaboration with R. Tate), "Lectures on non-perturbative canonical gravity", Advanced Series in Astrophysics and Cosmology Vol. 6, World Scientific, Singapore (1991).

[3] A. Ashtekar, "Mathematical problems of non-perturbative quantum gravity", to appear in Proceedings of the 1992 Les Houches Summer School on gravitation and quantization, editor B. Julia, North-Holland, Amsterdam (1995).

[4] J. Lewandowski, Class. Quan. Grav. **10**, 879 (1993).

[5] R. Gambini, A. Trias, Phys. Rev. **D23**, 553 (1981).

[6] J. Barrett, Int. J. Theor. Phys. **30**, 1171 (1991).

[7] K. Chen, Ann. Math. **97**, 217 (1973).

[8] S. Mandelstam, Ann. Phys. (NY) **19**, 1 (1962).

[9] S. Mandelstam, Phys. Rev. **175**, 1580 (1968).

[10] A. Migdal, Ann. Phys. (NY) **126**, 279 (1980).

[11] A. Migdal, Phys. Rep. **102**, 199 (1983).

[12] Yu. M. Makeenko, A. A. Migdal, Phys. Lett. **B88**, 135 (1979); Nucl. Phys. **B188**, 269 (1981).

[13] R. Gambini, A. Trias, Phys. Rev. **D27**, 2935 (1983).

[14] X. Fustero, R. Gambini, A. Trias, Phys. Rev. **D31**, 3144 (1985).

[15] R. Gambini "Teorías de Calibre en el espacio de ciclos", unpublished notes, Universidad Simón Bolivar (1986).

[16] M. Blencowe, Nucl. Phys. **B 341**, 213 (1990).

[17] R. Loll, Theor. Math. Phys. **93**, 1415 (1992).

[18] I. Aref'eva, Theor. Math. Phys **43**, 353 (1980) (Teor. Mat. Fiz. **43**, 111 (1980)).

[19] D. Marolf, Class. Quan. Grav. **10**, 2625 (1993).

[20] C. Di Bartolo, R. Gambini, J. Griego, Commun. Math. Phys. **158**, 217 (1993).

[21] C. Di Bartolo, R. Gambini, J. Griego, J. Pullin "The space of states of quantum gravity in terms of loops and extended loops: some remarks", to appear in J. Math. Phys (1995).

[22] This construction was first presented to us by Abhay Ashtekar (private communication).

[23] R. Loll, Class. Quan. Grav. **10**, 1471 (1993).

[24] R. Loll, Nuclear Physics **B350**,831 (1991). See also [26]

[25] R. Loll, Nucl. Phys. **B400**, 126 (1993).

[26] B. Brügmann, J. Pullin, Nucl. Phys. **B390**, 399 (1993).

[27] P. A. M. Dirac "Lectures in Quantum Mechanics", Yeshiva University, Belfer Graduate School of Science Monograph series, No. 2 (1964).

[28] M. Henneaux, Phys. Rep. **126**, 1 (1985).

[29] A. Hanson, T. Regge, C. Teitelboim, "Constrained Hamiltonian systems", Accademia Nazionale dei Lincei, Rome (1976).

[30] K. Sundermeyer, "Constrained dynamics", Lecture Notes in Physics, **169**, Springer-Verlag, Berlin (1982).

[31] R. Abraham, J. E. Marsden, "Foundations of mechanics", 2nd edition, Benjamin, Reading, MA (1978).

[32] V. Arnold, "Mathematical methods of classical mechanics", Springer-Verlag, New York (1978).

[33] V. W. Guillemin, "Symplectic techniques in physics", Cambridge University Press, Cambridge (1984).

[34] R. Gambini, A. Trias, Nucl. Phys. **B278**, 436 (1986).

[35] R. Giles, Phys. Rev. **D24**, 2160 (1981).

[36] A. Ashtekar, C. Isham, Class. Quan. Grav. **9**, 1069 (1992).

[37] J. Glimm, A. Jaffe, "Quantum physics. A functional integral point of view", 2nd edition, Springer Verlag, New York (1987).

[38] C. Rovelli, L. Smolin, Phys. Rev. Lett. **61**, 1155 (1988).

[39] C. Rovelli, L. Smolin, Nucl. Phys. **B331**, 80 (1990).

[40] A. Ashtekar, J. Lewandowski, in "Knots and quantum gravity", J. Baez editor, Oxford University Press, Oxford (1993).

[41] C. Isham, in "Relativity, groups and topology II", editors B.S. DeWitt and R. Stora, North-Holland, Amsterdam (1984), 1059–1290.

[42] A. Pressley, G. Segal, "Loop groups", Clarendon Press, Oxford (1986).

[43] J. N. Tavares, Int. J. Mod. Phys. **A9**, 4511 (1994).

[44] A. M. Polyakov, Phys. Lett. **B82**, 249 (1979); Nucl. Phys. **B164**, 171 (1979).

[45] E. Witten, Commun. Math. Phys **121**, 351 (1989).

[46] E. Witten, Nucl. Phys. **B311**, 46 (1988).

[47] C. Di Bartolo, J. Griego, Phys. Lett. **B317**, 540 (1993).

[48] K. Wilson, Phys. Rev. **D10**, 2445 (1974).

[49] J. Kogut, L. Susskind, Phys. Rev. **D11**, 395 (1975).

[50] G. t'Hooft, Nucl. Phys. **B153**, 141 (1979).

[51] A. Ashtekar, Phys. Rev. Lett. **57**, 2244 (1986); Phys. Rev. **D36**, 1587 (1987).

[52] B. Brügmann, R. Gambini, J. Pullin, Gen. Rel. Grav. **25**, 1 (1993).

[53] A. Ashtekar, "Lectures on non-perturbative quantum gravity", editors R. Kulkarni, J. Samuel, University of Poona press, Poona (1990).

[54] H. Kodama, Phys. Rev. **D42**, 2548 (1990).

[55] H. Kodama, Int. J. Mod. Phys. **D1**, 439 (1992).

[56] A. Ashtekar in "Conceptual problems of quantum gravity", editors A. Ashtekar, J. Stachel, Birkhauser, Boston (1993).

[57] K. Kuchař, in "Proceedings of the 4th Canadian meeting on Relativity and Relativistic Astrophysics", editors G. Kunstatter, D. Vincent, J. Williams, World Scientific, Singapore (1992).

[58] P. Peldán, " Large Diffeomorphisms in (2+1)-quantum gravity on the torus" preprint gr-qc@xxx.lanl.gov:9501020 (1995).

[59] R. Jackiw, in "Relativity, groups and topology II ", editors B. DeWitt and L. Stora, North Holland, Amsterdam (1984).

[60] J. Frieman, R. Sorkin, Phys. Rev. Lett. **44**, 1100 (1980); Gen. Rel. Grav. **14**, 615 (1982).

[61] A. Ashtekar, A. P. Balachandran, S. Jo, Int. J. Mod. Phys. **A4**, 1493 (1989).

[62] R. Gambini, A. Trias, Phys. Rev. **D22**, 1380 (1980).

[63] C. Di Bartolo, F. Nori, R. Gambini, A. Trias, Lett. Nuo. Cim. **38**, 497 (1983).

[64] G. Buendia, A. Trias, Phys. Rev. **D46**, 3649 (1992).

[65] A. Ashtekar, C. Rovelli, Class. Quan. Grav. **9**, 1121 (1992).

[66] J. Baez, in reference [218].

[67] A. Rendall, Class. Quan. Grav. **10**, 605 (1993).

[68] J. Goldberg, J. Lewandowski, C. Stornaiolo, Commun. Math. Phys. **148**, 377 (1992).

[69] C. Rovelli, Class. Quan. Grav. Class. Quan. Grav. **8**, 1613 (1991).

[70] V. Bargmann, Proc. Natl. Acad. Sci. (USA) **48**, 199 (1962).

[71] J. Bjorken, S. Drell, "Relativistic quantum fields", McGraw-Hill, New York (1965).

[72] D. Armand-Ugon, J. Griego, R. Gambini, L. Setaro, Phys. Rev. **D50**, 5352 (1994).

[73] A. Ashtekar, C. Isham, Phys. Lett. **B274**, 393 (1992).

[74] J. B. Kogut, Rev. Mod. Phys. **51**, 659 (1979).

[75] S. L. Glashow, Nucl. Phys. **22**, 579 (1961); A. Salam, J. C. Ward, Phys. Lett. **13**, 168 (1964); S. Weinberg, Phys. Rev. Lett. **18**, 507; **19**, 1264 (1967).

[76] C. N. Yang, R. L. Mills, Phys. Rev. **96**, 191 (1954).

[77] C. G. Bollini, J. J. Giambiagi, Phys. Lett. **40D**, 566 (1972).

[78] G. t'Hooft, Nucl. Phys. **B33**, 173; **B35**, 167 (1971); G. t'Hooft, M. Veltman, Nucl. Phys. **B44**, 189 (1972).

[79] R. Feynman, Acta Phys. Polonica **24**, 697 (1963).

[80] B. DeWitt, Phys. Rev. **162**, 195 (1967).

[81] L. Fadeev, V. Popov, Phys. Lett. **25B**, 29 (1967).

[82] A. M. Polyakov, Nucl. Phys. **82B**, 247 (1979).

[83] A. M. Polyakov, Nucl. Phys. **B164**, 171 (1979).

[84] A. M. Polyakov, "Gauge fields and strings", Harwood Academic Publishers, New York (1987).

[85] C.-H. Gu, L.-L. Wang, Phys. Rev. Lett. **25**, 2004 (1980).

[86] A. A. Migdal, Ann. Phys. **126**, 279 (1980).

[87] T. Eguchi, Phys. Lett. **B87**, 91 (1979); D. Foerster, Nucl. Phys. **B170**, 107 (1980); Phys. Lett. **B87**, 87 (1979).

[88] R. A. Brandt, A. Gocksch, M.-A. Sato, F. Neri, Phys. Rev. **D26**, 3611 (1982).

[89] R. Gambini, J. Griego, Phys. Lett. **B256**, 437 (1991).

[90] V. Kazakov, Nucl. Phys. **B179**, 283 (1981); V. Kazakov, V. Kostov, Nucl. Phys. **B176**, 199 (1980).

[91] A. A. Migdal, Zh. Eksp. Teor. Fiz. **69**, 810 (1975). (Sov. Phys. JETP **42**, 413 (1975).)

[92] A. A. Migdal, Nucl. Phys. Supplement Series **B4**, 59 (1988).

[93] E. Karjalainen, Nucl. Phys. **B413**, 84 (1994).

[94] J.-L. Gervais, A. Neveau, Phys. Lett. **B80**, 255 (1979); also in "Gauge theories in high energy physics : Les Houches, session XXXVII", editors M. Gaillard, R. Stora, North-Holland, Amsterdam (1983).

[95] J.M. Aroca-Farrerons, Ph. D. Thesis, Universitat Autònoma de Barcelona (unpublished) (1992).

[96] J. Baez, in reference [217].

[97] R. Gambini, J. Pullin, Phys. Rev. **D47**, R5214 (1993).

[98] M. Creutz, "Quarks, gluons and lattices", Cambridge University Press, Cambridge (1983).

[99] G. t'Hooft, Nucl. Phys. **B138**, 1 (1978).

[100] S. Mandelstam, Phys. Rev. **D19**, 2391 (1979).

[101] R. Gambini, A. Trias, Phys. Lett. **B141**, 403 (1984).

[102] J. Hetrick, Phys. Lett. **B230**, 88 (1994); Int. J. Mod. Phys. **A9** 3153 (1994).

[103] J. Hallin, Class. Quan. Grav. **11**, 1615 (1994).

[104] F. Wegner, J. Math. Phys. **12**, 2259 (1971).

[105] N. Metropolis, A. Rosembluth, M. Rosembluth, A. Teller, E. Teller, J. Chem. Phys. **21**, 1087 (1953).

[106] K. Wilson, Phys. Rep. **23C**, 331 (1976).

[107] M. Creutz, Phys. Rev. **D21**, 2308 (1980).

[108] R. Gambini, A. Trias, Phys. Rev. Lett. **53**, 25 (1984).

[109] B. Brügmann, Phys. Rev. **D43**, 566 (1991).

[110] R. Gambini, A. Leal, A. Trias, Phys. Rev. **D39**, 3127 (1989).

[111] S. Elitzur, Phys. Rev. **D12**, 3978 (1975).

[112] M. Creutz, Phys. Rev. **D15**, 1128 (1977).

[113] T. Schultz, D. Mattis, E. Lieb, Rev. Mod. Phys. **36**, 856 (1964).

[114] C. Di Bartolo, R. Gambini, A. Trias, Phys. Rev. **D39**, 3136 (1989).

[115] W. Furmanski, A. Kolawa, Nucl. Phys. **B291**, 594 (1987).

[116] R. Feynman, Nucl. Phys. **B188**, 479 (1981).

[117] A. Irving, C. Hamer, Z. Phys. **C27**, 307 (1985).

[118] A. Irbäck, C. Peterson, Phys. Lett. **B174**, 99 (1986).

[119] E. D'Hoker, Nucl. Phys. **180**, 341 (1981).

[120] L. Susskind, Phys. Rev. **D16**, 3031 (1977).

[121] R. Gambini, H. Fort, Phys. Rev. **D44**, 1257 (1991).

[122] R. Gambini, L. Setaro, "$SU(2)$ QCD in the path representation", to appear in Nucl. Phys. **B** (1995).

[123] R. M. Wald, "General relativity", University of Chicago Press, Chicago (1984).

[124] R. Arnowitt, S. Deser, C. Misner, in "Gravitation: an introduction to current research", editor L. Witten, Wiley, New York (1962).

[125] B. DeWitt, Phys. Rev. **150**, 1113 (1967).

[126] R. Utiyama, Phys. Rev. **101**, 1597 (1956).

[127] J. Kijowski, "On a purely affine formulation of general relativity", Springer Lecture Notes in Mathematics, **836**, Springer, New York (1980).

[128] A. Palatini, Rend. Circ. Mat. Palermo **43**, 203 (1917).

[129] G. Horowitz, Class. Quan. Grav. **8**, 587 (1991).

[130] A. Ashtekar, R. Tate, J. Math. Phys. **35**, 6434 (1994).

[131] K. Kuchař, in "General Relativity and Gravitation 1992", editors R. J. Gleiser, C. N. Kozameh, O. M. Moreschi, Institute of Physics Publishing, Bristol (1993).

[132] G. González, R. Tate, Class. Quan. Grav. **12**, 1287 (1995).

[133] A. Ashtekar, J. Romano, R. Tate, Phys. Rev. **D40**, 2572 (1989).

[134] T. Jacobson, L. Smolin, Nucl. Phys. **B299**, 295 (1988).

[135] O. Obregón, J. Pullin, M. Ryan, Phys. Rev. **D48**, 5642 (1993).

[136] V. Husain, Nucl. Phys. **B313**, 711 (1989).

[137] B. Brügmann, R. Gambini, J. Pullin, Nucl. Phys. **B385**, 587 (1992).

[138] R. Gambini, Phys. Lett. **B255**, 180 (1991).

[139] B. Brügmann, J. Pullin, Nucl. Phys. **390**, 399 (1993).

[140] C. Rovelli, L. Smolin, Phys. Rev. Lett. **72**, 446 (1994).

[141] R. Gambini, A. Garat, J. Pullin, "The constraint algebra of quantum gravity in the loop representation", Int. J. Mod. Phys. **D** (in press) (1995).

[142] N. Tsamis, R. Woodard, Phys. Rev. **D36**, 3691 (1987).

[143] J. Friedman, I. Jack, Phys. Rev. **D37**, 3495 (1987).

[144] C. Rovelli, "Outline of a generally covariant quantum field theory and a quantum field theory of gravity" preprint gr-qc@xxx.lanl.gov:9503067 (1995).

[145] L. Smolin, "Quantum gravity and cosmology. Proceedings, 22nd GIFT International Seminar on Theoretical Physics, Sant Feliu de Guixols, Spain", editors J. Perez-Mercader, J. Solá, E. Verdaguer, World Scientific, Singapore (1992).

[146] C. Rovelli, L. Smolin, Nucl. Phys. **B442**, 593 (1995).

[147] T. Jacobson, Class. Quan. Grav. **5**, L143 (1988).

[148] J. Samuel, Pramāna, **28**, L429 (1987).

[149] H. Morales-Técotl, C. Rovelli, Phys. Rev. Lett. **72**, 3642 (1994).

[150] S. Chakraborty, P. Peldán, Phys. Rev. Lett. **73**, 1195 (1994).

[151] G. Chapline, N. Manton, Phys. Lett. **B120**, 105 (1983).

[152] F. Wilczek, Phys. Rev. Lett. **40**, 279 (1978); S. Weinberg, Phys. Rev. Lett. **40**, 223 (1978). For a review of axions see, for example, J. Kim, Phys. Rep. **150**, 1 (1987).

[153] L. Smolin, in "Proceedings of the 12th Johns Hopkins meeting, Knots, Topology and Quantum Field theory", editor L. Lusanna, World Scientific, Singapore (1989).

[154] D. Freedman, P. Townsend, Nucl. Phys. **B177**, 287 (1981).

[155] P. Arias, C. Di Bartolo, X. Fustero, R. Gambini, A. Trias, Int. J. Mod. Phys. **A7**, 737 (1992).

[156] R. Gambini, Phys. Lett. **B131**, 251 (1986).

[157] A. Ashtekar, V. Husain, C. Rovelli, J. Samuel, L. Smolin, Class. Quan. Grav. **6**, 185 (1989).

[158] I. Bengtsson, Phys. Lett. **B220**, 51 (1989).

[159] F. Barbero, M. Varadarajan, Nucl. Phys. **B415**, 515 (1994).

[160] S. Martin, Nucl. Phys. **B**

[161] P. Peldán, "A Modular Invariant Quantum Theory From the Connection Formulation of (2+1)-Gravity on the Torus", preprint gr-qc@xxx.lanl.gov:9504014 (1995).

[162] J. Louko, D. Marolf, Class. Quan. Grav. **11**, 311 (1994).

[163] A. Ashtekar, R. Loll, Class. Quan. Grav. **11**, 2417 (1994).

[164] S. Carlip, in "Proceedings of the 5th Canadian conference on relativistic astrophysics", editors R. Mann, R. McLenaghan, World Scientific, Singapore (1994).

[165] A. Ashtekar, in "Strings '90", editor R. Arnowitt, World Scientific, Singapore (1990).

[166] C. Rovelli, Phys. Rev. **D47**, 1703 (1993).

[167] A. Ashtekar, C. Rovelli, L. Smolin, Phys. Rev. Lett. **69**, 237 (1992).

[168] J. Iwasaki, C. Rovelli, Int. J. Mod. Phys. **D1**, 533 (1993); Class. Quan. Grav. **11**, 1653 (1994).

[169] V. Husain, Class. Quan. Grav. **10**, L233 (1993).

[170] C. Rovelli, Nucl. Phys. **B405**, 1797 (1993).

[171] M. Kalb, P. Ramond, Phys. Rev. **D9**, 2273 (1974).

[172] L. Smolin, Phys. Rev. **D49**, 4028 (1994).

[173] L. Garay, Int. J. Mod. Phys. **A10** 145 (1995).

[174] J. C. Maxwell, "A treatise on electricity and magnetism," Volume II, Oxford (1873).

[175] S. Majid, J. Math. Phys. **31**, 925 (1990).

[176] P. G. Tait, "Scientific Papers", Cambridge (1898).

[177] L. Kauffman, "On knots", Annals of Mathematics Studies, Princeton University Press, Princeton (1987).

[178] L. Kauffman "Knots and physics", World Scientific Series on Knots and Everything **1**, World Scientific, Singapore (1991).

[179] E. Guadagnini, in "Nonperturbative methods in low dimensional quantum field theories, proceedings of the Johns Hopkins Workshop on Current Problems in Particle Theory 14, Debrecen, 1990 (August 27-30)" editors G. Domokos, Z. Horvath, S. Kovesi-Domokos, World Scientific, Singapore (1991).

[180] E. Guadagnini, "The link invariants of the Chern-Simons field theory, new developments in topological quantum field theory", De Gruyter expositions in mathematics, **10**, W. De Gruyter, New York (1993).

[181] J. Baez, J. Muniain, "Gauge fields, knots and gravity", World Scientific Series on Knots and Everything, **4**, World Scientific, Singapore (1995).

[182] C. Adams, "The knot book, an elementary introduction to the mathematical theory of knots", W. H. Freeman, San Francisco (1994).

[183] D. Armand-Ugon, R. Gambini, P. Mora, Phys. Lett. **B305**, 214 (1993).

[184] B. Brügmann, Int. J. Theor. Phys. **34**, 145 (1995).

[185] P. Cotta-Ramusino, E. Guadagnini, M. Martellini, M. Mintchev Nucl. Phys. **B330**, 557 (1990).

[186] L. Smolin, Mod. Phys. Lett. **A4** 1091 (1989).

[187] E. Guadagnini, M. Martellini, M. Mintchev, Phys. Lett. **B227**, 111 (1989).

[188] V. Vassiliev, Adv. in Sov. Math, **1**, AMS, Providence, RI (1990).

[189] L. Kauffman, Enseign. Math. **36**, 1 (1990).

[190] A. Balachandran, M. Bourdeau, S. Jo, Int. J. Mod. Phys. **A5**, 2423 (1990).

[191] J. Leinaas, J. Myrheim, Nuo. Cim. **B37**, 1 (1977).

[192] F. Wilczek, Phys. Rev. Lett. **49**, 957 (1982).

[193] Y. Wu, A. Zee, Phys. Lett. **B147**, 325 (1994).

[194] J. White, Amer. J. Math. **91**, 693 (1969).

[195] W. Bauer, F. Creek, J. White, Sci. Am. **243**, 118 (1980).

[196] A. Polyakov, Mod. Phys. Lett. **A3**, 325 (1988).

[197] J. Alexander, Trans. Am. Math. Soc. **20**, 275 (1923).

[198] J. Conway, in "Computational Problems in Abstract Algebra," editor J. Leech, Pergamon Press, London (1969).

[199] V. Jones, Bull. Am. Math. Soc. **129**, 103 (1985).

[200] J. Hoste, A. Ocneanu, K. Millet, P. Freyd, W. Lickorish, D. Yetter, Bull. Am. Math. Soc. **129**, 239 (1985).

[201] P. Kulish, N. Reshetikhin, E. Sklyanin, Lett. Math. Phys. **5**, 393 (1981).

[202] V. Turaev, Invent. Math. **92**, 527 (1988).

[203] A. Ashtekar, J. Lewandowski, D. Marolf, J. Mourão, T. Thiemann, "Coherent state transforms for spaces of connections", preprint gr-qc@xxx.lanl.gov:9412014, to appear in J. Funct. Anal. (1995).

[204] A. Ashtekar, D. Marolf and J. Mourão, in "The Proceedings of the Lanczos International Centenary Conference", editors J. Brown *et al.* SIAM, Philadelphia (1994).

[205] E. Guadagnini, M. Martellini, M. Mintchev, Nucl. Phys. **B330**, 575 (1990).

[206] R. Gambini, J. Pullin, in "Knots and quantum gravity", editor J. Baez, Oxford University Press, Oxford (1993).

[207] A. Ashtekar, C. Rovelli, L. Smolin, Phys. Rev. **D44**, 1740 (1991).

[208] L. Smolin, Class. Quan. Grav. **9**, 883 (1992).

[209] B. Brügmann, R. Gambini, J. Pullin, Phys. Rev. Lett. **68**, 431 (1992).

[210] J. Griego, private communication.

[211] G. Hall, J. Funct. Anal. **122**, 103 (1994).

[212] A. Ashtekar, J. Lewandowsky, D. Marolf, J. Mourão, T. Thiemann, in "Geometry of constrained dynamical systems", editor J. Charap, Publications of the Newton Institute, series editor P. Goddard, Cambridge University Press, Cambridge (1995).

[213] R. Loll, "Chromodynamics and gravity as theories in loop space" preprint hep-th@xxx.lanl.gov:9309056 (1993).

[214] J. Watson, in "QCD 94: proceedings of QCD 94, Montpellier, France, 7-13 July 1994", editors S. Narison, Nucl. Phys. Proceedings supplements **B39**, North-Holland, Amsterdam (1995).

[215] J. Aroca-Farrerons, H. Fort, Phys. Lett. **B325**, 166 (1994).

[216] J. Aroca-Farrerons, M. Baig, H. Fort, Phys. Lett. **B336**, 54 (1994).

[217] J. Baez, "Knots and quantum gravity", Oxford University Press, Oxford (1994).

[218] "Conference on Quantum Topology, 1993, Kansas State University. Proceedings of the Conference on Quantum Topology, Kansas State University, Manhattan, Kansas, 24-28 March 1993", editor D. Yetter, World Scientific, Singapore (1994).

[219] R. Baadhio, L. Kauffman, "Quantum Topology", World Scientific Series on Knots and Everything, **3**, World Scientific, Singapore (1993).

[220] C. Rovelli, Phys. Rev. **D48**, 2702 (1993).

[221] R. Gambini, P. Mora, "Intrinsic Time and Evolving Hilbert Spaces in Relational Dynamical Systems and Quantum Gravity", preprint hep-th@xxx.lanl.gov:9404169 (1994).

[222] T. Schilling, "Noncovariance of the extended holonomy: examples" preprint gr-qc@xxx.lanl.gov:9503064 (1995).

[223] C. Torre, Class. Quan. Grav. **8**, 1895 (1991).

[224] C. Di Bartolo, R. Gambini, J. Griego, J. Pullin, Phys. Rev. Lett. **72**, 3638 (1994).

[225] C. Di Bartolo, R. Gambini, J. Griego, Phys. Rev. **D51**, 502 (1995).

Index

Printed in the United States
By Bookmasters